**Principles of measurement
and instrumentation**

Principles of measurement and instrumentation

Second edition

Alan S. Morris

Prentice Hall

NEW YORK · LONDON · TORONTO · SYDNEY · TOKYO · SINGAPORE

First published 1993 by
Prentice Hall International (UK) Ltd
Campus 400, Maylands Avenue
Hemel Hempstead
Hertfordshire, HP2 7EZ
A division of
Simon & Schuster International Group

© Prentice Hall International (UK) Ltd, 1993

Typeset in 10/12pt Century Old Style
by Keyset Composition, Colchester, Essex CO1 2LP

Printed and bound in Great Britain by
Redwood Books Ltd, Trowbridge, Wiltshire

Library of Congress Cataloging-in-Publication Data

Morris, Alan S., 1948–
    Principles of measurement and instrumentation / by Alan S.
Morris. — 2nd ed.
        p.   cm.
    Includes bibliographical references and index.   Po 4079
    ISBN 0-13-489709-9
      1. Engineering instruments.   2. Engineering—Measurement.
I. Title.
TA165.M638   1993
681′.2—dc20   93-3286
                                                            CIP

British Library Cataloguing in Publication Data

A catalogue record for this book is available from the British Library

ISBN 0-13-489709-9 (pbk)

1 2 3 4 5   97 96 95 94 93

To Jane, Nicola and Julia

Contents

Preface to the second edition

Measurement has been of great relevance to humankind since the earliest days of human civilization, when it was first used as a means of quantifying the exchange of goods in barter trade systems. Today, measurement systems, and the instruments and transducers used in them, are of immense importance in a wide variety of domestic and industrial activities. The growth in the number and sophistication of instruments used in industry has been particularly significant over the last two decades as automation schemes have been developed. A similar rapid expansion in their use has also been evident in military and medical applications over the same period.

The aim of this book is to present the subject of instrumentation, and its use within measurement systems, as an integrated and coherent subject. The text is divided into two parts, with Part One covering the principles of measurement, and Part Two presenting the range of instruments available for measuring various physical quantities. This order of coverage has been chosen so that the general characteristics of measuring instruments and their behaviour in different operating environments are well established before the reader is introduced to the various types of instrument available. This approach means that the reader will have gained a thorough understanding of the theoretical considerations which govern the choice of a suitable instrument in any particular measurement situation before the properties of any particular instrument are discussed, and so will be properly equipped to appreciate and appraise critically the various merits and characteristics of the ranges of instrument presented.

Whilst the motivation for producing this book has been to provide a supporting text for introductory courses in measurement and instrumentation, the depth of coverage in many areas extends beyond the minimum level required for such courses and enables the book to be used also in support of courses which are somewhat more advanced. The amount of mathematics within the text has been minimized as far as possible to make the book suitable as a course text at all levels of further and higher education.

Besides this role as a student course text, the book is also of relevance to the practising instrumentation technologist who wishes to upgrade his/her knowledge about the latest developments in the theory of measurement. The catalogue of instrument ranges contained in Part Two is particularly recommended as a valuable source of reference to all practising engineers whose duties involve the design, specification or maintenance of instruments.

The opportunity has been taken in this revised and expanded edition to modify the order of presentation of certain material as well as to add new material, in response to comments from existing users of the text. This is particularly evident in Chapter 3, where the treatment of both random and sequential errors is now presented in a more cohesive and comprehensive way compared with the previous edition. The revised Chapter 7 now incorporates the theory of electrical meters within the general coverage of electrical parameter measurement. The section on signal measurement/recording has also been combined with the discussion on data presentation techniques in the revised Chapter 8.

All previous chapters in this second edition of the book have been expanded with new material and five new chapters have been added. For the benefit of previous readers of the first edition of the text, the following summary highlights the major additions in this revised edition. An introductory discussion on measurement errors has been included in Chapter 1, and a new section setting out selection criteria for instruments has been added here. Chapter 2 has been expanded with a new section on measurement precision, and the relationship of the latter to measurement accuracy has been carefully explained. A discussion of manufacturing tolerances and the use of error function tables in the analysis of random errors has been added to Chapter 3. The previous brief coverage of instrument calibration in the chapter on instrument characteristics in the first edition has been substantially expanded in the new Chapter 4 devoted to calibration principles and practice. Chapter 5 has been strengthened by the expansion of the section on signal amplification and by the addition of new sections on signal manipulation techniques and signal transmission. Additional material on power measurement has been included in Chapter 7 and a new section on phase measurement has been added. Chapter 9 is a new chapter which has been added in recognition of the growing importance of fiber optic systems. It explains the principles of fiber optics and goes on to describe the use of fiber optics in sensors, instrumentation networks and data transmission systems. Chapter 11 is another new chapter which explains some of the technology used in large-scale instrumentation and computer networks. Dimension measurement is a subject of some importance which was not previously covered but is now discussed in Chapter 16. Finally, Chapter 22 describes measurement techniques for various other miscellaneous quantities which were not previously covered in the book.

In addition to the above inclusions and expansions, the number of worked examples throughout the text and the number of sample problems at the ends of chapters have been approximately doubled.

Part One
Principles of measurement

1 Introduction to measurement

The last decade has seen a large and rapid growth in new industrial technology, encouraged by developments in electronics in general and computers in particular. This decade of rapid growth, often referred to as the 'electronics revolution', represents a step improvement in production techniques of a magnitude similar to that brought about by the Industrial Revolution in the last century. At the forefront of this thrust forward has been the digital computer – an essential component in both hard production automation schemes and in the even more advanced flexible manufacturing systems which are just beginning to emerge.

The massive growth in the application of computers to process control and monitoring tasks has spawned a parallel growth in the requirement for instruments to measure, record and control process variables. As modern production techniques dictate working to tighter and tighter limits of accuracy, and as economic forces limiting production costs become more severe, so the requirement for instruments to be both accurate and cheap becomes ever harder to satisfy. This latter problem is at the focal point of the research and development efforts of all instrument manufacturers. In the past few years, the most cost-effective means of improving instrument accuracy has been found in many cases to be the inclusion of digital computing power within the instruments themselves. These intelligent instruments therefore feature prominently in current instrument manufacturers' catalogues.

Such intelligent instruments represent the latest stage in the present era of measurement technology. This era goes back to the start of the Industrial Revolution in the nineteenth century when measuring instruments first began to be developed to satisfy the needs of industrialized production techniques. However, the complete history of measurement techniques goes back much further, in fact by thousands of years to the very start of human civilization. As humans evolved from their ape-like ancestors, they ceased to rely on using caves for shelter and to hunt and forage whatever they could for food, and started instead to build their own shelters and produce food by planting seed, rearing animals and farming in an organized manner. Initially, civilized humans lived in family communities and one assumes that such groupings were able to live in reasonable harmony without arguing about who was working hardest and who was consuming the most. With the natural diversity of human talents, however, particular family groups developed particular specializations. Some communities might excel at farming, perhaps because of the arable quality of the area of land they occupied, whilst other groups might be particularly proficient at building houses. This inevitably led to family communities producing an excess of some things and a deficit in others. Trade of excess production between family communities therefore developed naturally, and followed a barter system whereby produce or work of one sort would be exchanged for produce or work of another.

Clearly, this required a system of measurement to quantify the amounts being exchanged and to establish clear rules about the relative values of different commodities. Such early systems of measurement were based on whatever was available as a measuring unit. For the purposes of measuring length, the human torso was a convenient tool, and gave us units of the hand, the foot and the cubit. Although generally adequate for barter trade systems, such measurement units are of course imprecise, varying as they do from one person to the next; modern measurement systems are therefore based on units which are defined much more accurately. The standard measurement units which have become established are discussed in Chapter 2.

Today, the techniques of measurement are of immense importance in most facets of human civilization. Present-day applications of measuring instruments can be classified into three major areas. The first of these is their use in regulating trade, and includes instruments which measure physical quantities such as length, volume and mass in terms of standard units. The particular instruments and transducers employed in such applications are included in the general description of instruments presented in Part Two of this book.

The second area for the application of measuring instruments is in monitoring functions. These provide information which enables human beings to take some prescribed action accordingly. The gardener uses a thermometer to determine whether to turn the heat on in the greenhouse or open the windows if it is too hot. Regular study of a barometer allows us to decide whether we should take our umbrellas if we are planning to go out for a few hours. Whilst there are thus many uses of instrumentation in our normal domestic lives, the majority of monitoring functions exist to provide the information necessary to allow a human being to control some industrial operation or process. In a chemical process for instance, the progress of chemical reactions is indicated by the measurement of temperatures and pressures at various points, and such measurements allow the operator to take correct decisions regarding the electrical supply to heaters, cooling water flows, valve positions, etc. One other important use of monitoring instruments is in calibrating the instruments used in the automatic process control systems described below.

Use as part of automatic control systems forms the third area for the application of measurement systems. Figure 1.1 shows a functional block diagram of a simple control system where some output variable y of a controlled process P is maintained at a reference value r. The value of the controlled variable y, as determined by a measuring instrument M, is compared with the reference value r, and the difference e is applied as an error signal to the correcting unit C. The correcting unit then modifies the process output such that the output variable is given by $y = r$. A simple example of such a process is a shunt-wound d.c. servomotor with fixed field excitation as shown in Figure 1.2. The output variable requiring control is the motor speed, and

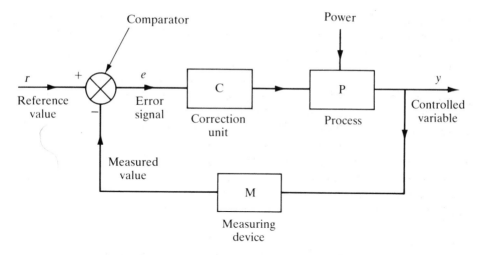

Figure 1.1 Elements of a simple closed-loop control system

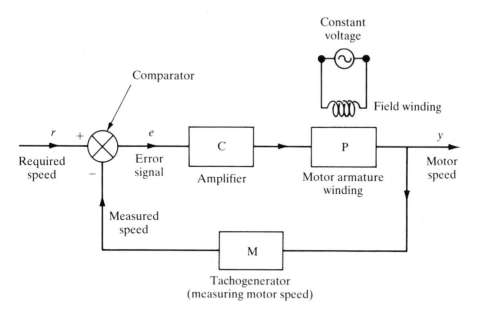

Figure 1.2 Control of a shunt-wound d.c. servomotor

this is achieved by varying the armature output according to the difference between the measured speed and the demanded reference speed r. The characteristics of measuring instruments in such feedback control systems are of fundamental importance to the quality of control achieved. The accuracy and resolution with which an output variable of a process is controlled can never be better than the accuracy and resolution of the measuring instruments used. This is a very important principle, but one which is often inadequately discussed in many texts on automatic control systems. Such texts explore the theoretical aspects of control system design in considerable depth, but fail to give sufficient emphasis to the fact that all gain and phase margin performance calculations, etc., are entirely dependent on the quality of the process measurements obtained.

A measuring instrument exists to provide information about the physical value of some variable being measured. In simple cases, an instrument consists of a single unit which gives an output reading or signal according to the magnitude of the unknown variable applied to it. However, in more complex measurement situations, a measuring instrument may consist of several separate elements as shown in Figure 1.3. These components might be contained within one or more boxes, and the boxes holding individual measurement elements might be either close together or physically separate. Because of the modular nature of the elements within it, a measuring instrument is commonly referred to as a measurement system, and this term is used extensively throughout this book to emphasize this modular nature.

Common to any measuring instrument is the primary transducer: this gives an output which is a function of the measurand (the input applied to it). For most but not all transducers, this function is at least approximately linear. Some examples of primary transducers are a liquid-in-glass thermometer, a thermocouple and a strain gauge. In the case of a mercury-in-glass thermometer, the output reading is given in terms of the level of the mercury, and so this particular primary transducer is also a complete measurement system in itself. In general, however, the primary transducer is only part of a measurement system. The types of primary transducers available for measuring a wide range of physical quantities are presented in Part Two of this book.

The output variable of a primary transducer is often in an inconvenient form and has to be converted to a more convenient one. For instance, the displacement-measuring strain gauge has an output in the form of a varying resistance. This is converted to a change in voltage by a **bridge circuit**, which is a typical example of the variable conversion element shown in Figure 1.3.

Signal processing elements exist to improve the quality of the output of a measurement system in some way. A very common type of signal processing element is the electronic amplifier, which amplifies the output of the primary transducer or variable conversion element, thus improving the sensitivity and resolution of measurement. This element of a measuring system is particularly important where the primary transducer has a low output. For example, thermocouples have a typical output of only a few millivolts. Other types of signal processing element are those which filter out induced noise and remove mean levels, etc.

The observation or application point of the output of a measurement system is often

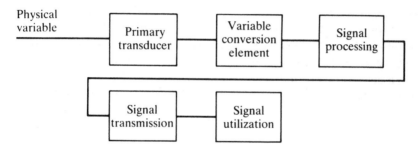

Figure 1.3 Elements of a measuring instrument

some physical distance away from the site of the primary transducer which is measuring a physical quantity, and some mechanism of transmitting the measured signal between these points is necessary. Sometimes, this separation is made solely for purposes of convenience, but more often it follows from the physical inaccessibility or environmental unsuitability of the site of the primary transducer for mounting the signal presentation/recording unit. The signal transmission element has traditionally consisted of single- or multi-cored cable, which is often screened to minimize signal corruption by induced electrical noise. Now, optical fiber cables are being used in ever increasing numbers in modern installations, in part because of their low transmission loss and imperviousness to the effects of electrical and magnetic fields.

The final element in a measurement system is the point where the measured signal is utilized. In some cases, this element is omitted altogether because the measurement is used as part of an automatic control scheme, and the transmitted signal is fed directly into the control system. In other cases, this element takes the form of either a signal presentation unit or a signal recording unit. These take many forms according to the requirements of the particular measurement application, and the range of possible units is discussed more fully in Chapter 8.

Before closing this introductory chapter, mention must also be made of measurement errors. Measurement system errors can be divided into two categories, systematic and random errors. Various mechanisms exist for reducing these two types of error.

A characteristic feature of all systematic errors is that they produce errors which are consistently on the same side of the true value, i.e. either all the errors are positive or they are all negative. Systematic errors arise from many causes, which are discussed fully in Chapter 3. These include system disturbance due to measurement, environmental changes (modifying inputs) and drift in instrument characteristics. Large errors due to instrument characteristic drift are avoided by recalibrating instruments at suitable intervals (see Chapter 4). In the case of other sources of systematic error, a good measurement technician can largely eliminate errors by calculating their effect and correcting the measurements. This is done automatically by intelligent instruments (Chapter 10).

Random errors are in many ways easier to deal with because they consist generally of small perturbations of the measurement either side of the correct value, i.e. positive errors and negative errors occur in approximately equal numbers for a series of measurements made of the same quantity. Therefore, random errors can be largely eliminated by averaging a few measurements of the same quantity. Unfortunately, averaging a number of measurements cannot be guaranteed to produce a value close to the true one because random errors occasionally cause large perturbations from the true value. Thus, it is necessary to describe measurements subject to random errors in probabilistic terms, typically that there is a 95% probability that the measurement error is within boundaries of ±1% from the true value. This is also discussed more fully in Chapter 3.

2 Instrument classification and characteristics

2.1 Instrument classification

Instruments can be subdivided into separate classes according to several criteria. These subclassifications are useful in broadly establishing several attributes of particular instruments such as accuracy, cost and general applicability to different applications.

2.1.1 Active/passive instruments

Instruments are either active or passive according to whether the instrument output is entirely produced by the quantity being measured or whether the quantity being measured simply modulates the magnitude of some external power source. This can be illustrated by examples.

An example of a passive instrument is the pressure measuring device shown in Figure 2.1. The pressure of the fluid is translated into movement of a pointer against a scale. The energy expended in moving the pointer is derived entirely from the change in pressure measured: there are no other energy inputs to the system.

An example of an active instrument is a float-type petrol-tank level indicator as sketched in Figure 2.2. Here, the change in petrol level moves a potentiometer arm, and the output signal consists of a proportion of the external voltage source applied across the two ends of the potentiometer. The energy in the output signal comes from the external power source; the primary transducer float system is merely modulating the value of the voltage from this external power source.

In active instruments, the external power source is usually electrical in form, but in some cases it can be in other forms of energy such as pneumatic or hydraulic.

One very important difference between active and passive instruments is the level of measurement resolution which can be obtained. With the simple pressure gauge shown, the amount of movement made by the pointer for a particular pressure change is closely defined by the nature of the instrument. Whilst it is possible to increase the measurement resolution by making the pointer longer, so that the pointer tip moves through a longer arc, the scope for such improvement is clearly restricted by the practical limit of how long the pointer can conveniently be. In an active instrument, however, adjustment of the magnitude of the external energy input allows much greater control over measurement resolution. Whilst the scope for improving this

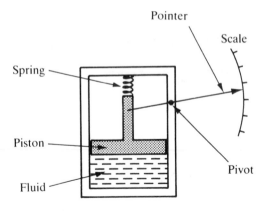

Figure 2.1 Passive pressure gauge

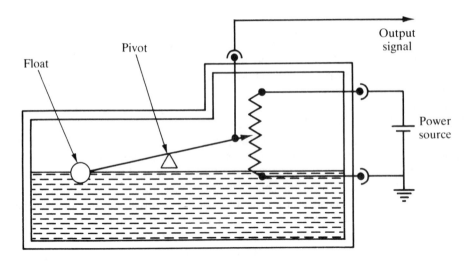

Figure 2.2 Petrol-tank level indicator

resolution is much greater incidentally, it is not infinite because of the limitations placed on the magnitude of the external energy input, in considering heating effects and for safety reasons.

In terms of cost, passive instruments are normally of a simpler construction than active ones and are therefore cheaper to manufacture. The choice between active and passive instruments for a particular application therefore involves carefully balancing the measurement resolution requirements against cost.

2.1.2 **Null/deflection-type instruments**

The pressure gauge just mentioned is a good example of a deflection-type instrument, where the value of the quantity being measured is displayed in terms of the amount of movement of a pointer. An alternative type of pressure gauge is the dead-weight gauge shown in Figure 2.3 which is a null-type instrument. Here, weights are put on top of the piston until the downward force balances the fluid pressure. Weights are added until the piston reaches a datum level, known as the null point. Pressure measurement is made in terms of the value of the weights needed to reach this null position.

The accuracy of these two instruments depends on different things. For the first one it depends on the linearity and calibration of the spring, whilst for the second it relies on the calibration of the weights. As calibration of weights is much easier than careful choice and calibration of a linear-characteristic spring, this means that the second type of instrument will normally be the more accurate. This is in accordance with the general rule that null-type instruments are more accurate than deflection types.

In terms of usage, the deflection-type instrument is clearly more convenient. It is far simpler to read the position of a pointer against a scale than to add and subtract weights until a null point is reached. A deflection-type instrument is therefore the one that would normally be used in the workplace. However, for calibration duties, the null-type instrument is preferable because of its superior accuracy. The extra effort required to use such an instrument is perfectly acceptable in this case because of the infrequent nature of calibration operations.

2.1.3 **Monitoring/control instruments**

An important distinction between different instruments is whether they are suitable only for monitoring functions or whether their output is in a form that can be directly

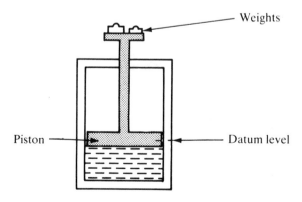

Figure 2.3 Dead-weight pressure gauge

included as part of an automatic control system. Instruments which only give an audio or visual indication of the magnitude of the physical quantity measured, such as a liquid-in-glass thermometer, are only suitable for monitoring purposes. This class normally includes all null-type instruments and most passive transducers.

For an instrument to be suitable for inclusion in an automatic control system, its output must be in a suitable form for direct input to the controller. Usually, this means that an instrument with an electrical output is required, although other forms of output such as optical or pneumatic signals are used in some systems.

2.1.4 Analog/digital instruments

An analog instrument gives an output which varies continuously as the quantity being measured changes. The output can have an infinite number of values within the range that the instrument is designed to measure. The deflection type of pressure gauge described earlier in this chapter (Figure 2.1) is a good example of an analog instrument. As the input value changes, the pointer moves with a smooth continuous motion. Whilst the pointer can therefore be in an infinite number of positions within its range of movement, the number of different positions which the eye can discriminate between is strictly limited, this discrimination being dependent upon how large the scale is and how finely it is divided.

A digital instrument has an output which varies in discrete steps and so can only have a finite number of values. The rev-counter sketched in Figure 2.4 is an example of a digital instrument. A cam is attached to the revolving body whose motion is being measured, and on each revolution the cam opens and closes a switch. The switching operations are counted by an electronic counter. This system can only count whole revolutions and cannot discriminate any motion which is less than a full revolution.

The distinction between analog and digital instruments has become particularly important with the rapid growth in the application of microcomputers to automatic control systems. Any digital computer system, of which the microcomputer is but one

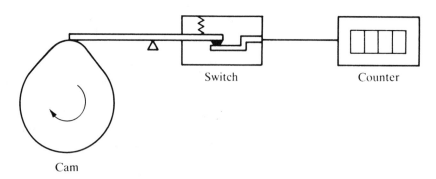

Figure 2.4 Rev-counter

example, performs its computations in digital form. An instrument whose output is in digital form is therefore particularly advantageous in such applications, as it can be interfaced directly to the control computer. Analog instruments must be interfaced to the microcomputer by an analog-to-digital (A/D) converter, which converts the analog output signal from the instrument into an equivalent digital quantity which can be read into the computer. This conversion has several disadvantages. First, the A/D converter adds a significant cost to the system. Secondly, a finite time is involved in the process of converting an analog signal to a digital quantity, and this time can be critical in the control of fast processes where the accuracy of control depends on the speed of the controlling computer. Degrading the speed of operation of the control computer by imposing a requirement for A/D conversion thus impairs the accuracy by which the process is controlled.

2.2 **Standards**

Referring back to our earlier comments in the last chapter about the development of measurement systems, we stated that the earliest units of measurement used were convenient parts of the human torso such as the foot and hand. Such a system allowed an approximate level of equivalence to be established about the relative value of quantities of different commodities. The length and breadth of constructional timber for instance could be measured in units of feet, and cloth could be similarly measured in units of feet squared. This allowed a basis for determining the relative values of timber and cloth for barter trade purposes. However, such a system was clearly unfair when a person with large hands exchanged timber for cloth from a person with small hands (assuming that they both used their own hands to measure the commodity being exchanged).

There was thus an urgent need to establish units of measurement which were based on non-varying quantities, and the first of these was a unit of length (the metre) defined as 10^{-7} times the polar quadrant of the earth. A platinum bar made to this length was established as a standard of length in the early part of the nineteenth century. This was superseded by a superior-quality standard bar in 1889, manufactured from a platinum–iridium alloy. Since that time, technological research has enabled further improvements to be made in the standard used for defining length. First, in 1960, a standard metre was redefined in terms of $1.650\,763\,73 \times 10^{6}$ wavelengths of the radiation from krypton-86 in vacuum. More recently, in 1983, the metre was redefined yet again as the length of path travelled by light in an interval of $1/299\,792\,458$ seconds.

In a similar fashion, standard units for the measurement of other physical quantities have been defined and progressively improved over the years. The latest standards for defining the units used for measuring a range of physical variables are given in Table 2.1.

The early establishment of standards for the measurement of physical quantities proceeded in several countries at broadly parallel times, and in consequence, several

Table 2.1 Definitions of standard units

Physical quantity	Standard unit	Definition
Length	metre	The length of path travelled by light in an interval of 1/299 792 458 seconds
Mass	kilogram	The mass of a platinum–iridium cylinder kept in the International Bureau of Weights and Measures, Sèvres, Paris
Time	second	$9.192\,631\,770 \times 10^9$ cycles of radiation from vaporized caesium-133 (an accuracy of 1 in 10^{12} or 1 second in 36 000 years)
Temperature	kelvin	The temperature difference between absolute zero and the triple point of water is defined as 273.16° kelvin
Current	ampere	One ampere is the current flowing through two infinitely long parallel conductors of negligible cross-section placed 1 metre apart in a vacuum and producing a force of 2×10^{-7} newtons per metre length of conductor
Luminous intensity	candela	One candela is the luminous intensity in a given direction from a source emitting monochromatic radiation at a frequency of 540 terahertz ($Hz \times 10^{12}$) and with a radiant density in that direction of 1.4641 mW/ steradian. (1 steradian is the solid angle which, having its vertex at the centre of a sphere, cuts off an area of the sphere surface equal to that of a square with sides of length equal to the sphere radius)
Matter	mole	The number of atoms in a 0.012 kg mass of carbon-12

sets of units emerged for measuring the same physical variable. For instance, length can be measured in yards or metres or several other units. Apart from the major units of length, subdivisions of standard units exist such as feet, inches, centimetres and millimetres, with a fixed relationship between each fundamental unit and its subdivisions.

Yards, feet and inches belong to the Imperial System of units, which is characterized by having varying and cumbersome multiplication factors relating fundamental units to subdivisions such as 1760 (miles to yards), 3 (yards to feet) and

12 (feet to inches). This system is the one which was used predominantly in Britain until a few years ago.

The metric system is an alternative set of units which includes for instance the unit of the metre and its centimetre and millimetre subdivisions for measuring length. All multiples and subdivisions of basic metric units are related to the base by factors of ten and such units are therefore much easier to use than Imperial units. However, in the case of derived units such as velocity, the number of alternative ways in which these can be expressed in the metric system can lead to confusion.

As a result of this, an internationally agreed set of standard units (SI units or Systèmes Internationales d'Unités) has been defined, and strong efforts are being made to encourage the adoption of this system throughout the world. In support of this effort, SI units will be used exclusively in this book.

The full range of fundamental SI measuring units and the further set of units derived from them are given in Table 2.2. Two supplementary units for measuring angles are also included within this table. Following this, conversion tables relating common Imperial and metric units to their equivalent SI units are to be found in Appendix 1. Apart from the establishment of a common set of measurement units (the SI system), international agreements have also established several other common conventions. One such agreement is the convention used to mark component values and tolerances on electric circuit components such as resistors and capacitors, as detailed in Appendix 2. Further examples are internationally accepted codes (G-codes) for programming numerically controlled machine tools, and a system of standard symbols for representing components in electric circuit drawings.

2.3 Static characteristics of instruments

If we have a thermometer in a room and its reading shows a temperature of 20 °C, then it does not really matter whether the true temperature of the room is 19.5 °C or 20.5 °C. Such small variations around 20 °C are too small to affect whether we feel warm enough or not. Our bodies cannot discriminate between such close levels of temperature and therefore a thermometer with an accuracy of ±0.5 °C is perfectly adequate. If we had to measure the temperature of certain chemical processes, however, a variation of 0.5 °C might have a significant effect on the rate of reaction or even the products of a process. A measurement accuracy much better than ±0.5 °C is therefore clearly required.

Accuracy of measurement is thus one consideration in the choice of instrument for a particular application. Other parameters such as sensitivity, linearity and the reaction to ambient temperature changes are further considerations. These attributes are collectively known as the static characteristics of instruments, and are given in the data sheet for a particular instrument. It is important to note that the values quoted for instrument characteristics in such a data sheet only apply when the instrument is used under specified standard calibration conditions. Due allowance must be made for variations in the characteristics when the instrument is used under other conditions.

Table 2.2 Fundamental and derived SI units

(a) Fundamental units

Quantity	Standard unit	Symbol
Length	metre	m
Mass	kilogram	kg
Time	second	s
Electric current	ampere	A
Temperature	kelvin	K
Luminous intensity	candela	cd
Matter	mole	mol

(b) Supplementary fundamental units

Quantity	Standard unit	Symbol
Plane angle	radian	rad
Solid angle	steradian	sr

(c) Derived units

Quantity	Standard unit	Symbol	Derivation formula
Area	square metre	m^2	
Volume	cubic metre	m^3	
Velocity	metre per second	m/s	
Acceleration	metre per second squared	m/s^2	
Angular velocity	radian per second	rad/s	
Angular acceleration	radian per second squared	rad/s^2	
Density	kilogram per cubic metre	kg/m^3	
Specific volume	cubic metre per kilogram	m^3/kg	
Mass flow rate	kilogram per second	kg/s	
Volume flow rate	cubic metre per second	m^3/s	
Force	newton	N	$kg\,m/s^2$
Pressure	newton per square metre	N/m^2	
Torque	newton metre	N m	
Momentum	kilogram metre per second	$kg\,m/s$	
Moment of inertia	kilogram metre squared	$kg\,m^2$	
Kinematic viscosity	square metre per second	m^2/s	
Dynamic viscosity	newton second per sq. metre	$N\,s/m^2$	
Work, energy, heat	joule	J	N m
Specific energy	joule per cubic metre	J/m^3	
Power	watt	W	J/s
Thermal conductivity	watt per metre kelvin	W/m K	

Quantity	Standard unit	Symbol	Derivation formula
Electric charge	coulomb	C	A s
Voltage, e.m.f., pot. diff.	volt	V	W/A
Electric field strength	volt per metre	V/m	
Electric resistance	ohm	Ω	V/A
Electric capacitance	farad	F	A s/V
Electric inductance	henry	H	V s/A
Electric conductance	siemen	S	A/V
Resistivity	ohm metre	Ω m	
Permittivity	farad per metre	F/m	
Permeability	henry per metre	H/m	
Current density	ampere per square metre	A/m^2	
Magnetic flux	weber	Wb	V s
Magnetic flux density	tesla	T	Wb/m^2
Magnetic field strength	ampere per metre	A/m	
Frequency	hertz	Hz	s^{-1}
Luminous flux	lumen	lm	cd sr
Luminance	candela per square metre	cd/m^2	
Illumination	lux	lx	lm/m^2
Molar volume	cubic metre per mole	m^3/mol	
Molarity	mole per kilogram	mol/kg	
Molar energy	joule per mole	J/mol	

The various static characteristics are defined in the following subsections. More formal definitions can be found in the documents referenced (BS 5233, BS 5532 and ISO 3534).

2.3.1 Accuracy

Accuracy is the extent to which a reading might be wrong, and is often quoted as a percentage of the full-scale reading of an instrument.

If, for example, a pressure gauge of range 0–10 bar has a quoted inaccuracy of $\pm 1.0\%$ f.s. ($\pm 1\%$ of full-scale reading), then the maximum error to be expected in any reading is 0.1 bar. This means that when the instrument is reading 1.0 bar, the possible error is 10% of this value. For this reason, it is an important system design rule that instruments are chosen such that their range is appropriate to the spread of values being measured, in order that the best possible accuracy is maintained in instrument readings. Thus, if we were measuring pressures with expected values between 0 and 1 bar, we would not use an instrument with a range of 0–10 bar.

2.3.2 Precision/repeatability/reproducibility

Precision is a term which describes an instrument's degree of freedom from random errors. If a large number of readings are taken of the same quantity by a high-precision instrument, then the spread of readings will be very small.

Precision is often, though incorrectly, confused with accuracy. High precision does not imply anything about measurement accuracy. A high-precision instrument may have a low accuracy. Low-accuracy measurements from a high-precision instrument are normally caused by a bias in the measurements, which is removable by recalibration.

The terms repeatability and reproducibility mean approximately the same but are applied in different contexts as given next. **Repeatability** describes the closeness of output readings when the same input is applied repetitively over a short period of time, with the same measurement conditions, same instrument and observer, same location and same conditions of use maintained throughout. **Reproducibility** describes the closeness of output readings for the same input when there are changes in the method of measurement, observer, measuring instrument, location, conditions of use and time of measurement. Both terms thus describe the spread of output readings for the same input. This spread is referred to as repeatability if the measurement conditions are constant and as reproducibility if the measurement conditions vary.

The degree of repeatability or reproducibility in measurements from an instrument is an alternative way of expressing its precision. Figure 2.5 illustrates this more clearly. The figure shows the results of tests on three industrial robots which were programmed to place components at a particular point on a table. The target point was at the centre of the concentric circles shown, and the black dots represent the points where each robot actually deposited components at each attempt. Both the accuracy and the precision of Robot 1 are shown to be low in this trial. Robot 2 consistently puts the component down at approximately the same place but this is the wrong point; therefore, it has high precision but low accuracy. Finally, Robot 3 has both high precision and high accuracy, because it consistently places the component at the correct target position.

2.3.3 Tolerance

Tolerance is a term which is closely related to accuracy and defines the maximum error which is to be expected in some value. Whilst it is not, strictly speaking, a static characteristic of measuring instruments, it is mentioned here because the accuracy of some instruments is sometimes quoted as a tolerance figure.

Tolerance, when used correctly, describes the maximum deviation of a manufactured component from some specified value. Crankshafts, for instance, are machined with a diameter tolerance quoted as so many microns (10^{-6} m), and electric circuit components such as resistors have tolerances of perhaps 5%. One resistor chosen at random from a batch having a nominal value 1000 Ω and tolerance 5% might have an actual value anywhere between 950 Ω and 1050 Ω.

2.3.4 Range or span

The range or span of an instrument defines the minimum and maximum values of a quantity that the instrument is designed to measure.

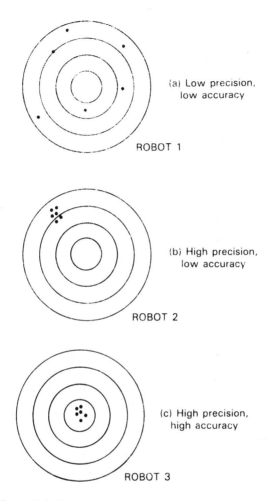

Figure 2.5 Comparison of accuracy and precision

2.3.5 **Bias**

Bias describes a constant error which exists over the full range of measurement of an instrument. This error is normally removable by calibration.

Bathroom scales are a common example of instruments which are prone to bias. It is quite usual to find that there is a reading of perhaps 1 kg with no one stood on the scales. If someone of known weight 70 kg were to get on the scales, the reading would be 71 kg, and if someone of known weight 100 kg were to get on the scales, the reading would be 101 kg. This constant bias of 1 kg can be removed by calibration: in the case of bathroom scales this normally means turning a thumbwheel with the scales unloaded until the reading is zero.

2.3.6 Linearity

It is normally desirable that the output reading of an instrument is linearly proportional to the quantity being measured. The crosses marked on Figure 2.6 show a plot of the typical output readings of an instrument when a sequence of input quantities is applied to it. Normal procedure is to draw a good-fit straight line through the crosses, as shown in the figure. (Whilst this can often be done with reasonable accuracy by eye, it is always preferable to apply a mathematical least-squares line-fitting technique, as described in Chapter 8.) The non-linearity is then defined as the maximum deviation of any of the output readings marked with a cross from this straight line. Non-linearity is usually expressed as a percentage of full-scale reading.

2.3.7 Sensitivity of measurement

The sensitivity of measurement is a measure of the change in instrument output which occurs when the quantity being measured changes by a given amount. Sensitivity is thus the ratio:

$$\frac{\text{scale deflection}}{\text{value of measurand causing deflection}}$$

The sensitivity of measurement is therefore the slope of the straight line drawn on Figure 2.6. If, for example, a pressure of 2 bar produces a deflection of 10 degrees in a pressure transducer, the sensitivity of the instrument is 5 degrees/bar (assuming that the deflection is zero with zero pressure applied).

Example 2.1
The following resistance values of a platinum resistance thermometer were measured at a range of temperatures. Determine the measurement sensitivity of the instrument in $\Omega/°C$.

Resistance (Ω)	Temperature (°C)
307	200
314	230
321	260
328	290

Solution
If these values are plotted on a graph, the straight-line relationship between resistance change and temperature change is obvious.

For a change in temperature of 30 °C, the change in resistance is 7 Ω. Hence the measurement sensitivity is equal to 7/30 = 0.233 $\Omega/°C$.

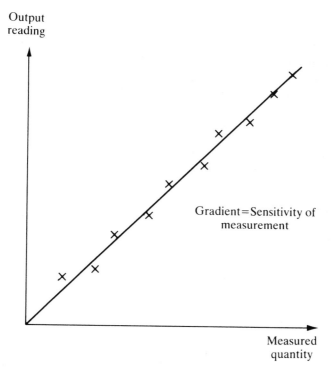

Figure 2.6 Instrument output characteristic

2.3.8 **Sensitivity to disturbance**

All calibrations and specifications of an instrument are only valid under controlled conditions of temperature, pressure, etc. These standard ambient conditions are usually defined in the instrument specification. As variations occur in the ambient temperature, etc., certain static instrument characteristics change, and the sensitivity to disturbance is a measure of the magnitude of this change. Such environmental changes affect instruments in two main ways, known as zero drift and sensitivity drift.

Zero drift describes the effect where the zero reading of an instrument is modified by a change in ambient conditions. Typical units by which zero drift is measured are volts/°C, in the case of a voltmeter affected by ambient temperature changes. This is often called the zero drift coefficient related to temperature changes. If the characteristic of an instrument is sensitive to several environmental parameters, then it will have several zero drift coefficients, one for each environmental parameter. The effect of zero drift is to impose a bias in the instrument output readings; this is normally removable by recalibration in the usual way. A typical change in the output characteristic of a pressure gauge subject to zero drift is shown in Figure 2.7(a).

Sensitivity drift (also known as **scale factor drift**) defines the amount by which an instrument's sensitivity of measurement varies as ambient conditions change. It is

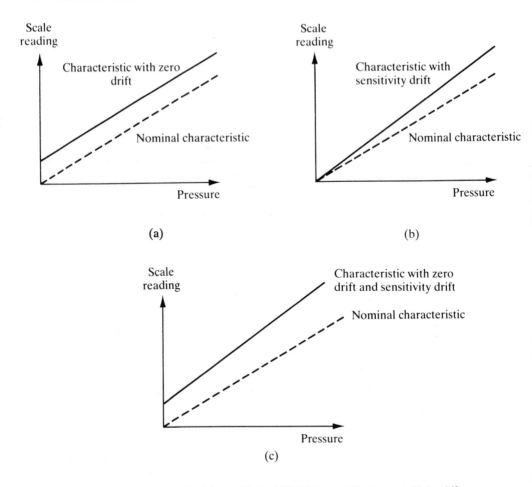

Figure 2.7 (a) Zero drift; (b) sensitivity drift; (c) zero drift plus sensitivity drift

quantified by sensitivity drift coefficients which define how much drift there is for a unit change in each environmental parameter that the instrument characteristics are sensitive to.

Many components within an instrument are affected by environmental fluctuations, such as temperature changes: for instance, the modulus of elasticity of a spring is temperature dependent. Figure 2.7(b) shows what effect sensitivity drift can have on the output characteristic of an instrument. Sensitivity drift is measured in units of the form (angular degree/bar)/°C.

If an instrument suffers both zero drift and sensitivity drift at the same time, then the typical modification of the output characteristic is as shown in Figure 2.7(c).

Example 2.2

A spring balance is calibrated in an environment at a temperature of 20 °C and has the following deflection/load characteristic:

Load (kg)	0	1	2	3
Deflection (mm)	0	20	40	60

It is then used in an environment at a temperature of 30 °C and the following deflection/load characteristic is measured:

Load (kg)	0	1	2	3
Deflection (mm)	5	27	49	71

Determine the zero drift and sensitivity drift per °C change in ambient temperature.

Solution

At 20 °C, deflection/load characteristic is a straight line. Sensitivity = 20 mm/kg
At 30 °C, deflection/load characteristic is still a straight line. Sensitivity = 22 mm/kg
Bias (zero drift) = 5 mm (the no-load deflection)
Sensitivity drift = 2 mm/kg
Zero drift/°C = 5/10 = 0.5 mm/°C
Sensitivity drift/°C = 2/10 = 0.2 (mm per kg)/°C.

2.3.9 Hysteresis

Figure 2.8 illustrates the output characteristic of an instrument which exhibits hysteresis. If the input measured quantity to the instrument is steadily increased from a negative value, the output reading varies in the manner shown in curve A. If the input variable is then steadily decreased, the output varies in the manner shown in curve B. The non-coincidence between these loading and unloading curves is known as hysteresis.

The two quantities of maximum input hysteresis and maximum output hysteresis are defined as shown in Figure 2.8. They are normally expressed as a percentage of the full-scale input or output reading respectively.

2.3.10 Dead space

Dead space is defined as the range of different input values over which there is no change in output value.

Any instrument which exhibits hysteresis also displays dead space, as marked on Figure 2.8. Some instruments which do not suffer from any significant hysteresis can still exhibit a dead space in their output characteristics, however. Backlash in gears is a typical cause of dead space, and results in the sort of instrument output characteristic shown in Figure 2.9.

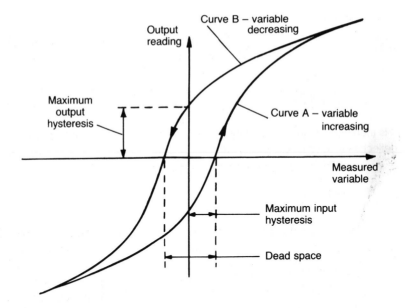

Figure 2.8 Instrument characteristic with hysteresis

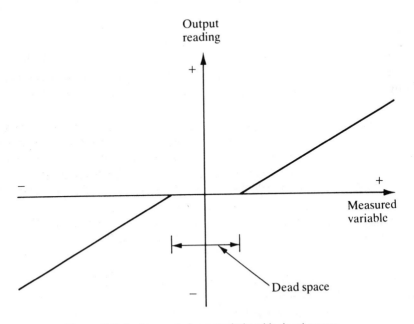

Figure 2.9 Instrument characteristic with dead space

2.3.11 **Threshold**

If the input to an instrument is gradually increased from zero, the input will have to reach a certain minimum level before the change in the instrument output reading is of a large enough magnitude to be detectable. This minimum level of input is known as the threshold of the instrument. Manufacturers vary in the way that they specify threshold for instruments. Some quote absolute values, whereas others quote threshold as a percentage of full-scale readings.

As an illustration, a car speedometer typically has a threshold of about 15 km/h. This means that, if the vehicle starts from rest and accelerates, no output reading is observed on the speedometer until the speed reaches 15 km/h.

2.3.12 **Resolution**

When an instrument is showing a particular output reading, there is a lower limit on the magnitude of the change in the input measured quantity which produces an observable change in the instrument output. Like threshold, resolution is sometimes specified as an absolute value and sometimes as a percentage of f.s. deflection.

One of the major factors influencing the resolution of an instrument is how finely its output scale is subdivided. Using a car speedometer again as an example, this has subdivisions of typically 20 km/h. This means that when the needle is between the scale markings, we cannot estimate speed more accurately than to the nearest 5 km/h. This figure of 5 km/h thus represents the resolution of the instrument.

2.4 **Dynamic characteristics of instruments**

The static characteristics of measuring instruments are concerned only with the steady-state reading that the instrument settles down to, such as the accuracy of the reading, etc.

The dynamic characteristics of a measuring instrument describe its behaviour between the time a measured quantity changes value and the time when the instrument output attains a steady value in response. As with static characteristics, any values for dynamic characteristics quoted in instrument data sheets only apply when the instrument is used under specified environmental conditions. Outside these calibration conditions, some variation in the dynamic parameters can be expected.

In any linear, time-invariant measuring system, the following general relation can be written between input and output for time t greater than zero:

$$a_n \frac{\mathrm{d}_n q_0}{\mathrm{d}t_n} + a_{n-1} \frac{\mathrm{d}_{n-1} q_0}{\mathrm{d}t_{n-1}} + \ldots + a_1 \frac{\mathrm{d}q_0}{\mathrm{d}t} + a_0 q_0 = b_m \frac{\mathrm{d}_m q_\mathrm{i}}{\mathrm{d}t_m}$$

$$(2.1)$$

$$+ b_{m-1} \frac{\mathrm{d}_{m-1} q_\mathrm{i}}{\mathrm{d}t_{m-1}} + \ldots + b_1 \frac{\mathrm{d}q_\mathrm{i}}{\mathrm{d}t} + b_0 q_\mathrm{i}$$

where q_i is the measured quantity, q_0 is the output reading and $a_0 \ldots a_n$, $b_0 \ldots b_m$ are constants.

The reader whose mathematical background is such that the above equation appears daunting should not worry unduly, as only certain special, simplified cases of it are applicable in normal measurement situations. The important point is to have a practical appreciation of the manner in which various different types of instrument respond when the measurand applied to them varies.

If we limit consideration to that of step changes in the measured quantity only, then Equation (2.1) reduces to:

$$a_n \frac{d_n q_0}{dt_n} + a_{n-1} \frac{d_{n-1} q_0}{dt_{n-1}} + \ldots + a_1 \frac{dq_0}{dt} + a_0 q_0 = b_0 q_i \qquad (2.2)$$

Further simplification can be made by taking certain special cases of Equation (2.2) which collectively apply to nearly all measurement systems.

2.4.1 Zero-order instrument

If all the coefficients $a_1 \ldots a_n$ other than a_0 in Equation (2.2) are assumed zero, then:

$$a_0 q_0 = b_0 q_i \quad \text{or} \quad q_0 = b_0 q_i / a_0 = K q_i \qquad (2.3)$$

where K is a constant known as the instrument sensitivity as defined earlier.

Any instrument which behaves according to Equation (2.3) is said to be of zero-order type. A potentiometer, which measures motion, is a good example of such an instrument, where the output voltage changes instantaneously as the slider is displaced along the potentiometer track.

2.4.2 First-order instrument

If all the coefficients $a_2 \ldots a_n$ except for a_0 and a_1 are assumed zero in equation (2.2) then:

$$a_1 \frac{dq_0}{dt} + a_0 q_0 = b_0 q_i \qquad (2.4)$$

Any instrument which behaves according to Equation (2.4) is known as a first-order instrument.

If d/dt is replaced by the D operator in Equation (2.4), we get:

$$a_1 D q_0 + a_0 q_0 = b_0 q_i$$

and rearranging:

$$q_0 = \frac{(b_0/a_0) q_i}{[1 + (a_1/a_0) D]} \qquad (2.5)$$

Defining $K = b_0/a_0$ as the static sensitivity and $\tau = a_1/a_0$ as the time constant of the system, Equation (2.5) becomes:

$$q_0 = \frac{Kq_i}{1 + \tau D} \qquad (2.6)$$

If Equation (2.6) is solved analytically, the output quantity q_0 in response to a step change in q_i varies with time in the manner shown in Figure 2.9. The time constant τ of the step response is the time taken for the output quantity q_0 to reach 63% of its final value.

The thermocouple (see Chapter 15) is a good example of a first-order instrument. It is well known that, if a thermocouple at room temperature is plunged into boiling water, the output e.m.f. does not rise instantaneously to a level indicating 100 °C, but instead approaches a reading indicating 100 °C in a manner similar to that shown in Figure 2.10.

A large number of other instruments also belong to this first-order class. This is of particular importance in control systems where it is necessary to take account of the time lag that occurs between a measured quantity changing in value and the measuring instrument indicating the change. Fortunately, the time constant of many first-order instruments is small relative to the dynamics of the process being measured, and so no serious problems are created.

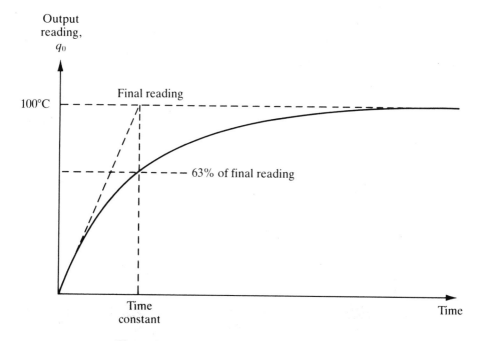

Figure 2.10 First-order instrument characteristic

Example 2.3

A balloon is equipped with temperature and altitude measuring instruments and has radio equipment which can transmit the output readings of these instruments back to ground. The balloon is initially anchored to the ground with the instrument output readings in steady state. The altitude measuring instrument is approximately zero order and the temperature transducer first order with a time constant of 15 seconds. The temperature on the ground, T_0, is 10 °C and the temperature T_x at an altitude of x metres is given by the relation $T_x = T_0 - 0.01x$.

(a) If the balloon is released at time zero, and thereafter rises upwards at a velocity of 5 metres/second, construct a table showing the temperature and altitude measurements reported at intervals of 10 seconds over the first 50 seconds of travel. Show also in the table the error in each temperature reading.

(b) What temperature does the balloon report at an altitude of 5000 metres?

Solution

In order to answer this question, it is assumed that the solution of a first-order differential equation has been presented to the reader in a mathematics course. If the reader is not so equipped, the following solution will be difficult to follow.

Let the temperature reported by the balloon at some general time t be T_r. Then T_x is related to T_r by the relation:

$$T_r = \frac{T_x}{1 + \tau D} = \frac{T_0 - 0.01x}{1 + \tau D} = \frac{10 - 0.01x}{1 + 15D}$$

It is given that $x = 5t$. Thus:

$$T_r = \frac{10 - 0.05t}{1 + 15D}$$

The transient or complementary function part of the solution ($T_x = 0$) is given by:

$$T_{r_{cf}} = C \exp^{(-t/15)}$$

The particular integral part of the solution is given by:

$$T_{r_{pi}} = 10 - 0.05(t - 15)$$

Thus the whole solution is given by:

$$T_r = T_{r_{cf}} + T_{r_{pi}} = C \exp^{(-t/15)} + 10 - 0.05(t - 15)$$

Applying initial conditions: At $t = 0$, $T_r = 10$, i.e.

$$10 = C \exp^{(-0)} + 10 - 0.05(-15)$$

Thus $C = -0.75$ and the solution can be written as:

$$T_r = 10 - 0.75\exp^{(-t/15)} - 0.05(t - 15)$$

(a) Using the above expression to calculate T_r for various values of t, the following table can be constructed:

Time	Altitude	Temp. reading	Temp. error
0	0	10	0
10	50	9.86	0.36
20	100	9.55	0.55
30	150	9.15	0.65
40	200	8.70	0.70
50	250	8.22	0.72

(b) At 5000 M, $t = 1000$ seconds.
   Calculating T_r from the above expression:

$$T_r = 10 - 0.75\exp^{(-1000/15)} - 0.05(1000 - 15)$$

The exponential term approximates to zero and so T_r can be written as:

$$T_r \approx 10 - 0.05(985) = -39.25\,°C$$

This result might have been inferred from the table above where it can be seen that the error is converging towards a value of 0.75. For large values of t, the transducer reading lags behind the true temperature value by a period equal to the time constant of 15 seconds. In this time, the balloon travels a distance of 75 metres and the temperature falls by 0.75 °C. Thus for large values of t, the output reading is always 0.75 °C less than it should be.

2.4.3 Second-order instrument

If all coefficients $a_3 \ldots a_n$ other than a_0, a_1 and a_2 in Equation (2.2) are assumed zero, then we get:

$$a_2\frac{d^2q_0}{dt^2} + a_1\frac{dq_0}{dt} + a_0q_0 = b_0q_i \tag{2.7}$$

Applying the D operator again:

$$a_2D^2q_0 + a_1Dq_0 + a_0q_0 = b_0q_i$$

and rearranging:

$$q_0 = \frac{b_0 q_i}{a_0 + a_1 D + a_2 D^2} \tag{2.8}$$

It is convenient to re-express the variables a_0, a_1, a_2 and b_0 in Equation (2.8) in terms of three parameters K (static sensitivity), ω (undamped natural frequency) and ε (damping ratio), where:

$$K = b_0/a_0$$

$$\omega = a_0/a_2$$

$$\varepsilon = a_1/2a_0 a_2$$

Re-expressing Equation (2.8) in terms of K, ω and ε we get:

$$\frac{q_0}{q_i} = \frac{K}{D^2/\omega^2 + 2\varepsilon D/\omega + 1} \tag{2.9}$$

This is the standard equation for a second-order system and any instrument whose response can be described by it is known as a second-order instrument.

If Equation (2.9) is solved analytically, the shape of the step response obtained depends on the value of the damping ratio parameter ε. The output responses of a second-order instrument for various values of ε are shown in Figure 2.11. For case A where $\varepsilon = 0$, there is no damping and the instrument output exhibits constant amplitude oscillations when disturbed by any change in the physical quantity measured. For light damping of $\varepsilon = 0.2$, represented by case B, the response to a step change in input is still oscillatory but the oscillations gradually die down. A further increase in the value of ε reduces oscillations and overshoot still more, as shown by curves C and D, and finally the response becomes very overdamped as shown by curve E where the output reading creeps up slowly towards the correct reading. Clearly, the extreme response curves A and E are grossly unsuitable for any measuring instrument. If an instrument were to be only ever subjected to step inputs, then the design strategy would be to aim towards a damping ratio of 0.707, which gives the critically damped response C. Unfortunately, most of the physical quantities which instruments are required to measure do not change in the mathematically convenient form of steps, but rather in the form of ramps of varying slopes. As the form of the input variable changes, so the best value for ε varies, and the choice of ε becomes a compromise between those values that are best for each type of input variable behaviour anticipated. Commercial second-order instruments, of which the accelerometer is a common example, are generally designed to have a damping ratio (ε) somewhere in the range of 0.6–0.8.

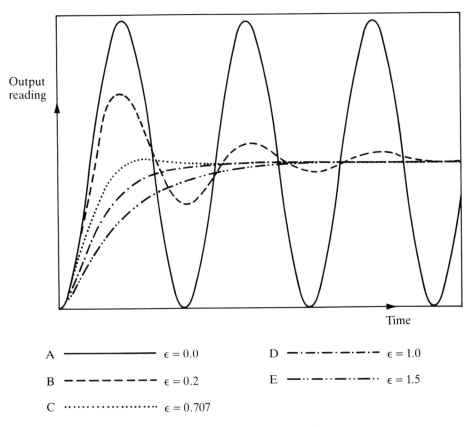

A	———————————	$\epsilon = 0.0$	D	—·——·——·——·—	$\epsilon = 1.0$
B	— — — — — — —	$\epsilon = 0.2$	E	—··——··——··—··	$\epsilon = 1.5$
C	··················	$\epsilon = 0.707$			

Figure 2.11 Responses of second-order instruments

2.5 **Calibration**

The foregoing discussion has described the static and dynamic characteristics of measuring instruments in some detail. An important qualification which has been omitted from this discussion, however, is the statement that an instrument only conforms to stated static and dynamic patterns of behaviour after it has been calibrated. It can normally be assumed that a new instrument will have been calibrated when it is obtained from an instrument manufacturer, and will therefore initially behave according to the characteristics stated in the specifications. During use, however, its behaviour will gradually diverge from the stated specification for a variety of reasons. Such reasons include mechanical wear and the effects of dirt, dust, fumes and chemicals in the operating environment. The rate of divergence from standard specifications varies according to the type of instrument, the frequency of usage and the severity of the operating conditions. However, there will come a time, determined by practical knowledge, when the characteristics of the instrument will have drifted from the standard specification by an unacceptable amount. When this situation is reached, it is necessary to recalibrate the instrument back to the standard specifications. Such recalibration is performed by adjusting the instrument at each point in its output range until its output readings are the same as those of a second standard instrument to which the same inputs are applied. This second instrument is one kept solely for calibration purposes and whose specifications are accurately known. Calibration procedures are discussed more fully in Chapter 4.

2.6 **Choice of instruments**

The starting point in choosing the most suitable instrument to use for measurement of a particular quantity in a manufacturing plant or other system is the specification of the instrument characteristics required, especially parameters such as the desired measurement accuracy, resolution and sensitivity. It is also essential to know the environmental conditions that the instrument will be subjected to, as certain conditions will immediately eliminate the possibility of using some types of instrument. Provision of this type of information usually requires the expert knowledge of personnel who are intimately acquainted with the operation of the manufacturing plant or system in question. Then, a skilled instrument engineer, having knowledge of all the instruments that are available for measuring the quantity in question, will be able to evaluate the possible list of instruments in terms of their accuracy, cost and suitability for the environmental conditions and thus to choose the most appropriate instrument. As far as possible, measurement systems and instruments should be chosen which are as insensitive as possible to the operating environment, although this requirement is often difficult to meet because of cost and other performance considerations.

Published literature is of considerable help in the choice of a suitable instrument for a particular measurement situation. Many books are available which give valuable assistance in the necessary evaluation by providing lists and data about all the instruments available for measuring a range of physical quantities (e.g. Part Two of this text). However, new techniques and instruments are being developed all the time, and therefore a good instrumentation engineer must keep abreast of the latest developments by reading the appropriate technical journals regularly.

The instrument characteristics discussed earlier in this chapter are those features which form the technical basis for a comparison between the relative merits of different instruments. Generally, the better the characteristics, the higher the cost. However, in comparing the cost and relative suitability of different instruments for a particular measurement situation, considerations of durability, maintainability and constancy of performance are also very important because the instrument chosen will often have to be capable of operating for long periods without performance degradation and a requirement for costly maintenance. As a consequence, the initial cost of an instrument often has a low weighting in the evaluation exercise.

Cost is very strongly correlated with the performance of an instrument, as measured by its static characteristics. Increasing the accuracy or resolution of an instrument, for example, can only be done at a penalty of increasing its manufacturing cost. Instrument choice therefore proceeds by specifying the minimum characteristics required by a measurement situation and then searching manufacturers' catalogues to find an instrument whose characteristics match those required. To select an instrument with characteristics superior to those required would only mean paying more than necessary for a level of performance greater than that needed.

As well as purchase cost, other important factors in the assessment exercise are instrument durability and the maintenance requirements. Assuming that one had £10 000 to spend, one would not spend £8000 on a new motor car whose projected life was 5 years if a car of equivalent specification with a projected life of 10 years was available for £10 000. Likewise, durability is an important consideration in the choice of instruments. The projected life of an instrument often depends on the conditions in which the instrument will have to operate. Maintenance requirements must also be taken into account, as they also have cost implications.

As a general rule, a good assessment criterion is obtained if the total purchase cost and estimated maintenance costs of an instrument over its life are divided by the period of its expected life. The figure obtained is thus a cost per year. However, this rule is modified where instruments are being installed on a process whose life is expected to be limited, perhaps in the manufacture of a particular model of car. Then, the total costs can only be divided by the period of time an instrument is expected to be used, unless an alternative use for the instrument is envisaged at the end of this period.

To summarize, instrument choice is therefore a compromise between performance characteristics, ruggedness and durability, maintenance requirements and purchase cost. To carry out such an evaluation properly, the instrument engineer must have a wide knowledge of the range of instruments available for measuring particular physical quantities, and he/she must also have a deep understanding of how instrument characteristics are affected by particular measurement situations and operating conditions.

2.7 **Exercises**

2.1 Explain what is meant by:
    (a) active instruments
    (b) passive instruments.
    Give examples of each and discuss the relative merits of these two classes of instruments.

2.2 Discuss the advantages and disadvantages of null and deflection types of measuring instrument. What are null types of instrument mainly used for and why?

2.3 Briefly define and explain all the static characteristics of measuring instruments.

2.4 Explain the difference between accuracy and precision in an instrument.

2.5 A tungsten/5% rhenium–tungsten/26% rhenium thermocouple has an output e.m.f.

as shown in the following table when its hot (measuring) junction is at the temperatures shown. Determine the sensitivity of measurement for the thermocouple in mV/°C.

mV	4.37	8.74	13.11	17.48
°C	250	500	750	1000

2.6 Define sensitivity drift and zero drift. What factors can cause sensitivity drift and zero drift in instrument characteristics?

2.7 (a) An instrument is calibrated in an environment at a temperature of 20 °C and the following output readings y are obtained for various input values x:

y:	13.1	26.2	39.3	52.4	65.5	78.6
x:	5	10	15	20	25	30

Determine the measurement sensitivity, expressed as the ratio y/x.

(b) When the instrument is subsequently used in an environment at a temperature of 50 °C, the input/output characteristic changes to the following:

y:	14.7	29.4	44.1	58.8	73.5	88.2
x:	5	10	15	20	25	30

Determine the new measurement sensitivity. Hence determine the sensitivity drift due to the change in ambient temperature of 30 °C.

2.8 A load cell is calibrated in an environment at a temperature of 21 °C and has the following deflection/load characteristic:

Load (kg):	0	50	100	150	200
Deflection (mm):	0.0	1.0	2.0	3.0	4.0

When used in an environment at 35 °C, its characteristic changes to the following:

Load (kg):	0	50	100	150	200
Deflection (mm):	0.2	1.3	2.4	3.5	4.6

(a) Determine the sensitivity at 21 °C and 35 °C.
(b) Calculate the total zero drift and sensitivity drift at 35 °C.
(c) Hence determine the zero drift and sensitivity drift coefficients (in units of μm/°C and (μm per kg)/°C).

2.9 An unmanned submarine is equipped with temperature and depth measuring instruments and has radio equipment which can transmit the output readings of these instruments back to the surface. The submarine is initially floating on the surface of the sea with the instrument output readings in steady state. The depth measuring instrument is approximately zero order and the temperature transducer first order with a time constant of 50 seconds. The water temperature on the sea surface, T_0, is 20 °C and the temperature T_x at a depth of x metres is given by the relation:

$$T_x = T_0 - 0.01x$$

(a) If the submarine starts diving at time zero, and thereafter goes down at a velocity of 0.5 metres/second, construct a table showing the temperature and depth measurements reported at intervals of 100 seconds over the first 500 seconds of travel. Show also in the table the error in each temperature reading.
(b) What temperature does the submarine report at a depth of 1000 metres?

2.10 Write down the general differential equation describing the dynamic response of a second-order measuring instrument and state the expressions relating the static sensitivity, undamped natural frequency and damping ratio to the parameters in this differential equation. Sketch the instrument response for the cases of heavy damping, critical damping and light damping, and state which of these is the usual target when a second-order instrument is being designed.

References and further reading

BS 5233 (1986) Glossary of terms used in metrology (incorporating BS 2643), British Standards Institution, London.
BS 5532 (1978) Statistical terminology, British Standards Institution, London.
ISO 3534 (1977) Statistics – Vocabulary and symbols, International Organization for Standardization, Geneva.

3 Measurement system errors

3.1 Introduction

This chapter is concerned with identifying the various errors which exist in a measurement system, and suggesting mechanisms for reducing their magnitude and effect. A discussion is also included about the way in which the separate error components are combined together in order to calculate the overall measurement system error level.

It is extremely important in any measurement system to reduce errors in instrument output readings to the minimum possible level and to quantify the maximum error which may exist in any output reading. This requires a detailed analysis of the sources of error which exist in the system as a prerequisite in this process of reducing the total measurement error level. These errors in measurement data can be divided into two groups, known as systematic errors and random errors.

Systematic errors describe errors in the output readings of a measurement system which are consistently on one side of the correct reading, i.e. either all the errors are positive or they are all negative. Two major sources of systematic errors are system disturbance during measurement and the effect of modifying inputs, as discussed in sections 3.2.1 and 3.2.2. Other sources of systematic error include bent meter needles, the use of uncalibrated instruments, poor cabling practices and the generation of thermal e.m.f.s. The last two sources are considered in section 3.2.3. Even when systematic errors due to the above factors have been reduced or eliminated, some errors remain which are inherent in the manufacture of an instrument. These are quantified by the accuracy figure quoted in the published specifications contained in the instrument data sheet.

Random errors are perturbations of the measurement either side of the true value caused by random and unpredictable effects, such that positive errors and negative errors occur in approximately equal numbers for a series of measurements made of the same quantity. Such perturbations are mainly small, but large perturbations occur from time to time, again unpredictably. Random errors often arise when measurements are taken by human observation of an analog meter, especially where this involves interpolation between scale points. Electrical noise can also be a source of random errors. To a large extent, random errors can be overcome by taking the same measurement a number of times and extracting a value by averaging or other statistical techniques, as discussed in section 3.3.1. However, any quantification of the measurement value and statement of error bounds remains a statistical quantity. Because of the nature of random errors and the fact that large perturbations in the measured quantity occur from time to time, the best that we can do is to express measurements in probabilistic terms: we may be able to assign a 95% or even 99% confidence level that the measurement is a certain value within error bounds of, say, ±1%, but we can never attach a 100% probability to measurement values which are subject to random errors.

Finally, a word must be said about the distinction between systematic and random errors. Error sources in the measurement system must be examined carefully to determine what type of error is present, systematic or random, and to apply the appropriate treatment. In the case of manual data measurements, a human observer may make a different observation at each attempt, but it is often reasonable to assume that the errors are random and that the mean of these readings is likely to be close to the correct value. However, this is only true as long as the human observer is not introducing a systematic parallax-induced error as well by persistently reading the position of a needle against the scale of an analog meter from the same side rather than from directly above. In that case, correction would have to be made for this systematic error (bias) in the measurements before statistical techniques were applied to reduce the effect of random errors.

3.2 Systematic errors

Systematic errors in the output of many instruments are due to factors inherent in the manufacture of the instrument arising out of tolerances in the components of the instrument. They can also arise from wear in the instrument's components over a period of time. In other cases, systematic errors are introduced either by the effect of environmental disturbances or through the disturbance of the measured system by the act of measurement. These various sources of systematic error, and the ways in which the magnitude of the errors can be reduced, are discussed below.

3.2.1 System disturbance due to measurement

Disturbance of the measured system by the act of measurement is one source of systematic error. If we were to start with a beaker of hot water and wished to measure its temperature with a mercury-in-glass thermometer, then we would take the thermometer, which would be initially at room temperature, and plunge it into the water. In so doing, we would be introducing a relatively cold mass (the thermometer) into the hot water and a heat transfer would take place between the water and the thermometer. This heat transfer would lower the temperature of the water. Whilst in this case the reduction in temperature would be so small as to be undetectable by the limited measurement resolution of such a thermometer, the effect is finite and clearly establishes the principle that, in nearly all measurement situations, the process of measurement disturbs the system and alters the values of the physical quantities being measured.

Another example is that of measuring car tyre pressures with the type of pressure gauge commonly obtainable from car accessory shops. Measurement is made by pushing one end of the pressure gauge on to the valve of the tyre and reading the displacement of the other end of the gauge against a scale. As the gauge is used, a quantity of air flows from the tyre into the gauge. This air does not subsequently flow back into the tyre after measurement, and so the tyre has been disturbed and the air pressure inside it has been permanently reduced.

Thus, as a general rule, the process of measurement always disturbs the system being measured. The magnitude of the disturbance varies from one measurement system to the next and is affected particularly by the type of instrument used for measurement. Ways of minimizing the disturbance of measured systems are an important consideration in instrument design. A prerequisite for this, however, is an accurate understanding of the mechanisms of system disturbance.

Measurements in electric circuits

In analyzing system disturbance during measurements in electric circuits, Thévenin's theorem (see Appendix 3) is often of great assistance.

Consider for instance the circuit shown in Figure 3.1(a) in which the voltage across resistor R_5 is to be measured by a voltmeter with resistance R_m. Here, R_m acts as a shunt resistance across R_5, decreasing the resistance between points AB and so disturbing the circuit. The voltage E_m measured by the meter is not therefore the value of the voltage E_0 that existed prior to measurement. The extent of the disturbance can be assessed by calculating the open-circuit voltage E_0 and comparing it with E_m.

Thévenin's theorem allows the circuit of Figure 3.1(a), comprising two voltage sources and five resistors, to be replaced by an equivalent circuit containing a single resistance and one voltage source, as shown in Figure 3.1(b). For the purpose of defining the equivalent single resistance of a circuit by Thévenin's theorem, all voltage sources are represented just by their internal resistance, which can be approximated to zero, as shown in Figure 3.1(c). Analysis proceeds by calculating the equivalent resistances of sections of the circuit and building these up until the required equivalent resistance of the whole of the circuit is obtained. So in (c), starting at C and D, the circuit to the left of C and D consists of a series pair of resistances (R_1 and R_2) in parallel with R_3, and the equivalent resistance can be written as:

$$\frac{1}{R_{CD}} = \frac{1}{R_1 + R_2} + \frac{1}{R_3} \quad \text{or} \quad R_{CD} = \frac{(R_1 + R_2)R_3}{R_1 + R_2 + R_3}$$

Moving now to A and B, the circuit to the left consists of a pair of series resistances (R_{CD} and R_4) in parallel with R_5. The equivalent circuit resistance R_{AB} can thus be written as:

$$\frac{1}{R_{AB}} = \frac{1}{R_{CD} + R_4} + \frac{1}{R_5} \quad \text{or} \quad R_{AB} = \frac{(R_4 + R_{CD})R_5}{R_4 + R_{CD} + R_5}$$

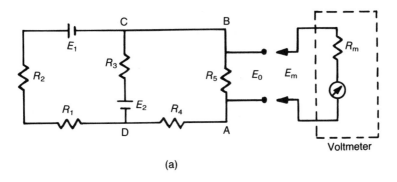

(a)

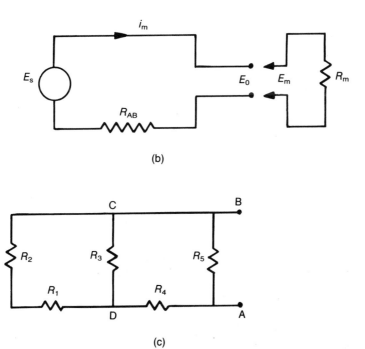

(b)

(c)

Figure 3.1 (a) A circuit in which the voltage across R_5 is to be measured; (b) equivalent circuit by Thévenin's theorem; (c) the circuit used to find the equivalent single resistance R_{AB}

Substituting for R_{CD} using the previously derived expression, we obtain:

$$R_{AB} = \frac{\left[\dfrac{(R_1+R_2)R_3}{R_1+R_2+R_3} + R_4\right]R_5}{\dfrac{(R_1+R_2)R_3}{R_1+R_2+R_3} + R_4 + R_5}$$

(3.1)

Defining I as the current flowing in the circuit when the measuring instrument is connected to it, we can write:

$$I = \frac{E_0}{R_{AB}+R_m}$$

and the voltage measured by the meter is given by:

$$E_m = \frac{R_m E_0}{R_{AB}+R_m}$$

In the absence of the measuring instrument and its resistance R_m, the voltage across AB would be the equivalent circuit voltage source whose value is E_0. The effect of measurement is therefore to reduce the voltage across AB by the ratio given by:

$$\frac{E_m}{E_0} = \frac{R_m}{R_{AB}+R_m}$$

(3.2)

It is thus obvious that as R_m gets larger, the ratio E_m/E_0 gets closer to unity, showing that the design strategy should be to make R_m as high as possible to minimize disturbance of the measured system. (Note that we did not calculate the value of E_0, since this is not required in quantifying the effect of R_m.)

Example 3.1
Suppose that the components of the circuit shown in Figure 3.1(a) have the following values: $R_1 = 400\,\Omega$, $R_2 = 600\,\Omega$, $R_3 = 1000\,\Omega$, $R_4 = 500\,\Omega$, $R_5 = 1000\,\Omega$. The voltage across AB is measured by a voltmeter whose internal resistance is $9500\,\Omega$. What is the measurement error caused by the resistance of the measuring instrument?

Solution
Proceeding by applying Thévenin's theorem to find an equivalent circuit to that of

Figure 3.1(a) of the form shown in Figure 3.1(b), and substituting the given component values into Equation (3.1) for R_{AB}, we obtain:

$$R_{AB} = \frac{[(1000^2/2000) + 500]1000}{(1000^2/2000) + 500 + 1000} = \frac{1000^2}{2000} = 500\ \Omega$$

From Equation 3.2, we have:

$$\frac{E_m}{E_0} = \frac{R_m}{R_{AB} + R_m}$$

The measurement error is given by $(E_0 - E_m)$:

$$E_0 - E_m = E_0\left(1 - \frac{R_m}{R_{AB} + R_m}\right)$$

Substituting in values:

$$E_0 - E_m = E_0\left(1 - \frac{9500}{10\,000}\right) = 0.95E_0$$

Thus the error in the measured value is 5%.

At this point, it is interesting to note what constraints exist when practical attempts are made to achieve a high internal resistance in the design of a moving-coil voltmeter. Such an instrument consists of a coil carrying a pointer mounted in a fixed magnetic field. As current flows through the coil, the interaction between the field generated and the fixed field causes the pointer to turn in proportion to the applied current (see Chapter 7, especially Figure 7.2, for further details).

The simplest way of increasing the input impedance (the resistance) of the meter is either to increase the number of turns in the coil or to construct the same number of turns with a higher-resistance material. Either of these solutions decreases the current flowing in the coil, however, giving less magnetic torque and thus decreasing the measurement sensitivity of the instrument (i.e. for a given applied voltage, we get less deflection of the pointer). This problem can be overcome by changing the spring constant of the restraining springs of the instrument, such that less torque is required to turn the pointer by a given amount. This, however, reduces the ruggedness of the instrument and also demands better pivot design to reduce friction. This highlights a very important but tiresome principle in instrument design: any attempt to improve the performance of an instrument in one aspect generally decreases the performance in some other aspect. This is an inescapable fact of life with passive instruments such as the type of voltmeter mentioned, and is often the reason for the use of alternative active instruments such as digital voltmeters, where the inclusion of auxiliary power greatly improves performance.

Bridge circuits for measuring resistance values are a further example of the need for careful design of the measurement system. The impedance of the instrument measuring the bridge output voltage must be very large in comparison with the component resistances in the bridge circuit. Otherwise, the measuring instrument will load the circuit and draw current from it. This is discussed more fully in section 6.2.

3.2.2 Modifying inputs in measurement systems

The fact that the static and dynamic characteristics of measuring instruments are specified for particular environmental conditions (e.g. of temperature and pressure) has already been discussed at considerable length in Chapter 2. These specified conditions must be reproduced as closely as possible during calibration exercises because, away from the specified calibration conditions, the characteristics of measuring instruments vary to some extent. The magnitude of this variation is quantified by the two constants known as sensitivity drift and zero drift, both of which are generally included in the published specifications for an instrument. Such variations of environmental conditions away from the calibration conditions are described as modifying inputs to the system and are a further source of systematic error. The environmental variation is described as a measurement system input because the effect on the system output is the same as if the value of the real input (the measured quantity) had changed by a certain amount.

Without proper analysis, it is impossible to establish how much of an instrument's output is due to the real input and how much to one or more modifying inputs. This is illustrated by the following example.

Suppose that we have a small closed box weighing 0.1 kg empty which we think contains a rat or a mouse. If we put the box on to bathroom scales and observe a reading of 1.0 kg, this does not immediately tell us what is in the box because the reading may be due to one of three things:

1. A 0.9 kg rat in the box (real input).
2. An empty box with a 0.9 kg bias on the scales due to a temperature change (modifying input).
3. A 0.4 kg mouse in the box together with a 0.5 kg bias (real + modifying inputs).

Thus, the magnitude of any modifying input must be measured before the value of the measured quantity (the real input) can be determined from the output reading of an instrument.

In any general measurement situation, it is very difficult to avoid modifying inputs, because it is either impractical or impossible to control the environmental conditions surrounding the measurement system. System designers are therefore charged with the task of either reducing the susceptibility of measuring instruments to modifying inputs or alternatively quantifying the effect of modifying inputs and correcting for them in the instrument output reading.

The techniques used to deal with modifying inputs and to minimize their effect on the final output measurement follow a number of routes as discussed below.

Careful instrument design

Careful instrument design is the most useful weapon in the battle against modifying inputs, by reducing the sensitivity of an instrument to modifying inputs to as low a level as possible. In the design of strain gauges for instance, the element should be constructed from a material whose resistance has a very low temperature coefficient (i.e. the variation of the resistance with temperature is very small). For many instruments, however, it is not possible to reduce their sensitivity to modifying inputs to a satisfactory level by simple design adjustments, and other measures have to be taken.

Method of opposing inputs

The method of opposing inputs compensates for the effect of a modifying input in a measurement system by introducing an equal and opposite modifying input which cancels it out. One example of how this technique is applied is in the type of millivoltmeter shown in Figure 3.2. This consists of a coil suspended in a fixed magnetic field produced by a permanent magnet. When an unknown voltage is applied to the coil, the magnetic field due to the current interacts with the fixed field and causes the coil (and a pointer attached to the coil) to turn. If the coil resistance is

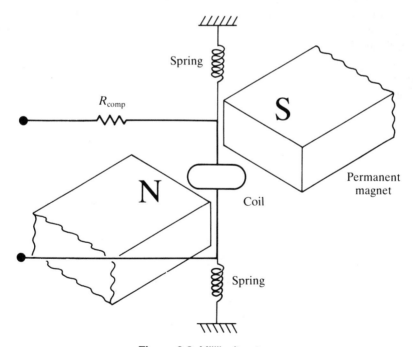

Figure 3.2 Milllivoltmeter

sensitive to temperature, then any modifying input to the system in the form of a temperature change will alter the value of the coil current for a given applied voltage and so alter the pointer output reading. Compensation for this is made by introducing a compensating resistance R_{comp} into the circuit, where R_{comp} has a temperature coefficient which is equal in magnitude but opposite in sign to that of the coil.

High-gain feedback

The benefit of adding high-gain feedback to many measurement systems is illustrated by considering the case of the voltage measuring instrument whose block diagram is shown in Figure 3.3. In this system, the unknown voltage E_i is applied to a motor of torque constant K_m, and the torque induced turns a pointer against the restraining action of a spring with spring constant K_s. The effect of modifying inputs on the motor and spring constants are represented by variables D_m and D_s.

In the absence of modifying inputs, the displacement of the pointer X_0 is given by:

$$X_0 = K_m K_s E_i$$

However, in the presence of modifying inputs, both K_m and K_s change and the relationship between X_0 and E_i can be affected greatly. It therefore becomes difficult or impossible to calculate E_i from the measured value of X_0.

Consider now what happens if the system is converted into a high-gain, closed-loop one, as shown in Figure 3.4, by adding an amplifier of gain constant K_a and a feedback device with gain constant K_f. Assume also that the effect of modifying inputs on the values of K_a and K_f are represented by D_a and D_f. The feedback device feeds back a voltage E_0 proportional to the pointer displacement X_0. This is compared with the unknown voltage E_i by a comparator and the error is amplified.

Writing down the equations of the system, we have:

$$E_0 = K_f X_0$$

$$X_0 = (E_i - E_0)K_a K_m K_s = (E_i - K_f X_0)K_a K_m K_s$$

Thus:

$$E_i K_a K_m K_s = (1 + K_f K_a K_m K_s)X_0$$

i.e.

$$X_0 = \frac{K_a K_m K_s}{1 + K_f K_a K_m K_s} E_i \qquad (3.3)$$

Because K_a is very large (it is a high-gain amplifier), $K_f K_a K_m K_s \gg 1$, and Equation (3.3) reduces to:

$$X_0 = E_i/K_f$$

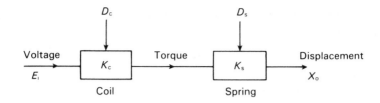

Figure 3.3 Block diagram for voltage measuring instrument

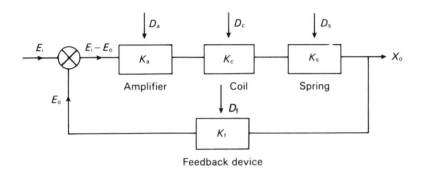

Figure 3.4 Block diagram of voltage measuring instrument with high-gain feedback

This is a highly important result because we have reduced the relationship between X_0 and E_i to one which involves K_f only. The sensitivity of the gain constants K_a, K_m and K_s to the modifying inputs D_a, D_m and D_s has thereby been rendered irrelevant and we only have to be concerned with one modifying input D_f. Conveniently, it is usually an easy matter to design a feedback device which is insensitive to modifying inputs: this is much easier than trying to make a motor or spring insensitive. Thus high-gain feedback techniques are often a very effective way of reducing a measurement system's sensitivity to modifying inputs. One potential problem which must be mentioned, however, is that there is a possibility that high-gain feedback will cause instability in the system. Any application of this method must therefore include careful stability analysis of the system.

Signal filtering

One frequent problem in measurement systems is corruption of the output reading by periodic noise, often at a frequency of 50 Hz caused by pick-up through the close proximity of the measurement system to apparatus or current-carrying cables operating on a mains supply. Periodic noise corruption at higher frequencies is also often introduced by mechanical oscillation or vibration within some component of a measurement system. The amplitude of all such noise components can be substantially attenuated by the inclusion of filtering of an appropriate form in the system, as discussed at great length in Chapter 5. Band-stop filters can be especially useful where corruption is of one particular known frequency, or, more generally, low-pass filters are employed to attenuate all noise in the range of 50 Hz and higher frequencies.

Measurement systems with a low-level output, such as a bridge circuit measuring a strain gauge resistance, are particularly prone to noise, and Figure 3.5(a) shows the typical corruption of a bridge output by 50 Hz pick-up. The beneficial effect of putting a simple passive RC low-pass filter across the output is shown in Figure 3.5(b).

3.2.3 Other sources of systematic error

Wear in instrument components

Systematic errors can frequently develop over a period of time because of wear in instrument components. Recalibration often provides a full solution to this problem.

Connecting leads

In connecting together the components of a measurement system, a common source of error is the failure to take proper account of the resistance of connecting leads (or pipes in the case of pneumatically or hydraulically actuated measurement systems). In typical applications of a resistance thermometer for instance, it is common to find the thermometer separated from other parts of the measurement system by perhaps 30 metres. The resistance of such a length of 7/0.0076 copper wire is 2.5 Ω and there is a further complication that such wire has a temperature coefficient of 1 mΩ/°C.

Careful consideration therefore needs to be given to the choice of connecting leads. Not only should they be of adequate cross-section so that their resistance is minimized, but also they should be adequately screened if they are thought likely to be subject to electrical or magnetic fields which could otherwise cause induced noise. Where screening is thought essential, then the routing of cables also needs careful planning. In one application in the author's experience involving the instrumentation of an electric-arc steelmaking furnace, screened signal-carrying cables between transducers on the arc furnace and a control room at the side of the furnace were initially corrupted by high-amplitude 50 Hz noise. However, by changing the route of the cables between the transducers and the control room, the magnitude of this induced noise was reduced by a factor of about ten.

Thermal e.m.f.s

Whenever metals of two different types are connected together, a thermal e.m.f. is generated according to the temperature of the joint. This is known as the **thermoelectric effect** and is the physical principle on which temperature-measuring thermocouples operate (see Chapter 12). Such thermal e.m.f.s are only a few

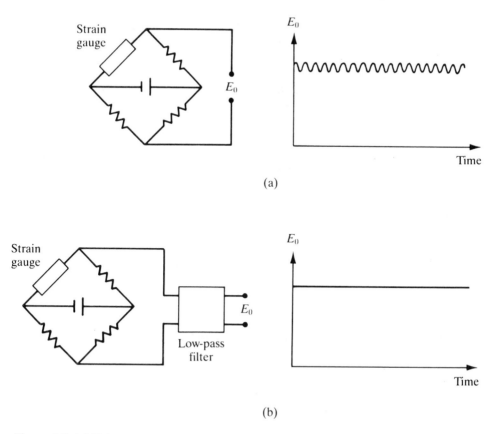

Figure 3.5 (a) Noise-corrupted output of bridge circuit measuring resistance; (b) effect of adding low-pass filter

millivolts in magnitude and so the effect is only significant when typical voltage output signals of a measurement system are of a similar low magnitude.

Such a situation is where one e.m.f.-measuring instrument is used to monitor the output of several thermocouples measuring the temperatures at different points in a process control system. This requires a means of automatically switching the output of each thermocouple to the measuring instrument in turn. Nickel–iron reed relays with copper connecting leads are commonly used to provide this switching function. This introduces a thermocouple effect of magnitude 40 μV/°C between the reed relay and the copper connecting leads. There is no problem if both ends of the reed relay are at the same temperature because then the thermal e.m.f.s will be equal and opposite and so cancel out. However, there are several recorded instances where, because of the lack of awareness of the problem, poor design had resulted in the two ends of a reed relay being at different temperatures and causing a net thermal e.m.f. The serious error that this introduces is clear. For a temperature difference between the two ends of only 2 °C, the thermal e.m.f. is 80 μV, which is very large compared with a typical thermocouple output level of 400 μV.

Another example of the difficulties that thermal e.m.f.s can create becomes apparent in considering the following problem which was reported in a current measuring system. This system had been designed such that the current flowing in a particular part of a circuit was calculated by applying it to an accurately calibrated wire-wound resistance of value 100 Ω and measuring the voltage drop across the resistance. In calibrating the system, a known current of 20 μA was applied to the resistance and a voltage of 2.20 mV was measured by an accurate high-impedance instrument. Simple application of Ohm's law reveals that such a voltage reading indicates a current value of 22 μA. What then was the explanation for this discrepancy? The answer once again is a thermal e.m.f. Because the designer was not aware of thermal e.m.f.s, the circuit had been constructed such that one side of the standard resistance was close to a power transistor, creating a difference in temperature of 2 °C between the two ends of the resistor. The thermal e.m.f. associated with this was sufficient to account for the 10% measurement error found.

3.2.4 Reduction of systematic errors using intelligent instruments

Intelligent instruments contain all the usual elements of a measurement system but are distinguished from normal, non-intelligent instruments by the inclusion of a microprocessor within the instrument. This enables them to apply preprogrammed signal processing and data manipulation algorithms to measurements (they are discussed in detail in Chapter 10). They can bring about a gross reduction in the level of output error in measurement systems which are subject to the types of error discussed earlier in this chapter. The ability of intelligent instruments to achieve this does, however, require that the following preconditions be satisfied:

1. The physical mechanism by which a measurement transducer is affected by changes in ambient conditions must be fully understood and all physical quantities which affect the transducer output must be identified.
2. The effect of each ambient variable on the output characteristic of the measurement transducer must be quantified.
3. Suitable secondary transducers for monitoring the value of all relevant ambient variables must be available for input to the intelligent instrument.

Condition 1 above means that the thermal expansion/contraction of all elements within a transducer must be considered in order to evaluate how it will respond to ambient temperature changes. Similarly, the transducer response, if any, to changes in ambient pressure, humidity, gravitational force or power supply level (active instruments) must be examined.

Quantification of the effect of each ambient variable on the characteristics of the measurement transducer is then necessary, as stated in condition (2). Analytic quantification of changes in ambient conditions from a purely theoretical consideration of the construction of a transducer is usually extremely complex and so is normally avoided. Instead, the effect is quantified empirically in laboratory tests where the output characteristic of the transducer is observed as the ambient environmental conditions are changed in a controlled manner.

Once the ambient variables affecting a measurement transducer have been identified and their effect quantified, an intelligent instrument can be designed which includes secondary transducers to monitor the value of the ambient variables. Suitable transducers which will operate satisfactorily within the environmental conditions prevailing for the measurement situation must of course exist, as stated in condition 3.

In the case of electrical circuits which are disturbed by the loading effect of the measuring instrument, an intelligent instrument can readily correct for measurement errors by applying equations such as (3.2) with the resistance of the measuring instrument inserted. A similar correction for the loading effect of the measuring instrument on the output of a bridge circuit is described in section 6.2.

Intelligent instruments are particularly useful for dealing with measurement errors due to modifying inputs. Secondary transducers are provided within the instrument to monitor the magnitude of the environmental conditions such as temperature and pressure which can affect the characteristics of the primary measurement transducer. The computer within the instrument then corrects the measurement obtained from the primary transducer according to the values read by the secondary transducers. This presupposes that all the modifying inputs in a measurement situation have been correctly identified and quantified, and that suitable secondary transducers exist to monitor these modifying inputs.

Suitable care must always be taken when introducing a microcomputer into a measurement system to avoid creating new sources of measurement noise. This is particularly so where one microcomputer is used to process the output of several transducers and is connected to them by signal wires. In such circumstances, the connections and connecting wires can create noise through electrochemical potentials, thermoelectric potentials, offset voltages introduced by common mode impedances, and a.c. noise at power, audio and radio frequencies. Recognition of all these possible noise sources allows them to be eliminated in most cases by employing good measurement system construction practices. All remaining noise sources are usually eliminated by the provision of a set of four earthing circuits within the interface fulfilling the following functions:

1. *Power earth:* provides a path for fault currents due to power faults.
2. *Logic earth:* provides a common line for all logic circuit potentials.
3. *Analog earth (ground):* provides a common reference for all analog signals.
4. *Safety earth:* is connected to all metal parts of equipment to protect personnel should power lines come into contact with metal enclosures.

3.3 **Random errors**

Random errors in measurements are caused by random, unpredictable variations in the measurement system and they can be largely eliminated by calculating the mean or median of the measurements. The degree of confidence in the calculated mean/median values can be quantified by calculating the standard deviation or variance of the data, these being parameters which describe how the measurements are distributed about the mean value/median. All of these terms are explained more fully in section 3.3.1.

Because of the unpredictability of random errors, any error bounds placed on measurements can only be quantified in probabilistic terms. Thus, if we say that the possible error in a measurement subject to random errors is ±2% of the measured value, we are only implying that this is probably true, i.e. there is, say, a 95% probability that the error level does not exceed ±2%.

The distribution of measurement data about the mean value can be displayed graphically by frequency distribution curves, as discussed in section 3.3.2. Calculation of the area under the frequency distribution curve gives the probability that the error will lie between any two chosen error levels.

3.3.1 **Statistical analysis of data**

Mean and median values

In any measurement situation subject to random errors, the normal technique is to take the same reading a number of times, ideally using different observers, and extract the most likely value from the measurement data set. For a set of n measurements $x_1, x_2, \ldots, x_n$, the most likely true value is the **mean** given by:

$$x_{mean} = \frac{x_1 + x_2 + \ldots + x_n}{n} \tag{3.4}$$

This is valid for all data sets where the measurement errors are distributed equally about the line of zero error, i.e. where the positive errors are balanced in quantity and magnitude by the negative errors.

When the number of values in the data set is large, however, calculation of the mean value is tedious, and it is more convenient to use the median value, which is a close approximation to the mean value. The **median** is given by the middle value when the measurements in the data set are written down in ascending order of magnitude.

For a set of n measurements $x_1, x_2, \ldots, x_n$ written down in ascending order of magnitude, the median value is given by:

$$x_{median} = x_{(n+1)/2}$$

Thus, for a set of nine measurements $x_1, x_2, \ldots, x_9$ the median value is x_5.

For an even number of data values, the median value is mid-way between the centre two values, i.e. for ten measurements $x_1, \ldots, x_{10}$, the median value is given by $(x_5 + x_6)/2$.

Suppose that, in a particular measurement situation, a mass is measured by a beam balance, and the following set of readings is obtained at a particular time by different observers:

$$81.6\ 81.1\ 81.4\ 80.9\ 81.1\ 80.5\ 81.3\ 80.8\ 81.2\ 81.8$$

$$81.1\ 81.5\ 81.0\ 81.3\ 81.1\ 80.8\ 81.3\ 81.6\ 81.1 \tag{3.5}$$

The mean value of this set of data is 81.18, calculated according to Equation (3.4). The median value is 81.1, which is the middle value if the data values are written down in ascending order, starting at 80.5 and ending at 81.8.

Standard deviation and variance

The probability that the mean or median value of a data set represents the true measurement value depends on how widely scattered the data values are. If the values in the last set of mass measurements (3.5) had ranged from 79 up to 83, our confidence in the mean value would be much less. The spread of values about the mean is analyzed by first calculating the deviation of each value from the mean. For any general value x_i, the deviation d_i is given by:

$$d_i = x_i - x_{\text{mean}}$$

The extent to which n measurement values are spread about the mean can now be expressed by the standard deviation σ, where σ is given by:

$$\sigma = \left(\frac{d_1^2 + d_2^2 + \ldots + d_n^2}{n - 1} \right)^{1/2} \tag{3.6}$$

This spread can alternatively be expressed by the variance V, which is the square of the standard deviation, i.e. $V = \sigma^2$. Mathematically minded readers may have observed that the expression for σ given above differs from the mathematical definition of the standard deviation, which has n instead of $n - 1$ in the denominator. This difference arises because the mathematical definition of σ is for an infinite data set, whereas in the case of measurements we are always concerned with only finite data sets. For a finite set of measurements $(d_i)_{i=1, n}$, the mean x_m will differ from the true mean μ of the infinite data set that the finite set d_i is part of. If somehow we knew the true mean μ of a set of measurements, then the deviations d_i could be calculated as the deviation of each data value from the true mean, and it would be correct to calculate σ using n instead of $n - 1$ in the expression for σ in Equation (3.6). However, in normal situations, using $n - 1$ in the denominator of Equation (3.6) produces a value of the standard deviation which is statistically closer to the correct value.

Example 3.2

The following measurements were taken of a current flowing in a circuit (the circuit was in steady state and therefore, although the measurements varied due to random errors, the current flowing was actually constant): 21.5, 22.1, 21.3, 21.7, 22.0, 22.2, 21.8, 21.4, 21.9 and 22.1 mA. Calculate the mean value, the deviations from the mean and the standard deviation.

Solution

Mean value $= \Sigma$(data values)/10 $=$ 218/10 $=$ 21.8 mA

Now construct a table of measurements and deviations:

Measurement	21.5	22.1	21.3	21.7	22.0	22.2	21.8	21.4	21.9	22.1
Deviation from mean	−0.3	+0.3	−0.5	−0.1	+0.2	+0.4	0.0	−0.4	+0.1	+0.3
(Deviations)^2	0.09	0.09	0.25	0.01	0.04	0.16	0.0	0.16	0.01	0.09

Σ(deviations)^2 $=$ 0.90

$n =$ number of measurements $=$ 10

Σ(deviations)$^2/(n-1) = \Sigma$(deviations)^2/9 $=$ 0.10

$[\Sigma$(deviations)^2/9]$^{1/2}$ $=$ 0.316

Thus, standard deviation $=$ 0.32 (the nature of the measurements does not justify expressing the standard deviation to any accuracy greater than two decimal places).

3.3.2 Frequency distributions

A further and very powerful way of analyzing the pattern in which measurements deviate from the mean value is to use graphical techniques. The simplest way of doing this is by means of a **histogram,** where bands of equal width across the range of measurement values are defined and the number of measurements within each band is counted. Figure 3.6 shows a histogram drawn from the set of mass measurement data in (3.5) by choosing bands 0.3 wide. There are for instance nine measurements in the range between 81.05 and 81.35, and so the height of the histogram at this point is nine units. (*Note:* The scaling of the bands was deliberately chosen so that no measurements fell on the boundary between different bands and caused ambiguity about which band to put them in.) Such a histogram has the characteristic shape shown by truly random data, with symmetry about the mean value of the measurements.

As the number of measurements increases, smaller bands can be defined for the histogram, which retains its basic shape but then consists of a larger number of smaller steps on each side of the peak. In the limit, as the number of measurements approaches infinity, the histogram becomes a smooth curve known as a frequency

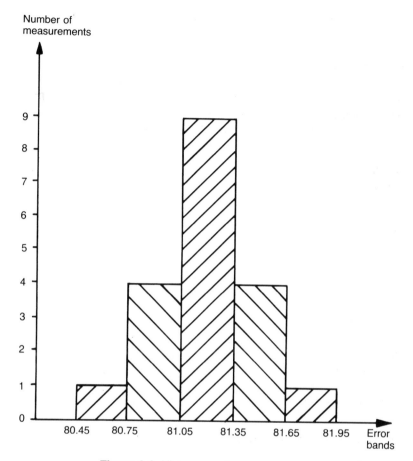

Figure 3.6 Histogram of measurements

distribution curve as shown in Figure 3.7. The ordinate of this curve is the frequency of occurrence of each measurement value, $F(X)$, and the abscissa is the magnitude, X.

The symmetry of Figures 3.6 and 3.7 about the mean data value is very useful for showing graphically that the measurement data has only random errors, but these figures cannot easily be used to quantify the magnitude and distribution of measurement errors. However, very similar graphical techniques can be used which do achieve this.

If the mean of the measurement data values is calculated first, then the error in each measurement in terms of its deviation from this mean value can be calculated. Error bands of equal width can then be defined and a histogram of errors drawn, as shown in Figure 3.8, according to the number of error values falling within each band. Provided that errors are only random, this histogram has symmetry about the line of zero error.

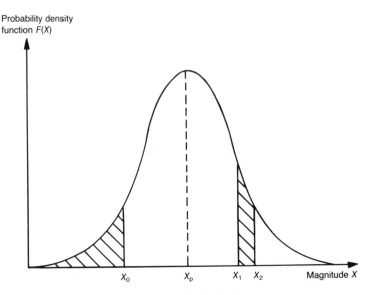

Figure 3.7 Frequency distribution curve

As the number of measurements increases, smaller error bands can be defined for the error histogram and in the limit, as the number of measurements approaches infinity, the histogram becomes a smooth curve as before. In this case, the curve is known as a **frequency distribution curve** as shown in Figure 3.9. The ordinate of this curve is the frequency of occurrence of each error level, $F(E)$, and the abscissa is the error magnitude, E.

If the height of the frequency distribution of errors curve is normalized such that the area under it is unity, then the curve is known as a **probability curve**, and the height $F(E)$ at any particular error magnitude E is known as the **probability density function** (p.d.f.). The condition that the area under the curve is unity can be expressed mathematically as:

$$\int_{-\infty}^{\infty} F(E)\, \mathrm{d}E = 1$$

The probability that the error in any one particular measurement lies between two levels E_1 and E_2 can be calculated by measuring the area under the curve contained between two vertical lines drawn through E_1 and E_2, as shown by the hatched area on the right of Figure 3.9. This can be expressed mathematically as:

$$P(E_1 \leqslant E \leqslant E_2) = \int_{E_1}^{E_2} F(E)\, \mathrm{d}E \tag{3.7}$$

This expression is often known as the **error function**.

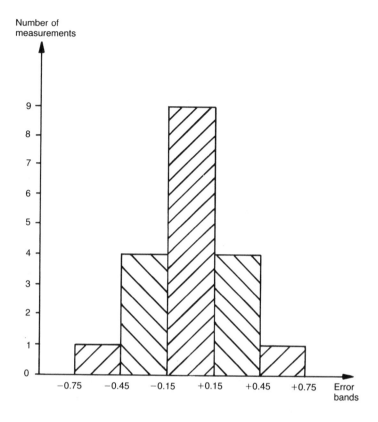

Figure 3.8 Histogram of errors

Of particular importance for assessing the maximum error likely in any one measurement is the **cumulative distribution function** (c.d.f.). This is defined as the probability of observing a value less than or equal to E_0, and is expressed mathematically as:

$$P(E \leqslant E_0) = \int_{-\infty}^{E_0} F(E)\, dE \tag{3.8}$$

Thus the c.d.f. is the area under the curve to the left of a vertical line drawn through E_0, as shown by the hatched area on the left of Figure 3.9.

Three special types of frequency distribution known as the Gaussian, binomial and Poisson distributions exist, and these are very important because most data sets approach closely to one or other of them. The distribution of relevance to data sets containing random measurement errors is the Gaussian one.

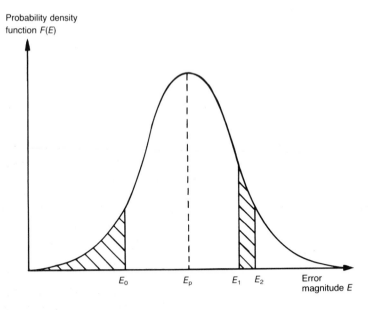

Probability density
function F(E)

E_0 E_p E_1 E_2 Error
magnitude E

Figure 3.9 Frequency distribution of errors

The error magnitude E_p corresponding to the peak of the frequency distribution curve (Figure 3.9) is the value of error which has the greatest probability. If the errors are entirely random in nature, then the value of E_p will equal zero. Any non-zero value of E_p indicates systematic errors in the data, in the form of a bias which is often removable by recalibration.

Gaussian distribution

A Gaussian curve is defined as a normalized frequency distribution where the frequency and magnitude of quantities are related by the expression:

$$F(x) = \frac{1}{\sigma(2\pi)^{1/2}} \exp[-(x - m)^2/2\sigma^2] \qquad (3.9)$$

where m is the mean value of the measurement set x and the other quantities are as defined before.

The Gaussian curve is only applicable to data which has only random errors, i.e. where no systematic errors exist. Equation (3.9) is particularly useful for analyzing a Gaussian set of measurements and predicting how many measurements lie within some particular defined range.

If the measurement errors E are calculated for all measurements such that

$E = x - m$, then the curve of error frequency $F(E)$ plotted against error magnitude E is a Gaussian curve known as the error frequency distribution curve. The mathematical relationship between $F(E)$ and E can then be derived by modifying Equation (3.9) to give:

$$F(E) = \frac{1}{\sigma(2\pi)^{1/2}} \exp(-E^2/2\sigma^2) \qquad (3.10)$$

Most measurement data sets such as the mass values above in (3.5) fit to a Gaussian distribution curve because, if errors are truly random, small deviations from the mean value occur much more often than large deviations, i.e. the number of small errors is much larger than the number of large ones. Alternative names for the Gaussian distribution are the **normal distribution** or **bell-shaped distribution.**

The Gaussian distribution curve is symmetrical about a line through the mean of the measurement values, which means that positive errors away from the mean value occur in equal quantities to negative errors in any data set containing measurements subject to random error. If the standard deviation is used as a unit of error, the curve can be used to determine what probability there is that the error in any particular measurement in a data set is greater than a certain value. By substituting the expression for $F(E)$ in (3.9) into the probability equation (3.7), the probability that the error lies in a band between error levels E_1 and E_2 can be expressed as:

$$P(E_1 \leqslant E \leqslant E_2) = \int_{E_1}^{E_2} \frac{1}{\sigma(2\pi)^{1/2}} \exp(-E^2/2\sigma^2)\, dE \qquad (3.11)$$

Equation (3.11) can be simplified by making the substitution

$$z = E/\sigma \qquad (3.12)$$

Then:

$$P(E_1 \leqslant E \leqslant E_2) = \int_{z_1}^{z_2} \frac{1}{\sigma(2\pi)^{1/2}} \exp(-z^2/2)\, dz \qquad (3.13)$$

Even after carrying out this simplification, Equation (3.13) still cannot be evaluated by the use of standard integrals. Instead, numerical integration has to be used. To simplify the burden involved in this, standard error function tables have been drawn up which evaluate the integral for various values of z.

Error function tables

Error function tables give values of $F(z)$ for various values of z and are included in this text in Appendix 5. $F(z)$ represents the proportion of data values which are less than or equal to z and is equal to the area under the normalized probability curve to the left of z. Study of the tables will show that $F(z) = 0.5$ for $z = 0$. This shows that, as expected, the number of data values less than or equal to zero is 50% of the total. This must be so if the data only has random errors.

Use of error function tables

It will be observed that the table in Appendix 5, in common with most published error function tables, only gives $F(z)$ for positive values of z. For negative values of z, we can make use of the following relationship because the frequency distribution curve is normalized:

$$F(-z) = 1 - F(z) \qquad (3.14)$$

($F(-z)$ is the area under the curve to the left of $-z$, i.e. it represents the proportion of data values less than or equal to $-z$).

Example 3.3
How many measurements in a data set subject to random errors lie outside the boundaries of $+\sigma$ and $-\sigma$, i.e. how many measurements have an error less than $|\sigma|$?

Solution
The required number is represented by the sum of the two shaded areas in Figure 3.10. This can be expressed mathematically as:

$$P(E < -\sigma \text{ or } E > +\sigma) = P(E < -\sigma) + P(E > +\sigma)$$

For $E = -\sigma$, $z = -1.0$ (from Equation (3.12)) and using error tables:

$$P(E < -\sigma) = F(-1) = 1 - F(1) = 1 - 0.8413 = 0.1587$$

Similarly, for $E = +\sigma$, $z = +1.0$, error tables give:

$$P(E > +\sigma) = 1 - P(E < +\sigma) = 1 - F(1) = 1 - 0.8413 = 0.1587$$

(This last step is valid because the frequency distribution curve is normalized such that the total area under it is unity.) Thus:

$$P(E < -\sigma) + P(E > +\sigma) = 0.1587 + 0.1587 = 0.3174 \approx 32\%$$

i.e. 32% of the measurements lie outside the $\pm\sigma$ boundaries; then 68% of the measurements lie inside.

The above analysis shows that, for Gaussian-distributed data values, 68% of the

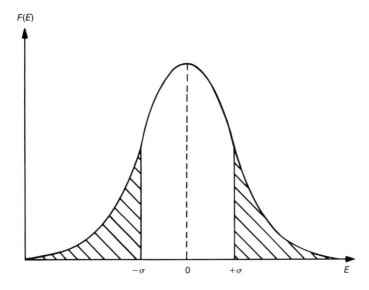

Figure 3.10 $\pm\sigma$ boundaries on measurement errors

measurements have errors which lie within the bounds of $\pm\sigma$. Similar analysis shows that the boundaries of $\pm2\sigma$ contain 95.4% of data points, and extending the boundaries to $\pm3\sigma$ encompasses 99.7% of data points. The probability of any data point lying outside particular error boundaries can therefore be expressed as in Table 3.1.

Table 3.1

Error boundaries	Data points within boundary (%)	Probability of any particular data point being outside boundary
$\pm\sigma$	68.0	34.0%
$\pm2\sigma$	95.4	4.6%
$\pm3\sigma$	99.7	0.3%

Distribution of manufacturing tolerances

The Gaussian distribution curve can be extended to analyze tolerances in manufactured components rather than errors in process measurements. In this form, it describes the frequency distribution of measurements as shown in Figure 3.11.

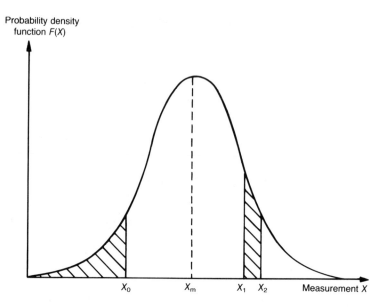

Figure 3.11 Frequency distribution of manufacturing tolerances

Here, $F(X)$ is the probability that the measurement has some particular value X. The most likely value of X, X_m, corresponds to the peak of the curve. If the measurements are Gaussian, the curve will be symmetrical about the line $X = X_m$ and X_m will represent the mean value of the measurements.

Equations similar to (3.7) and (3.8) can be written:

$$P(X_1 \leqslant X \leqslant X_2) = \int_{X_1}^{X_2} F(X)\, dX \tag{3.15}$$

$$P(X \leqslant X_0) = \int_{-\infty}^{X_0} F(X)\, dX \tag{3.16}$$

Also, by modifying Equation (3.13), we obtain:

$$P(X_1 \leqslant X \leqslant X_2) = \int_{X_1}^{X_2} \frac{1}{\sigma(2\pi)^{1/2}} \cdot \exp[-(X-\mu)^2/2\sigma^2]\, dX \tag{3.17}$$

where μ is the mean of the measurements ($=X_m$).

Having found the probability of any measurement chosen at random lying within the range of X_1 to X_2, the number of measurements N lying within the range of X_1 to X_2 can be calculated as follows:

$$N = \int_{X_1}^{X_2} \frac{n}{\sigma(2\pi)^{1/2}} \exp[-(X-\mu)^2/2\sigma^2]\, dX \qquad (3.18)$$

where n is the total number of measurements.

If the substitution $z = (X-\mu)/\sigma$ is made, Equation (3.18) simplifies to:

$$N = \int_{z_1}^{z_2} \frac{n}{\sigma(2\pi)^{1/2}} \exp(-z^2/2)\, dz \qquad (3.19)$$

This is now in a form which can be evaluated using standard error function tables (see Appendix 5).

Example 3.4

An integrated circuit chip contains 10^5 transistors. The transistors have a mean current gain of 20 and a standard deviation of 2. Calculate the following:
(a) The number of transistors with a current gain between 19.8 and 20.2.
(b) The number of transistors with a current gain greater than 17.

Solution
(a) The proportion of transistors with a gain between 19.8 and 20.2 is:

$$P(X<20.2) - P(X<19.8) = P(z<0.2) - P(z<-0.2)$$

where $z = (X-\mu/\sigma)$. For $X = 20.2$, $z = 0.1$, and for $X = 19.8$, $z = -0.1$. From tables, $P(z<0.1) = 0.5398$

$$P(z<-0.1) = 1 - P(z<0.1) = 1 - 0.5398 = 0.4602$$

Hence

$$P(z<0.1) - P(z<-0.1) = 0.5398 - 0.4602 = 0.0796$$

Thus $0.0796 \times 10^5 = 7960$ transistors have a current gain in the range from 19.8 to 20.2.
(b) The number of transistors with a gain greater than 17 is given by:

$$P(X>17) = 1 - P(X<17) = 1 - P(z<-1.5) = P(z<+1.5) = 0.9332$$

Thus, 93.32%, i.e. 93320 transistors, have a gain greater than 17.

Standard error of the mean

The foregoing analysis is only strictly true for measurement sets containing infinite populations. It is of course not possible to obtain an infinite number of data values, and some error must therefore be expected in the calculated mean value of the practical, finite data set available. If several subsets are taken from an infinite data population, then, by the central limit theorem, the means of the subsets will form a Gaussian distribution about the mean of the infinite data set. The error in the mean of a finite data set is usually expressed as the **standard error of the mean**, α, which is calculated as:

$$\alpha = \sigma/n^{1/2}$$

This tends towards zero as the number of measurements in the data set is expanded towards infinity. The value obtained from a set of n measurements $x_1, x_2, \ldots, x_n$ is then expressed as:

$$x = x_{\text{mean}} \pm \alpha$$

For the data set of mass measurements in (3.5), $n = 19$, $\sigma = 0.318$ and $\alpha = 0.073$. The mass can therefore be expressed as 81.18 ± 0.07 (68% confidence limit). However, it is more usual to express measurements with 95% confidence limits ($\pm 2\sigma$ boundaries). In this case, $2\sigma = 0.636$, $2\alpha = 0.146$ and the mass can be expressed as 81.18 ± 0.15 (95% confidence limits).

3.3.3 Application of intelligent instruments to reduce random errors

If a measurement system is known to be subject to random errors, intelligent instruments can be programmed to take the same measurement a number of times within a short space of time and perform simple averaging or other statistical techniques on the readings before displaying an output measurement. This is valid for all forms of random error, whether due to human observation deficiencies, electrical noise or any other random fluctuations.

3.4 Total measurement system errors

A measurement system often consists of several separate components, each of which is subject to systematic and/or random errors. Mechanisms have now been presented for quantifying the errors arising from each of these sources and therefore the total error at the output of each measurement system component can be calculated. What remains to be investigated is how the errors associated with each measurement system component combine together, so that a total error calculation can be made for the complete measurement system.

All four mathematical operations of addition, subtraction, multiplication and division may be performed on measurements derived from different instruments/transducers in a measurement system. Appropriate techniques for the various situations which arise are covered below.

3.4.1 **Error in a product**

If the outputs y and z of two measurement system components are multiplied together, the product can be written as:

$$P = yz$$

If the possible error in y is $\pm ay$ and in z is $\pm bz$, then the maximum and minimum values possible in P can be written as:

$$P_{max} = (y + ay)(z + bz)$$
$$= yz + ayz + byz + aybz$$

$$P_{min} = (y - ay)(z - bz)$$
$$= yz - ayz - byz + aybz$$

For typical measurement system components with output errors of up to 1% or 2% in magnitude, both a and b are very much less than one in magnitude and thus terms in $aybz$ are negligible compared with other terms. Therefore we have:

$$P_{max} = yz(1 + a + b) \quad P_{min} = yz(1 - a - b)$$

Thus the error in the product P is $\pm(a + b)$.

Example 3.5
If the power in a circuit is calculated from measurements of voltage and current in which the calculated maximum errors are respectively $\pm 1\%$ and $\pm 2\%$, then the possible error in the calculated power value is $\pm 3\%$.

3.4.2 **Error in a quotient**

If the output measurement y of one system component with possible error $\pm ay$ is divided by the output measurement z of another system component with possible error $\pm bz$, then the maximum and minimum possible values for the quotient can be written as:

$$Q_{max} = \frac{y + ay}{z - bz}$$

$$Q_{min} = \frac{y - ay}{z + bz}$$

$$= \frac{(y + ay)(z + bz)}{(z - bz)(z + bz)}$$

$$= \frac{(y - ay)(z - bz)}{(z + bz)(z - bz)}$$

$$= \frac{yz + ayz + byz + abyz}{z^2 - b^2 z^2}$$

$$= \frac{yz - ayz - byz + abyz}{z^2 - b^2 z^2}$$

For $a \ll 1$ and $b \ll 1$, terms in ab and b^2 are negligible compared with the other terms. Hence:

$$Q_{max} = \frac{yz(1 + a + b)}{z^2} \qquad Q_{min} = \frac{yz(1 - a - b)}{z^2}$$

i.e.

$$Q = \frac{y}{z} \pm \frac{y}{z}(a + b)$$

Thus the error in the quotient is $\pm(a + b)$.

Example 3.6
If the resistance in a circuit is calculated from measurements of voltage and current where the respective errors are $\pm 1\%$ and $\pm 0.3\%$, the likely error in the resistance value is $\pm 1.3\%$.

3.4.3 Error in a sum

If the two outputs y and z of separate measurement system components are to be added together, we can write the sum as:

$$S = y + z$$

If the maximum errors in y and z are $\pm ay$ and $\pm bz$ respectively, we can express the maximum and minimum possible values of S as:

$$S_{max} = y + ay + z + bz \qquad S_{min} = y - ay + z - bz$$

or:

$$S = y + z \pm (ay + bz)$$

This relationship for S is not convenient because in this form the error term cannot be expressed as a fraction or percentage of the calculated value for S. Fortunately, statistical analysis can be applied which expresses S in an alternative form such that the most probable maximum error in S is represented by a quantity e, where e is given by:

$$e = [(ay)^2 + (bz)^2]^{1/2} \tag{3.20}$$

Thus:

$$S = (y + z) \pm e$$

This can be expressed in the alternative form:

$$S = (y + z)(1 \pm f) \tag{3.21}$$

where $f = e/(y + z)$.

Example 3.7

A circuit requirement for a resistance of 550 Ω is satisfied by connecting together two resistors of nominal values 220 Ω and 330 Ω in series. If each resistor has a tolerance of $\pm 2\%$, the error in the sum calculated according to Equations (3.20) and (3.21) is given by:

$$e = [(0.02 \times 220)^2 + (0.02 \times 330)^2]^{1/2} = 7.93$$
$$f = 7.93/550 = 0.0144$$

Thus the total resistance S can be expressed as:

$$S = 550 \, \Omega \pm 7.93 \, \Omega$$

or $S = 550(1 \pm 0.0144) \, \Omega$, i.e. $S = 550 \, \Omega \pm 1.4\%$.

3.4.4 Error in a difference

If the two outputs y and z of separate measurement systems are to be subtracted from one another, and the possible errors are $\pm ay$ and $\pm bz$, then the difference S can be expressed as:

$$S = (y - z) \pm e \quad \text{or} \quad S = (y - z)(1 \pm f)$$

where e is calculated as above (Equation (3.20)) and $f = e/(y - z)$.

Example 3.8

A fluid flow rate is calculated from the difference in pressure measured on both sides of an orifice plate. If the pressure measurements are 10.0 bar and 9.5 bar and the error in the pressure measuring instruments is specified as $\pm 0.1\%$, then values for e and f can be calculated as:

$$e = [(0.001 \times 10)^2 + (0.001 \times 9.5)^2]^{1/2} = 0.0138$$
$$f = 0.0138/0.5 = 0.0276$$

Thus the pressure difference can be expressed as 0.5 bar $\pm 2.8\%$.

This example illustrates very poignantly the relatively large error which can arise when calculations are based on the difference between two measurements.

3.4.5 **Total error when combining multiple measurements**

The final case to be covered is where the final measurement is calculated from several measurements which are combined together in a way which involves more than one type of arithmetic operation. For example, the density of a rectangular-sided solid block of material can be calculated from measurements of its mass divided by the product of measurements of its length, height and width. The errors involved in each stage of arithmetic are cumulative, and so the total measurement error can be calculated by adding together the two error values associated with the two multiplication stages involved in calculating the volume and then calculating the error in the final arithmetic operation when the mass is divided by the volume.

> **Example 3.9**
> A rectangular-sided block has edges of lengths a, b and c, and its mass is m. If the values and possible errors in quantities a, b, c and m are $a = 100$ mm $\pm 1\%$, $b = 200$ mm $\pm 1\%$, $c = 300$ mm $\pm 1\%$ and $m = 20$ kg $\pm 0.5\%$, calculate the value of the density and the possible error in this value.
>
> **Solution**
> Value of $ab = 0.02$ m^2 $\pm 2\%$ (possible error $= 1\% + 1\% = 2\%$)
> Value of $(ab)c = 0.006$ m^3 $\pm 3\%$ (possible error $= 2\% + 1\% = 3\%$)
> Value of $m/(abc) = 20/0.006 = 3330$ kg/m^3 $\pm 3.5\%$ (possible
> error $= 3\% + 0.5\% = 3.5\%$).

3.5 **Exercises**

3.1 Explain the difference between systematic and random errors. What are the typical sources of these two types of error?

3.2 In what ways can the act of measurement cause a disturbance in the system being measured?

3.3 Suppose that the components in the circuit shown in Figure 3.1(a) have the following values: $R_1 = 330$ Ω, $R_2 = 1000$ Ω, $R_3 = 1200$ Ω, $R_4 = 220$ Ω, $R_5 = 270$ Ω. If the instrument measuring the output voltage across AB has a resistance of 5000 Ω, what is the measurement error caused by the loading effect of this instrument?

3.4 Instruments are normally calibrated and their characteristics defined for particular standard ambient conditions. What procedures are normally taken to avoid measurement errors when using instruments which are subjected to changing ambient conditions?

3.5 The voltage across a resistance R_5 in the circuit of Figure 3.12 is to be measured by a voltmeter connected across it.
   (a) If the voltmeter has an internal resistance (R_m) of 4750 Ω, what is the measurement error?
   (b) What value would the voltmeter internal resistance need to be in order to reduce the measurement error to 1%?

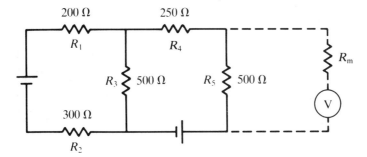

Figure 3.12 Circuit for Exercise 3.5

3.6 In the circuit shown in Figure 3.13, the current flowing between A and B is measured by an ammeter whose internal resistance is 100 Ω. What is the measurement error caused by the resistance of the measuring instrument?

3.7 What steps can be taken to reduce the effect of modifying inputs in measurement systems?

3.8 The output of a potentiometer is measured by a voltmeter having a resistance R_m, as shown in Figure 3.14. R_t is the resistance of the total length X_t of the potentiometer and R_i is the resistance between the wiper and common point C for a general wiper position X_i. Show that the measurement error due to the resistance R_m of the measuring instrument is given by:

$$\text{Error} = E\left(\frac{R_i^2(R_t - R_i)}{R_t(R_i R_t + R_m R_t - R_i^2)}\right)$$

Hence show that the maximum error occurs when X_i is approximately equal to $2X_t/3$. (*Hint:* Differentiate the error expression with respect to R_i and set to zero. Note that maximum error does not occur exactly at $X_i = 2X_t/3$, but this value is very close to the position where the maximum error occurs.)

3.9 In a survey of 15 owners of a certain model of car, the following figures for average petrol consumption were reported:

25.5 30.3 31.1 29.6 32.4 39.4 28.9 30.0

33.3 31.4 29.5 30.5 31.7 33.0 29.2

Calculate the mean value, the median value and the standard deviation of the data set.

3.10 (a) What do you understand by the term probability density function?

(b) Write down an expression for a Gaussian probability density function of given mean value μ and standard deviation σ and show how you would obtain the best estimate of these two quantities from a sample of population n.

(c) The following ten measurements are made of the output voltage from a high-gain amplifier which is contaminated due to noise fluctuations: 1.53, 1.57, 1.54, 1.54, 1.50, 1.51, 1.55, 1.54, 1.56 and 1.53. Determine the mean value and

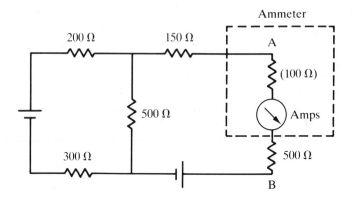

Ammeter

200 Ω   150 Ω

(100 Ω)

500 Ω   Amps

300 Ω

500 Ω

B

Figure 3.13 Circuit for Exercise 3.6

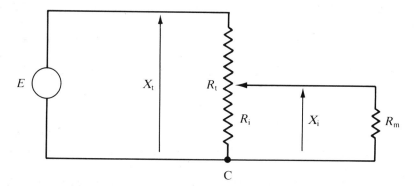

Figure 3.14 Circuit for Exercise 3.8

standard deviation. Hence estimate the accuracy to which the mean value is determined from these ten measurements. If 1000 measurements were taken, instead of ten, but σ remained the same, by how much would the accuracy of the calculated mean value be improved?

3.11 The measurements in a data set are subject to random errors but it is known that the data set fits a Gaussian distribution. Use error function tables to determine the percentage of measurements which lie within the boundaries of $\pm1.5\sigma$, where σ is the standard deviation of the measurements.

3.12 The thickness of a set of gaskets varies because of random manufacturing disturbances but the thickness values measured belong to a Gaussian distribution. If the mean thickness is 3 mm and the standard deviation is 0.25, calculate the percentage of gaskets which have a thickness greater than 2.5 mm.

3.13 A 3 volt d.c. power source required for a circuit is obtained by connecting together two 1.5 volt batteries in series. If the error in the voltage output of each battery is specified as $\pm 1\%$, calculate the possible error in the 3 volt power source which they make up.

3.14 In order to calculate the heat loss through the wall of a building, it is necessary to know the temperature difference between the inside and outside walls. If temperatures of 5 °C and 20 °C are measured on each side of the wall by mercury-in-glass thermometers with a range of -25 °C to $+25$ °C and a quoted accuracy figure of $\pm 1\%$ of full-scale reading, calculate the possible error in the calculated figure for the temperature difference.

3.15 The power dissipated in a car headlight is calculated by measuring the d.c. voltage drop across it and the current flowing through it ($P = V \times I$). If the possible errors in the measured voltage and current values are $\pm 1\%$ and $\pm 2\%$ respectively, calculate the possible error in the power value deduced.

3.16 The resistance of a carbon resistor is measured by applying a d.c. voltage across it and measuring the current flowing ($R = V/I$). If the voltage and current values are measured as 10 ± 0.1 V and 214 ± 5 mA respectively, calculate the value of the carbon resistor.

3.17 The density d of a liquid is calculated by measuring its depth c in a calibrated rectangular tank and then emptying it into a mass measuring system. The length and width of the tank are a and b respectively and thus the density is given by:

$$d = m/(a \times b \times c)$$

where m is the measured mass of the liquid emptied out. If the possible errors in the measurements of a, b, c and m are 1%, 1%, 2% and 0.5% respectively, determine the possible error in the calculated value of the density d.

3.18 The volume flow rate of a liquid is calculated by allowing the liquid to flow into a cylindrical tank (stood on its flat end) and measuring the height of the liquid surface before and after the liquid has flowed for 10 minutes. The volume collected after 10 minutes is given by:

$$\text{Volume} = (h_2 - h_1)\,\pi(d/2)^2$$

where h_1 and h_2 are the starting and finishing surface heights and d is the measured diameter of the tank.

(a) If $h_1 = 2$ m, $h_2 = 3$ m and $d = 2$ m, calculate the volume flow rate in m³/min.

(b) If the possible error in each measurement h_1, h_2 and d is $\pm 1\%$, estimate the possible error in the calculated value of volume flow rate.

4 Instrument calibration

4.1 Introduction

Instrument calibration consists of comparing the output of the instrument under test against the output of an instrument of known accuracy when the same input is applied to both instruments. This procedure is carried out for a range of inputs covering the whole measurement range of the instrument. Calibration ensures that the measuring accuracy of all instruments used in a measurement system is known over the whole measurement range, provided that the instruments are used in environmental conditions which are the same as those under which they were calibrated. For the use of instruments under different environmental conditions, appropriate correction has to be made for the ensuing modifying inputs, as described in Chapter 3.

Instrument calibration has to be repeated at prescribed intervals because the characteristics of any instrument change over a period of time. Changes in instrument characteristics are brought about by such factors as mechanical wear and the effects of dirt, dust, fumes and chemicals in the operating environment. To a great extent, the magnitude of the drift in characteristics depends on the amount of use an instrument receives and hence on the amount of wear and the length of time that it is subjected to the operating environment. However, some drift also occurs even in storage, as a result of ageing effects in components within the instrument.

The reasons for this drift in characteristics were discussed in detail in Chapter 3. It is sufficient here to accept that such drift does occur and that the rate at which characteristics change with time varies according to the type of instrument used, the frequency of use and the prevailing environmental conditions.

Because the rate of change of instrument characteristics is influenced by so many factors, it is difficult or even impossible to determine the required frequency of instrument recalibration from theoretical considerations. Instead, practical experimentation is applied to determine the rate of such changes. Once the maximum permissible measurement error has been defined, knowledge of the rate at which the characteristics of an instrument change allows a time interval to be calculated which represents the moment in time when an instrument will have reached the bounds of its acceptable performance level. The instrument must be recalibrated either at this time or earlier. This measurement error level which an instrument reaches just before recalibration is the error bound which must be quoted in the documented specifications for the instrument.

4.2 **Process instrument calibration**

Calibration consists of comparing the output of the process instrument being calibrated against the output of a standard instrument of known accuracy, when the same input (measured quantity) is applied to both instruments. During this calibration process, the instrument is tested over its whole range by repeating the comparison procedure for a range of inputs.

The instrument used as a standard for this procedure must be one which is kept solely for calibration duties. It must never be used for other purposes. Most particularly, it must not be regarded as a spare instrument which can be used for process measurements if the instrument normally used for that purpose breaks down. Proper provision for process instrument failures must be made by keeping a spare set of process instruments. Standard calibration instruments must be totally separate.

To ensure that these conditions are met, the calibration function must be managed and executed in a professional manner. This will normally mean setting aside a particular place within the instrumentation department of a company where all calibration operations take place and where all instruments used for calibration are kept. As far as possible this should take the form of a separate room, rather than a sectioned-off area in a room used for other purposes as well. This will enable better environmental control to be applied in the calibration area and will also offer better protection against unauthorized handling or use of the calibration instruments. The level of environmental control required during calibration should be considered carefully with due regard to what level of accuracy is required in the calibration procedure, but should not be overspecified as this will lead to unnecessary expense. Full air conditioning is not normally required for calibration at this level, as it is very expensive, but sensible precautions should be taken to guard the area from extremes of heat or cold, and also good standards of cleanliness should be maintained. Useful guidance on the operation of standards facilities can be found elsewhere (British Standards Society 1979).

Whilst it is desirable that all calibration functions are performed in this carefully controlled environment, it is not always practical to achieve this. Sometimes, it is not convenient or possible to remove instruments from process plant, and in these cases it is standard practice to calibrate them *in situ*. In these circumstances, appropriate corrections must be made for the deviation in the calibration environmental conditions away from those specified. This practice does not obviate the need to protect calibration instruments and maintain them in constant conditions in a calibration laboratory at all times other than when they are involved in such calibration duties on plant.

Apart from the precautions taken to preserve the accuracy of instruments used for calibration, by treating them carefully and reserving them only for calibration duties, they are often chosen to be of a greater inherent accuracy than the process instruments that they are used to calibrate. Where instruments are only used for calibration purposes, greater accuracy can often be achieved by specifying a type of instrument which would be unsuitable for normal process measurements. Ruggedness for instance is not a requirement, and freedom from this constraint opens up a much wider range of possible instruments. In practice, high-accuracy, null-type instruments are very commonly used for calibration duties, because their requirement for a human operator is not a problem in these circumstances.

As far as management of calibration procedures is concerned, it is important that the performance of all calibration operations is assigned as the clear responsibility of just one person. That person should have total control over the calibration function and be able to limit access to the calibration laboratory to designated approved personnel only. Only by giving this appointed person total control over the calibration function can the function be expected to operate efficiently and effectively. Lack of such definite management can only lead to unintentional neglect of the calibration system, resulting in the use of equipment in an out-of-date state of calibration and subsequent loss of traceability to reference standards. Professional management is essential so that the customer can be assured that an efficient calibration system is in operation and that the accuracy of measurements is guaranteed.

Calibration procedures which relate in any way to measurements used for quality control functions are controlled by British Standard BS 5750 (ISO 9000 is the international equivalent of this standard). One of the clauses in BS 5750 requires that all persons using calibration equipment be adequately trained. The manager in charge of the calibration function is clearly responsible for ensuring that this condition is met. Training must be adequate and targeted at the particular needs of the calibration systems involved. People must understand what they need to know and especially why they must have this information. Successful completion of training courses should be marked by the award of qualification certificates. These attest to the proficiency of personnel involved in calibration duties and are a convenient way of demonstrating that the BS 5750 training requirement has been satisfied.

The calibration facilities provided within the instrumentation department of a company provide the first link in the calibration chain. Instruments used for calibration at this level are known as working standards. A fundamental responsibility in the supervision of this calibration function is to establish the frequency at which the various shop-floor instruments should be calibrated and to ensure that calibration is carried out at the appropriate times.

Determination of the frequency at which instruments should be calibrated is dependent upon several factors which require specialist knowledge. If an instrument is required to measure some quantity to an accuracy of $\pm 2\%$, then a certain amount of performance degradation can be allowed if its accuracy immediately after recalibration is $\pm 1\%$. What is important is that the pattern of performance degradation be quantified, such that the instrument can be recalibrated before its accuracy has reduced to the limit defined by the application.

The quantities which cause the deterioration in the performance of instruments over a period of time are mechanical wear, dust, dirt, ambient temperature and frequency of usage. Susceptibility to these factors varies according to the type of instrument involved. The effect of these quantities on the accuracy and other characteristics of an instrument can only be quantified by possessing an in-depth knowledge of the mechanical construction and other features involved in the instrument. Some form of practical experimentation is normally required to determine the required calibration frequency precisely in the typical operating conditions for the instrument. Further discussion on the means of quantifying the rate of change of instrument characteristics was given in Chapter 3.

A proper course of action must be defined which describes the procedures to be followed when an instrument is found to be out of calibration, i.e. when its output is different to that of the calibration instrument when the same input is applied. The required action depends very much upon the nature of the discrepancy and the type of instrument involved. In many cases, deviations in the form of a simple output bias can be corrected by a small adjustment to the instrument (following which the adjustment screws must be sealed to prevent tampering). In other cases, the output scale of the instrument may have to be redrawn, or scaling factors altered where the instrument output is part of some automatic control or inspection system. In extreme cases, where the calibration procedure shows up signs of instrument damage, it may be necessary to send the instrument for repair or even to scrap it.

Whatever system and frequency of calibration is established, it is important to review this from time to time to ensure that the system remains effective and efficient. It may happen that a cheaper (but equally effective) method of calibration becomes available with the passage of time, and such an alternative system must clearly be adopted in the interests of cost efficiency. However, the main item under scrutiny in this review is normally whether the calibration interval is still appropriate. Records of the calibration history of the instrument will be the primary basis on which this review is made. It may happen that an instrument starts to go out of calibration more quickly after a period of time, either because of ageing factors within the instrument or because of changes in the operating environment. The conditions or mode of usage of the instrument may also be subject to change. As the environmental and usage conditions of an instrument may change beneficially as well as adversely, there is the possibility that the recommended calibration interval may decrease as well as increase.

Maintaining proper records is an important part of fulfilling this calibration function. A separate record, similar to that shown in Figure 4.1, should be kept for every instrument in the factory, whether it is in use or kept as a spare. This record should start by giving a description of the instrument and follow this by stating what the required calibration frequency is. Each occasion when the instrument is calibrated should be recorded in this record, and every such calibration log should show the status of the instrument in terms of the deviation from its required specification and the action taken to correct it. The calibration record is also very useful in providing feedback which shows whether the calibration frequency has been chosen correctly or not.

Type of instrument:	Company serial number:
Manufacturer's part number:	Manufacturer's serial number:
Measurement limit:	Date introduced:
Location:	
Instructions for use:	
Calibration frequency:	Signature of person responsible for calibration:

CALIBRATION RECORD

Calibration date	Calibration results	Calibrated by

Figure 4.1 Typical format for instrument record sheets

4.3 Standards laboratories

We have established so far that process instruments which are used to make quality-related measurements must be calibrated from time to time against a working standard instrument. As this working standard instrument is one which is kept by the instrumentation department of a company for calibration duties, and for no other purpose, then it can be assumed that it will maintain its accuracy over a reasonable period of time because use-related deterioration in accuracy is largely eliminated. However, over the longer term, the characteristics of even such a standard instrument will drift, mainly as a result of ageing effects in its components. Over this longer term, therefore, a programme must be instituted for calibrating this working standard instrument against one of yet higher accuracy at appropriate intervals of time. The instrument used for calibrating working standard instruments is known as a secondary reference standard. This must obviously be a very well-engineered instrument which gives high accuracy and is stabilized against drift in its performance with time. This implies that it will be an expensive instrument to buy. It also requires that the environmental conditions in which it is used are carefully controlled in respect of ambient temperature, humidity, etc.

When the working standard instrument has been calibrated by an authorized standards laboratory, a calibration certificate will be issued (see NAMAS Document B 5103 (1985)). This will contain at least the following information:

- The identification of the equipment calibrated.
- The calibration results obtained.
- The measurement uncertainty.
- Any use limitations on the equipment calibrated.
- The date of calibration.
- The authority under which the certificate is issued.

The establishment of a company standards laboratory to provide a calibration facility of the required quality is economically viable only in the case of very large companies where large numbers of instruments need to be calibrated across several factories. In the case of small- to medium-sized companies, the cost of buying and maintaining such equipment is not justified. Instead, they would normally use the calibration service provided by various companies which specialize in offering a standards laboratory. What these specialist calibration companies effectively do is to share out the high cost of providing this highly accurate but infrequently used calibration service over a large number of companies. Such standards laboratories are closely monitored by national standards organizations (see ISO Guide 25 (1982), BS 6460 (1983)).

4.4 Validation of standards laboratories

In the United Kingdom, the appropriate national standards organization for validating standards laboratories is the National Physical Laboratory (in the United States, the equivalent body is the National Bureau of Standards). This has established a National Measurement Accreditation Service (NAMAS) which monitors both instrument calibration and mechanical testing laboratories. The formal structure for accrediting instrument calibration standards laboratories is known as the British Calibration Service (BCS), and that for accrediting testing facilities is known as the National Testing Laboratory Accreditation Scheme (NATLAS).

Although each different country has its own structure for the maintenance of standards, each of these different frameworks tends to be equivalent in its effect. To achieve confidence in the goods and services which move across national boundaries, international agreements have established the equivalence of the different accreditation schemes in existence. As a result, NAMAS and the similar schemes operated by France, Germany, Italy, the United States, Australia and New Zealand enjoy mutual recognition.

BCS lays down strict conditions which a standards laboratory has to meet before it is approved. These conditions control laboratory management, environment, equipment and documentation. The person appointed as head of the laboratory must be suitably qualified, and independence of operation of the laboratory must be guaranteed. The management structure must be such that any pressure to rush or skip calibration procedures for production reasons can be resisted. As far as the laboratory environment is concerned, proper temperature and humidity control must be provided, and high standards of cleanliness and housekeeping must be maintained. All equipment used for calibration purposes must be maintained to reference standards and supported by calibration certificates which establish this traceability. Finally, full documentation must be maintained. This should describe all calibration procedures, maintain an index system for recalibration of equipment, and include a full inventory of apparatus and traceability schedules. Having met these conditions, a standards laboratory becomes an accredited laboratory for providing calibration services and issuing calibration certificates. This accreditation is reviewed at approximately 12 monthly intervals to ensure that the laboratory is continuing to satisfy the conditions for approval laid down.

4.5 Primary reference standards

Primary reference standards, as listed in Table 2.1, describe the highest level of accuracy that is achievable in the measurement of any particular physical quantity. All items of equipment used in standards laboratories as secondary reference standards have themselves to be calibrated against primary reference standards at appropriate intervals of time. This procedure is acknowledged by the issue of a calibration certificate in the standard way. National standards organizations maintain suitable facilities for this calibration, which in the case of the United Kingdom are at the National Physical Laboratory. The equivalent organization in the United States is the National Bureau of Standards. In certain cases, such primary reference standards can be located outside national standards organizations. For instance, the primary reference standard for dimension measurement is defined by the wavelength of the orange–red line of krypton light, and it can therefore be realized in any laboratory equipped with an interferometer.

In certain cases (e.g. the measurement of viscosity), such primary reference standards are not available and reference standards for calibration are achieved by collaboration between several national standards organizations who perform measurements on identical samples under controlled conditions (see BS 5497 (1987), ISO 5725 (1986)).

4.6 Traceability

What has emerged from the foregoing discussion is that calibration has a chain-like structure in which every instrument in the chain is calibrated against a more accurate instrument immediately above it in the chain, as shown in Figure 4.2. All of the elements in the calibration chain must be known so that the calibration of process instruments at the bottom of the chain is traceable to the fundamental measurement standards.

This knowledge of the full chain of instruments involved in the calibration procedure is known as traceability, and is specified as a mandatory requirement in satisfying standards such as BS 5750. Documentation must exist which shows that process instruments are calibrated by standard instruments which are linked by a chain of increasing accuracy back to national reference standards. There must be clear evidence to show that there is no break in this chain.

To illustrate a typical calibration chain, consider the calibration of micrometers (Figure 4.3). A typical shop-floor micrometer has an uncertainty (inaccuracy) of less than 1 in 10^4. These micrometers would normally be calibrated in the instrumentation department or standards laboratory of a company against laboratory standard gauge blocks with a typical uncertainty of less than 1 in 10^5. A specialist calibration service company would provide facilities for calibrating these blocks against reference-grade gauge blocks with a typical uncertainty of less than 1 in 10^6. More accurate calibration equipment is still provided by national standards organizations. The National Physical Laboratory maintains two sets of standards for this type of calibration, a working standard and a primary standard. Spectral lamps are used to provide a working reference standard with an uncertainty of less than 1 in 10^7. The primary standard is provided by an iodine-stabilized helium–neon laser which has a specified uncertainty of less than 1 in 10^9. All of the links in this calibration chain must be shown in any documentation which describes the use of micrometers in making quality-related measurements.

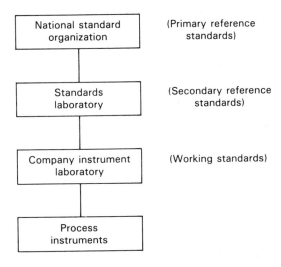

Figure 4.2 Instrument calibration chain

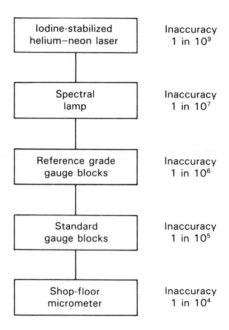

Figure 4.3 Typical calibration chain for micrometers

4.7 **Documentation in the workplace**

An essential element in the maintenance of measurement systems and the operation of calibration procedures is the provision of full documentation. This must give a full description of the measurement requirements throughout the workplace, the instruments used, and the calibration system and procedures operated. Individual calibration records for each instrument must be included within this. The documentation is a necessary part of the quality manual, although it may physically exist as a separate volume if this is more convenient. An overriding constraint on the style in which the documentation is presented is that it should be simple and easy to read. This is often greatly facilitated by the copious use of appendices.

The starting point in the documentation must be a statement of what measurement limits have been defined for each measurement system documented. Such limits are established by balancing the costs of improved accuracy against customer requirements, and also with regard to what overall quality level has been specified in *The Quality Manual*. The technical procedures required for this, which involve assessing the type and magnitude of relevant measurement errors, are described in Chapter 3. It is customary to express the final measurement limit calculated as ±2 standard deviations, i.e. within 95% confidence limits (see Chapter 3 for an explanation of these terms).

The instruments specified for each measurement situation must be listed next. This list must be accompanied by full instructions for the proper use of the instruments concerned. These instructions will include details of any environmental control or other special precautions which must be taken to ensure that the instruments provide measurements of sufficient accuracy to meet the measurement limits defined. The proper training courses appropriate to plant personnel who will use the instruments must be specified.

Having disposed of the question about what instruments are used, the documentation must go on to cover the subject of calibration. Full calibration is not applied to every measuring instrument used in a workplace because BS 5750 acknowledges that formal calibration procedures are not necessary for some equipment where it is uneconomic or technically unnecessary because the accuracy of the measurement involved has an insignificant effect on the overall quality target for a product. However, any equipment which is excluded from calibration procedures in this manner must be specified as such in the documentation. Identification of equipment which is in this category is a matter of informed judgement.

For instruments which are the subject of formal calibration, the documentation must specify what standard instruments are to be used for the purpose and define a formal procedure of calibration. This procedure must include instructions for the storage and handling of standard calibration instruments and specify the required environmental conditions under which calibration is to be performed. Where a calibration procedure for a particular instrument uses published standard practices, it is sufficient to include reference to that standard procedure in the documentation rather than to reproduce the whole procedure. Whatever calibration system is established, a formal review procedure must be defined in the documentation which ensures its continued effectiveness at regular intervals. The results of each review must also be documented in a formal way.

A standard format for recording calibration results should be defined in the documentation. A separate record must be kept for every instrument present in the workplace which includes details of the instrument's description, the required calibration frequency, the date of each calibration and the calibration results on each occasion. Where appropriate, the documentation must also define the manner in which calibration results are to be recorded on the instruments themselves.

The documentation must specify procedures which are to be followed if an instrument is found to be outside the calibration limits. This may involve adjustment, redrawing its scale or withdrawing it, depending upon the nature of the discrepancy and the type of instrument involved. Instruments withdrawn will be either repaired or scrapped. In the case of withdrawn instruments, a formal procedure for marking them as such must be defined to prevent them being accidentally put back into use.

Two other items must also be covered by the calibration document. The traceability of the calibration system back to national reference standards must be defined and supported by calibration certificates (see section 4.3). Training procedures must also be documented, specifying the particular training courses to be attended by various personnel and what, if any, refresher courses are required.

All aspects of these documented calibration procedures will be given consideration as part of the periodic audit of the quality control system which calibration procedures are instigated to support. Whilst the basic responsibility for choosing a suitable interval between calibration checks rests with the engineers responsible for the instruments concerned, the quality system auditor will require to see the results of tests which show that the calibration interval has been chosen correctly and that instruments are not going outside allowable measurement uncertainty limits between calibrations. Particularly important in such audits will be the existence of procedures which are instigated in response to instruments found to be out of calibration. Evidence that such procedures are effective in avoiding degradation of the quality assurance function will also be required.

References and further reading

British Standards Society (1979) 'The operation of a company standards department', British Standards Society, London.

BS 5497 (1987) Guide for the determination of repeatability and reproducibility for a standard test method by inter-laboratory tests, British Standards Institution, London.

BS 6460 (1983) Accreditation of testing laboratories, British Standards Institution, London.

ISO Guide 25 (1982) General requirements for the technical competence of testing laboratories, International Organization for Standardization, Geneva.

ISO 5725 (1986) Precision of test methods – determination of repeatability and reproducibility by inter-laboratory tests, International Organization for Standardization, Geneva.

NAMAS Document B 5103 (1985) Certificates of calibration, NAMAS Executive, National Physical Laboratory, Teddington, Middlesex.

5 Signal processing, manipulation and transmission

Signal processing is concerned with improving the quality of the reading or signal at the output of a measurement system. The form which signal processing takes depends on the nature of the raw output signal from a measurement transducer. Procedures of signal amplification, signal attenuation, signal linearization, bias removal and signal filtering are applied according to the form of correction required in the raw signal. Noise-free transmission of the signal between remote transducers and the usage point of signals is also an important target in measurement systems, and appropriate means for this are discussed in the final section of this chapter.

The implementation of signal processing procedures can be carried out either by analog techniques or by digital computation. For the purposes of explaining the procedures involved, this chapter mainly describes analog signal processing and concludes with a relatively brief discussion of the equivalent digital signal processing techniques. One particular reason for this method of treatment is that some prior analog signal conditioning is often necessary even when the major part of the signal processing is carried out digitally.

The choice between analog and digital signal processing is largely determined by the degree of accuracy required in the signal processing procedure. Digital processing is very much more accurate than analog processing but the cost of the equipment involved is much greater and the processing time is much longer. Normal practice, therefore, is to use analog techniques for all signal processing tasks except where the accuracy of this is insufficient. It should be noted also that in some measurement situations where a physical quantity is measured by an inherently inaccurate transducer, the extra accuracy provided by digital signal processing is insignificant and therefore such expensive and slow processing techniques are inappropriate.

5.1 Signal amplification

Signal amplification is carried out when the typical signal output level of a measurement transducer is considered to be too low. Amplification by analog means is carried out by an operational amplifier. This is normally required to have a high input impedance so that its loading effect on the transducer output signal is minimized. In some circumstances, such as when amplifying the output signal from accelerometers and some optical detectors, the amplifier must also have a high-frequency response, to avoid distortion of the output reading.

The operational amplifier is an electronic device which has two input terminals and one output terminal, the two inputs being known as the inverting input and non-inverting input respectively. When used for signal processing duties, it is connected in the configuration shown in Figure 5.1. The raw (unprocessed) signal V_i is connected to the inverting input through a resistor R_1 and the non-inverting input is connected to ground. A feedback path is provided from the output terminal through a resistor R_2 to the inverting input terminal. The processed signal V_0 at the output terminal is then related to the voltage V_i at the input terminal by the expression (assuming ideal operational amplifier characteristics):

$$V_0 = -\frac{R_2 V_i}{R_1} \tag{5.1}$$

The amount of signal amplification is therefore defined by the relative values of R_1 and R_2. This ratio between R_1 and R_2 in the amplifier configuration is often known as the amplifier gain. If, for instance, $R_1 = 1\,\text{M}\Omega$ and $R_2 = 10\,\text{M}\Omega$, an amplification factor of ten is obtained (i.e. gain = 10). It is important to note that, in this standard way of connecting the operational amplifier, the sign of the processed signal is inverted. This can be corrected for if necessary by feeding the signal through a further amplifier set up for unity gain ($R_1 = R_2$). This inverts the signal again and returns it to its original sign.

5.1.1 Instrumentation amplifier

For some applications requiring the amplification of very low-level signals, a special type of amplifier known as an **instrumentation amplifier** is used. This consists of a circuit containing three standard operational amplifiers, as shown in Figure 5.2. The advantage of the instrumentation amplifier compared with a standard operational amplifier is that its differential input impedance is much higher. In consequence, its common mode rejection capability* is much better. This means that if a twisted wire pair is used to connect a transducer to the differential inputs of the amplifier, any induced noise will contaminate each wire equally and will be rejected by the common mode rejection capacity of the amplifier.

5.2 Signal attenuation

One method of attenuating signals by analog means is to use a potentiometer connected in a voltage-dividing circuit, as shown in Figure 5.3. For the potentiometer wiper positioned a distance of X_w along the resistance element of total length X_t, the

---

*Common mode rejection describes the ability of the amplifier to reject equal-magnitude signals that appear on both of its inputs.

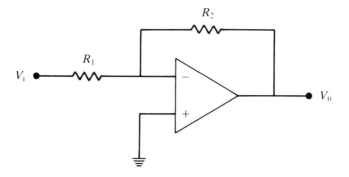

Figure 5.1 Operational amplifier connected for signal amplification

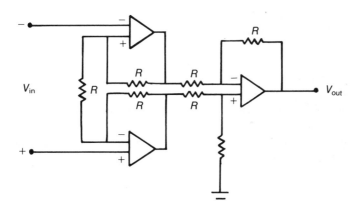

Figure 5.2 Instrumentation amplifier

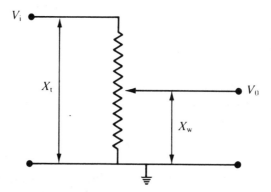

Figure 5.3 Potentiometer in voltage-dividing circuit

voltage level of the processed signal V_0 is related to the voltage level of the raw signal V_i by the expression:

$$V_0 = \frac{X_w V_i}{X_t}$$

An alternative device to the potentiometer for signal attenuation is the operational amplifier. This is connected in exactly the same way as for an amplifier as shown in Figure 5.1, but R_1 is chosen to be greater than R_2. Equation (5.1) still holds and therefore, if R_1 is chosen to be 10 MΩ and R_2 as 1 MΩ, an attenuation factor of ten is achieved (gain = 0.1). Use of an operational amplifier as an attenuating device is a more expensive solution than using a potentiometer, but it has advantages in terms of its smaller size and low power consumption.

5.3 Signal linearization

Several types of transducer used in measuring instruments have an output which is a non-linear function of the measured quantity input. In many cases, this non-linear signal can be converted to a linear one by special operational amplifier configurations which have an equal and opposite non-linear relationship between the amplifier input and output terminals.

For example, light intensity transducers typically have an exponential relationship between the output signal and the input light intensity, i.e.:

$$V_0 = K \exp(-\alpha Q) \tag{5.2}$$

where Q is the light intensity, V_0 is the voltage level of the output signal, and K and α are constants.

If a diode is placed in the feedback path between the input and output terminals of the amplifier as shown in Figure 5.4, the relationship between the amplifier output voltage V_2 and input voltage V_1 is given by:

$$V_2 = C \log_e(V_1) \tag{5.3}$$

If the output of the light transducer with characteristic given by Equation (5.2) is conditioned by an amplifier of characteristic given by Equation (5.3), the voltage level of the processed signal is given by:

$$V_2 = C \log_e(K) - \alpha C Q \tag{5.4}$$

Expression (5.4) shows that the output signal now varies linearly with light intensity Q but with an offset of $C \log_e(K)$. This offset would normally be removed by further signal conditioning, as described below.

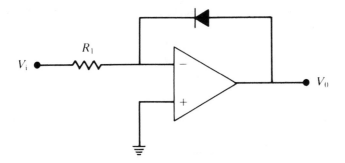

Figure 5.4 Operational amplifier connected for signal linearization

5.4 **Bias removal**

Sometimes, either because of the nature of the measurement transducer itself, or as a result of other signal conditioning operations (see section 5.3), a bias exists in the output signal. This can be expressed mathematically for a physical quantity x and measurement signal y as:

$$y = Kx + C \qquad (5.5)$$

where C represents a bias in the output signal which needs to be removed by signal processing. Analog processing consists of using an operational amplifier connected in a differential amplification mode, as shown in Figure 5.5. Referring to this circuit, for $R_1 = R_2$ and $R_3 = R_4$, the output V_0 is given by:

$$V_0 = (R_3/R_1)(V_c - V_i) \qquad (5.6)$$

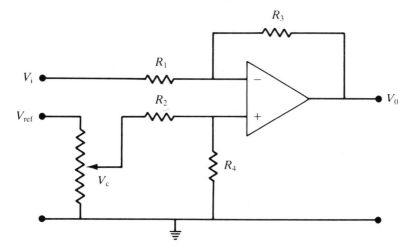

Figure 5.5 Operational amplifier connected in differential amplification mode

where V_i is the unprocessed signal y equal to $Kx + C$ and V_c is the output voltage from a potentiometer supplied by a known reference voltage V_{ref}, which is set such that $V_c = C$. Now, substituting these values for V_i and V_c into Equation (5.6) and referring the quantities back into Equation (5.5), we find that:

$$y = K'x \tag{5.7}$$

where the new constant K' is related to K by the amplifier gain factor R_3/R_1. It is clear that a straight-line relationship now exists between the measurement signal y and the measured quantity x and that the unwanted bias has been removed.

5.5 Signal filtering

Signal filtering consists of processing a signal to remove a certain band of frequencies within it. The band of frequencies removed can be at the low-frequency end of the frequency spectrum, at the high-frequency end, at both ends, or in the middle of the spectrum. Filters to perform each of these operations are known respectively as low-pass filters, high-pass filters, band-pass filters and band-stop filters. All such filtering operations can be carried out by either analog or digital methods.

The result of filtering can be readily understood if the analogy with a procedure such as sieving soil particles is considered. Suppose that a sample of soil A is passed through a system of two sieves of differing meshes such that the soil is divided into three parts, B, C and D, consisting of large, medium and small particles, as shown in Figure 5.6. Suppose that the system also has a mechanism for delivering one or more of the separated parts, B, C and D, as the system output. If the graded soil output consists of parts C and D, the system is behaving as a low-pass filter (rejecting large particles), whereas if it consists of parts B and C, the system is behaving as a high-pass filter (rejecting small particles). Other options are to deliver just part C (band-pass filter mode) or parts B and D together (band-stop filter mode). As any gardener knows, however, such perfect sieving is not achieved in practice and any form of graded soil output always contains a few particles of the wrong size.

Signal filtering consists of selectively passing or rejecting low-, medium- and high-frequency signals from the frequency spectrum of a general signal. The range of frequencies passed by a filter is known as the **pass band**, the range not passed is known as the **stop band**, and the boundary between the two ranges is known as the **cut-off frequency**. To illustrate this, consider a signal whose frequency spectrum is such that all frequency components in the frequency range from zero to infinity have equal magnitude. If this signal is applied to an ideal filter, then the outputs for a low-pass filter, high-pass filter, band-pass filter and band-stop filter respectively are as shown in Figure 5.7. Note that for the last two types, the bands are defined by a pair of frequencies rather than by a single cut-off frequency.

Just as in the case of the soil sieving analogy presented above, the signal filtering mechanism is not perfect, with unwanted frequency components not being erased

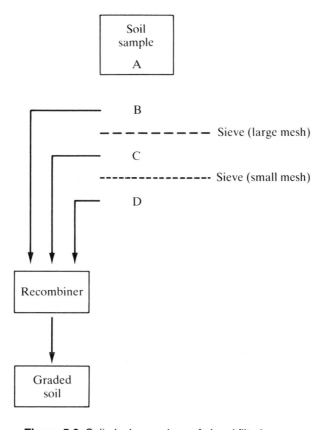

Figure 5.6 Soil-sieving analogy of signal filtering

completely but only attenuated by varying degrees instead, i.e. the filtered signal always retains some components (of a relatively low magnitude) in the unwanted frequency range. There is also a small amount of attenuation of frequencies within the pass band which increases as the cut-off frequency is approached. Figure 5.8 shows the typical output characteristics of a practical constant-k* filter designed respectively for high-pass, low-pass, band-pass and band-stop filtering. Filter design is concerned with trying to obtain frequency rejection characteristics which are as close to the ideal as possible. However, an improvement in characteristics is only achieved at the expense of greater complexity in the design. The filter chosen for any given situation is therefore a compromise between performance, complexity and cost.

In the majority of measurement situations, the physical quantity being measured has a value which is either constant or only changing slowly with time. In these

---

*'Constant-k' is a term used to describe a common class of passive filters, as discussed in the following section.

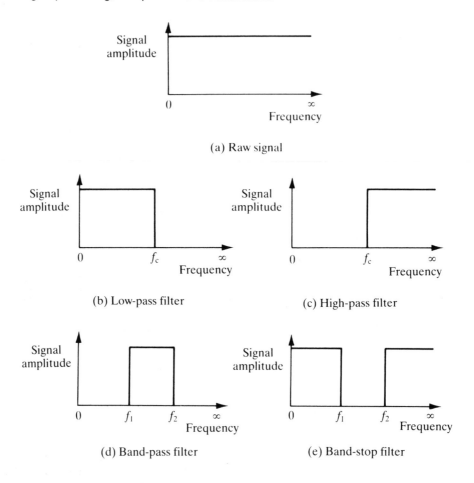

(a) Raw signal

(b) Low-pass filter

(c) High-pass filter

(d) Band-pass filter

(e) Band-stop filter

Figure 5.7 Outputs from ideal filters

circumstances, the most common types of signal corruption are high-frequency noise components, and the type of signal processing element required is a low-pass filter. In a few cases, the measured signal itself has a high frequency, for instance when mechanical vibrations are being monitored, and the signal processing required is the application of a high-pass filter to attenuate low-frequency noise components. Band-stop filters can be used where a measurement signal is corrupted by noise at a particular frequency. Such noise is frequently due to mechanical vibrations or the proximity of the measurement circuit to other electrical apparatus.

Both passive and active analog filter implementations of the four types of filter identified are considered below. The equivalent digital filters are discussed later in section 5.7.

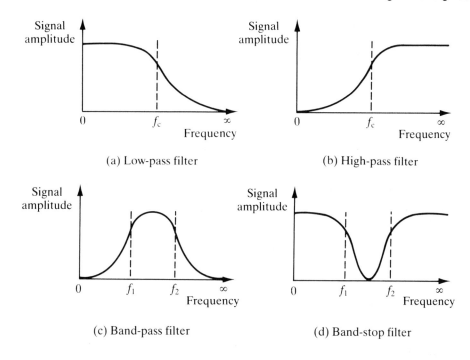

Figure 5.8 Outputs from practical constant-*k* filters

5.5.1 **Passive analog filters**

The detailed design of passive filters is a subject of some considerable complexity which is outside the scope of this book. In the following treatment, the major formulae appropriate to the design of filters are quoted without derivation. The derivations can be found elsewhere (Blinchikoff 1976, Skilling 1967, Williams 1963).

Simple passive filters consist of a network of impedances as shown in Figure 5.9(a). So that there is no dissipation of energy in the filter, these impedances should ideally be pure reactances (capacitors or resistance-less inductors). In practice, however, it is impossible to manufacture inductors that do not have a small resistive component and so this ideal cannot be achieved. Indeed, readers familiar with radio receiver design will be aware of the existence of filters consisting of pure resistors and capacitors. These only have a mild filtering effect which is useful in radio tone controls but not relevant to the signal processing requirements discussed in this chapter. Such filters are therefore not considered further.

Each element of the network shown in Figure 5.9(a) can be represented by either a T-section or π-section as shown in Figures 5.9(b) and 5.9(c) respectively. To obtain proper matching between filter sections, it is necessary for the input impedance of each section to be equal to the load impedance for that section. This value of

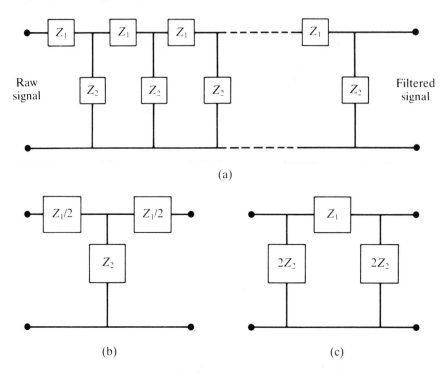

Figure 5.9 (a) Simple passive filter; (b) T-section; (c) π-section

impedance is known as the characteristic impedance (Z_0). For a T-section of filter, the characteristic impedance is calculated from:

$$Z_0 = \{Z_1 Z_2 [1 + (Z_1/4Z_2)]\}^{1/2} \qquad (5.8)$$

The frequency attenuation characteristics of the filter can be determined by inspecting this expression for Z_0. Frequency values for which Z_0 is real lie in the pass band and frequencies for which Z_0 is imaginary lie in the stop band.

Consider the case where $Z_1 = j\omega L$ and $Z_2 = 1/j\omega C$. Substituting these values into the expression for Z_0 above, we obtain:

$$Z_0 = [(L/C)(1 - 0.25\omega^2 LC)]^{1/2}$$

For frequencies where $\omega < (4/LC)^{1/2}$, Z_0 is real, and for higher frequencies, Z_0 is imaginary. These values of impedance therefore give a **low-pass** filter with cut-off frequency given by:

$$f_c = (\omega_c/2\pi) = 1/\omega(LC)^{1/2}$$

(*Note:* Parameters L and C are defined with respect to the filter components expressed in Figure 5.10.)

A **high-pass** filter can be synthesized with exactly the same cut-off frequency if the impedance values chosen are:

$$Z_1 = 1/j\omega C \quad \text{and} \quad Z_2 = j\omega L$$

A point worthy of note in both these last two examples is that the product $Z_1 Z_2$ could be represented by a constant k which is independent of frequency. Because of this, such filters are known by the name of constant-k filters.

(*Note:* Parameters L and C are defined with respect to the filter components expressed in Figure 5.10.)

A constant-k **band-pass** filter can be realized with the following choice of impedance values:

$$Z_1 = j\omega L + \frac{1}{j\omega C} \qquad Z_2 = \frac{(j\omega La)(a/j\omega C)}{j\omega La + (a/j\omega C)}$$

The frequencies f_1 and f_2 defining the end of the pass band are most easily

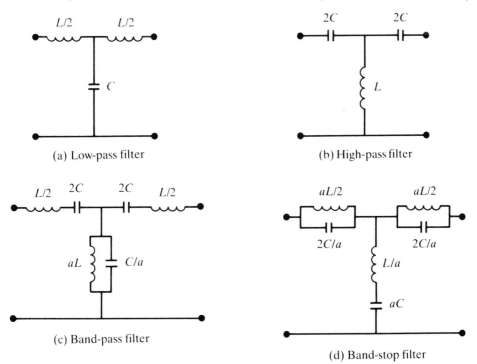

(a) Low-pass filter

(b) High-pass filter

(c) Band-pass filter

(d) Band-stop filter

Figure 5.10 Circuit components for passive filter T-sections

expressed in terms of a frequency f_0 in the centre of the pass band. The corresponding equations are:

$$f_0 = \frac{1}{2\pi(LC)^{1/2}} \quad f_1 = f_0[(1 + a)^{1/2} - a^{1/2}] \quad f_2 = f_0[(1 + a)^{1/2} + a^{1/2}]$$

(*Note:* Parameters L and C are defined with respect to the filter components expressed in Figure 5.10.)

For a constant-k **band-stop** filter, the appropriate impedance values are:

$$Z_1 = \frac{(j\omega La)(a/j\omega C)}{j\omega La + (a/j\omega C)} \quad Z_2 = \frac{1}{a}\left(j\omega L + \frac{1}{j\omega C}\right)$$

The frequencies defining the ends of the stop band are again normally defined in terms of the frequency f_0 in the centre of the stop band:

$$f_0 = \frac{1}{2\pi(LC)^{1/2}} \quad f_1 = f_0\left(1 - \frac{a}{4}\right) \quad f_2 = f_0\left(1 + \frac{a}{4}\right)$$

(*Note:* Parameters L and C are defined with respect to the filter components expressed in Figure 5.10.)

As has already been mentioned, a practical filter does not eliminate frequencies in the stop band but merely attenuates them by a certain amount. The attenuation, α, at a frequency in the stop band, f, for a single T-section of a low-pass filter is given by:

$$\alpha = 2\cosh^{-1}(f/f_c) \tag{5.9}$$

The relatively poor attenuation characteristics are obvious if we evaluate this expression for a value of frequency close to the cut-off frequency given by $f = 2f_c$. Then $\alpha = 2\cosh^{-1}(2) = 2.64$. Further away from the cut-off frequency, for $f = 20f_c$, $\alpha = 2\cosh^{-1}(20) = 7.38$.

Improved attenuation characteristics can be obtained by putting several T-sections in cascade. If perfect matching is assumed then two T-sections give twice the attenuation of one section, i.e. at frequencies of $2f_c$ and $20f_c$, α for two sections would have a value of 5.28 and 14.76 respectively.

The discussion so far has assumed resistance-less inductances and perfect matching between sections. Such conditions cannot be achieved in practice and this has several consequences.

Inspection of expression (5.8) for the characteristic impedance reveals frequency-dependent terms. Thus the condition that the load impedance is equal to the input impedance for a section is only satisfied at one particular frequency. The load impedance is normally chosen so that this frequency is 'well within the pass band'. (Zero frequency is usually chosen for a low-pass filter and infinite frequency for a high-pass one as the frequency where this is satisfied.) This is one of the reasons for the degree of attenuation in the pass band shown in the practical filter characteristics of Figure 5.8, the other reason being the presence of resistive components in the inductors of the filter. The effect of this in a practical filter is that the value of α at the cut-off frequency is 1.414 whereas the value predicted theoretically for an ideal filter (Equation (5.9)) is zero. Cascading filter sections together increases this attenuation in the pass band as well as increasing the attenuation of frequencies in the stop band.

This problem of matching successive sections in a cascaded filter seriously degrades the performance of constant-k filters and this has resulted in the development of other types such as m-derived and n-derived composite filters. These produce less attenuation within the pass band and greater attenuation outside it than constant-k filters, although this is only achieved at the expense of greater filter complexity and cost. The reader interested in further consideration of these filters is directed to consult one of the specialist texts mentioned in the References at the end of this chapter.

5.5.2 Active analog filters

In the foregoing discussion on passive filters, two main difficulties were encountered: one of obtaining resistance-less inductors, and one of achieving proper matching between signal source and load through the filter sections. Active filters overcome both of these problems and so are very popular for signal processing duties.

Active circuits to produce the four different types of filtering identified are illustrated in Figure 5.11. These particular circuits are all second-order filters because the relationship between filter input and output is described by a second-order differential equation. The major component in an active filter is an electronic amplifier, with the filter characteristics being defined by a network of amplifier input and feedback components consisting of resistors and capacitors. The fact that resistors are used instead of the more expensive and bulky inductors required by passive filters is a further advantage of the active type of filter.

As in the case of passive filters, the design of active filters is a subject of considerable complexity and is only considered here in simple terms. The reader requiring a deeper understanding is referred to a specialist text, such as the one by Hilburn and Johnson (1973).

The characteristics of the filter in terms of its attenuation behaviour in the pass and stop bands is determined by the choice of circuit components in Figure 5.11. A common set of design formulae is given below:

(a) *low-pass filter*

$$K = \frac{G}{R_1 R_2 C_1 C_2}$$

$$a = \frac{1-G}{R_2 C_2} + \frac{1}{R_1 C_1} + \frac{1}{R_2 C_1}$$

(b) *high-pass filter*

$$K = G$$

$$a = \frac{1-G}{R_1 C} + \frac{2}{R_2 C}$$

$$b = \frac{1}{R_1 R_2 C_1 C_2}$$

$$b = \frac{1}{R_1 R_2 C^2}$$

$$G = 1 + (R_4/R_3) = K/b$$

$$G = 1 + (R_4/R_3)$$

$$= \text{d.c. gain}$$

(c) *band-pass filter*

(d) *band-stop filter*

$$K = G/R_1 C$$

$$R_3 R_4 = 2 R_1 R_5$$

$$B = \omega_2 - \omega_1 = \frac{4 - G}{R_1 C}$$

$$B = \frac{2}{R_4 C}$$

$$\omega_0^2 = \frac{2}{R_1^2 C^2}$$

$$\omega_0^2 = \frac{1}{R_4 C^2}\left(\frac{1}{R_1} + \frac{1}{R_2}\right)$$

where $G = 1 + R_3/R_2$, ω_1 and ω_2 are frequencies at ends of pass band, ω_0 is centre frequency of pass band and gain at frequency ω_0 is K/b

where ω_1 and ω_2 are frequencies at ends of stop band, ω_0 is centre frequency of pass band and inverting gain magnitude is R_6/R_3

Design procedures using the formulae above involve the use of graphs in which gains and cut-off frequencies are plotted for a range of parameter values (see Hilburn and Johnson 1973). Suitable parameter values are then chosen by inspection.

Whilst the above formulae represent the form of active filter which is the most general-purpose one available, many other forms also exist with the same circuit structure but with different rules for defining component values. Butterworth filters, for instance, optimize the pass-band attenuation characteristics at the expense of stop-band performance. Another form, Chebyshev filters, have very good stop-band attenuation characteristics but poorer pass-band performance. These are considered in considerable detail in the text by Hilburn and Johnson (1973).

5.6 Signal manipulation

To complete the discussion on analog signal processing techniques, mention must also be made of certain other special-purpose devices and circuits used to manipulate signals. These are listed below.

5.6.1 Voltage-to-current conversion

Many process control systems use current to transmit signals rather than voltage. Hence, voltage-to-current conversion is important.

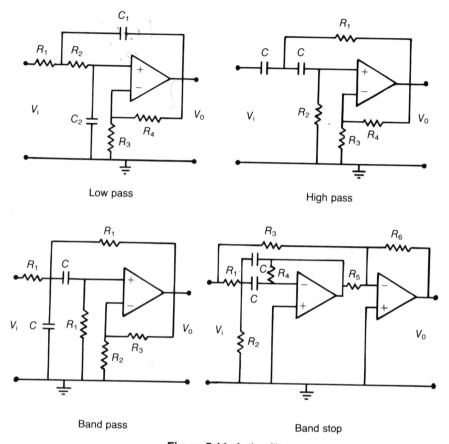

Low pass

High pass

Band pass

Band stop

Figure 5.11 Active filters

An operational amplifier circuit, as shown in Figure 5.12, is a suitable voltage-to-current converter in which the output current I is related to the input voltage V by the equation:

$$I = -\frac{R_2}{R_1 R_3} V$$

If required, the amplifier gain can be reduced to lower the output current level by either increasing R_1 and R_3 or reducing R_2.

5.6.2 **Current-to-voltage conversion**

Current-to-voltage conversion is often required at the termination of transmission lines in process control systems to change the transmitted currents back to voltages.

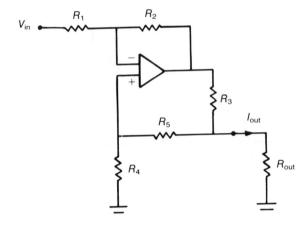

Figure 5.12 Operational amplifier connected for voltage-to-current conversion

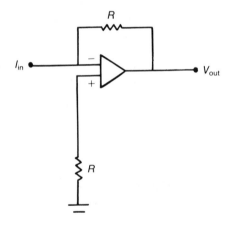

Figure 5.13 Operational amplifier connected for current-to-voltage conversion

Again, an operational amplifier, connected as shown in Figure 5.13, is suitable for this. The output voltage V is simply related to the input current I by:

$$V = IR$$

5.6.3 Signal integration

Connected in the configuration shown in Figure 5.14, an operational amplifier is able to integrate the input signal V_i such that the output signal V_o is given by:

$$V_0 = -\frac{1}{RC} \int V_i \, dt$$

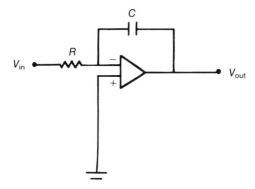

Figure 5.14 Operational amplifier connected as integrating element

This circuit is used whenever there is a requirement to integrate the output signal from a transducer.

5.6.4 Voltage follower (pre-amplifier)

The voltage follower, also known as a pre-amplifier, is a unity-gain amplifier circuit with a short circuit in the feedback path, as shown in Figure 5.15, such that:

$$V_0 = V_i$$

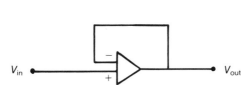

Figure 5.15 Operational amplifier connected as voltage follower (pre-amplifier)

It has a very high input impedance and its main application is to reduce the load on the measured system. It also has a very low output impedance which is very useful in some applications.

5.6.5 Voltage comparator

The output of a voltage comparator switches between positive and negative values according to whether the difference between the two input signals to it is positive or negative. An operational amplifier connected as shown in Figure 5.16 gives an output which switches between positive and negative saturation levels according to whether $V_1 - V_2$ is greater than or less than zero.

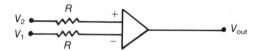

Figure 5.16 Comparison between two voltage signals

Alternatively, the voltage of a single input signal can be compared against positive and negative reference levels with the circuit shown in Figure 5.17.

In practice, operational amplifiers have drawbacks as voltage comparators for several reasons. These include non-compatibility between output voltage levels and industry-standard logic circuits, propagation delays and slow recovery. In consequence, various other special-purpose integrated circuits have been developed for voltage comparison.

5.6.6 **Phase-locked loop**

The phase-locked loop, described in detail in section 7.7, is primarily a circuit for measuring the frequency of a signal. However, because the output waveform is a pure (i.e. perfectly clean) square wave at the same frequency as the input signal, irrespective of the amount of noise, modulation or distortion on the input signal, the phase-locked loop also finds application as a signal processing element to clean up poor-quality signals.

5.6.7 **Signal addition**

The most common mechanism for summing two or more input signals is the use of an operational amplifier connected in signal-inversion mode, as shown in Figure 5.18. For input signal voltages V_1, V_2 and V_3, the output voltage V_0 is given by:

$$V_0 = - (V_1 + V_2 + V_3)$$

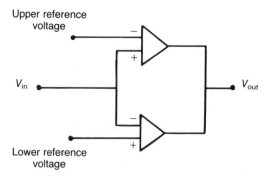

Figure 5.17 Comparison of input signal against reference value

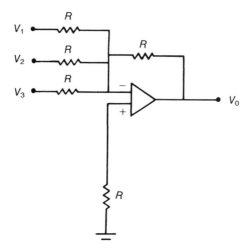

Figure 5.18 Operational amplifier connected for signal addition

5.6.8 **Signal multiplication**

Great care must be taken when choosing a signal multiplier because, whilst many circuits exist for multiplying two analog signals together, most of them are two-quadrant types which only work for signals of a single polarity, i.e. both positive or both negative. Such schemes are unsuitable for general analog signal processing, where the signals to be multiplied may be of changing polarity.

For analog signal processing, a four-quadrant multiplier is required. Two forms of such a multiplier are easily available, the Hall-effect multiplier and the translinear multiplier.

5.6.9 **Sample and hold circuit**

A sample and hold circuit is often an essential element at the interface between an analog instrument/transducer and an analog-to-digital converter. It holds the input signal at a constant level whilst the analog-to-digital conversion process is taking place and prevents the conversion errors which would probably result if variations in the measured signal were allowed to pass through to the converter. The operational amplifier circuit shown in Figure 5.19 provides this sample and hold function. After the input signal has been applied to the circuit for a very short time duration, the signal level is held until the circuit is reset when the next sample is required.

5.6.10 **Analog-to-digital conversion**

In most computer-controlled systems, there is a fundamental mismatch between the analog form of output data from instruments and transducers and the digital form of

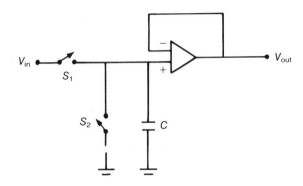

Figure 5.19 Operational amplifier connected as 'sample and hold' circuit

data required by a digital computer. This problem is solved by the provision of an analog-to-digital converter in the computer input interface. This is discussed in more detail in Chapter 10.

5.6.11 Digital-to-analog conversion

Similarly in computer-controlled systems, control actuators usually require the application of analog signals whereas the controlling computer has a digital output. This is overcome by providing a digital-to-analog converter in the computer output interface. This is also discussed further in Chapter 10.

5.7 Digital signal processing

Digital techniques achieve much greater levels of accuracy in signal processing than equivalent analog methods. However, the time taken to process a signal digitally is much longer than that required to carry out the same operation by analog techniques, and the equipment required is more expensive. Some care is needed therefore in making the correct choice between digital and analog methods in a particular signal processing application.

Whilst digital signal processing elements in a measurement system can exist as separate units, it is more usual to find them as an integral part of an intelligent instrument. However, their construction and mode of operation are the same irrespective of whether they exist physically as separate boxes or within an intelligent instrument.

The hardware aspect of a digital signal processing element consists of a digital computer and analog interface boards, which will be considered in detail in Chapter 10. The actual form which signal processing takes depends on the software program executed by the processor. However, before consideration is given to this, some theoretical aspects of signal sampling need to be discussed.

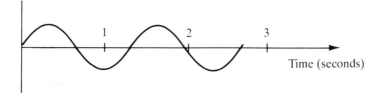

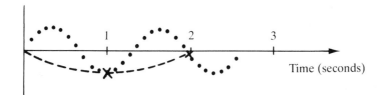

Figure 5.20 Conversion of continuous analog signal to discrete sampled signal

As mentioned earlier, digital computers require signals to be in digital form whereas most instrumentation transducers have an output signal in analog form. Analog-to-digital conversion is therefore required at the interface between analog transducers and the digital computer. The procedure followed is to sample the analog signal at a particular moment in time and then convert the analog value to an equivalent digital one. This conversion takes a certain finite time, during which the analog signal can be changing in value. The next sample of the analog signal cannot be taken until the conversion of the last sample to digital form is completed. The representation within a digital computer of a continuous analog signal is therefore a sequence of samples whose pattern only approximately follows the shape of the original signal. This pattern of samples taken at successive, equal intervals of time is known as a discrete signal. The process of conversion between a continuous analog signal and a discrete digital one is illustrated for a sine wave in Figure 5.20.

The raw analog signal in Figure 5.20 has a frequency of approximately 0.75 cycles per second. With the rate of sampling shown, which is approximately 11 samples per second, reconstruction of the samples matches the original analog signal very well. If the rate of sampling were decreased, the fit between the reconstructed samples and the original signal would be less good. If the rate of sampling was very much less than the frequency of the raw analog signal, such as 1 sample per second, only the samples marked with a cross in Figure 5.20 would be obtained. Fitting a line through these crosses incorrectly estimates a signal whose frequency is approximately 0.25 cycles per second. This phenomenon, whereby the process of sampling transmutes a high-frequency signal into a lower-frequency one, is known as **aliasing**. To avoid aliasing, it is necessary theoretically for the sampling rate to be at least twice the highest frequency in the analog signal sampled. In practice, sampling rates of between five and ten times the highest-frequency signal are normally chosen so that the discrete sampled signal is a close approximation to the original analog signal in amplitude as well as frequency.

Problems can arise in sampling when the raw analog signal is corrupted by high-frequency noise of unknown characteristics. It would be normal practice to choose the sampling interval as, say, a ten-times multiple of the frequency of the measurement component in the raw signal. If such a sampling interval is chosen, aliasing can in certain circumstances transmute high-frequency noise components into the same frequency range as the measurement component in the signal, thus giving erroneous results. This is one of the circumstances mentioned earlier, where prior analog signal conditioning in the form of a low-pass filter must be carried out before processing the signal digitally.

One further factor which affects the quality of a signal when it is converted from analog to digital form is **quantization**. Quantization describes the procedure whereby the continuous analog signal is converted into a number of discrete levels. At any particular value of the analog signal, the digital representation is either the discrete level immediately above this value or the discrete level immediately below it. If the difference between two successive discrete levels is represented by the parameter Q, then the maximum error in each digital sample of the raw analog signal is $\pm Q/2$. This error is known as the quantization error and is clearly proportional to the resolution of the analog-to-digital converter, i.e. to the number of bits used to represent the samples in digital form.

Once a satisfactory digital representation in discrete form of an analog signal has been obtained, the procedures of signal amplification, signal attenuation and bias removal become trivial. For signal amplification and attenuation, all samples have to be multiplied or divided by a fixed constant. Bias removal involves simply adding or subtracting a fixed constant from each sample of the signal.

Signal linearization requires **a priori** knowledge of the type of non-linearity involved, in the form of a mathematical equation which expresses the relationship between the output measurements from an instrument and the value of the physical quantity being measured. This can be obtained either theoretically through knowledge of the physical laws governing the system or empirically using input–output data obtained from the measurement system under controlled conditions. Once this relationship has been obtained, it is used to calculate the value of the measured physical quantity corresponding to each discrete sample of the measurement signal. Whilst the amount of computation involved in this is greater than for the trivial cases of signal amplification, etc., already mentioned, the computational burden is still relatively small in most measurement situations.

Digital signal processing can also perform all of the filtering functions mentioned earlier in respect of analog filters, i.e. low pass, high pass, band pass and band stop. However, the design of digital filters requires a level of theoretical knowledge, including the use of z-transform theory, which is outside the range of this book. The reader interested in digital filter design is therefore referred elsewhere (Lynn 1973, Huelsman 1970).

5.8 **Signal transmission**

There is a necessity in many measurement systems to transmit measurement signals over quite large distances from the point of measurement to the place where the signals are recorded and/or used in a process control system. This creates several problems for which a solution must be found. Of the difficulties associated with long-distance signal transmission, the most serious is contamination of the measurement signal by noise. Many sources of noise exist in industrial environments, such as radiated elctromagnetic fields from electrical machinery and power cables, induced fields through wiring loops and spikes on the a.c. power supply.

5.8.1 **Signal amplification**

The output signal levels from many types of measurement transducer are very low, and amplification of the signal prior to transmission is essential if a reasonable signal-to-noise ratio is to be obtained after transmission. Amplification at the input to the transmission system is also required to compensate for the attenuation of the signal which results from the resistance of the signal wires. The means of amplifying signals has already been discussed in section 5.1.

5.8.2 **Shielding**

Shielding consists of surrounding the signal wires in a cable with a braided metal shield which is connected to earth. This provides a high degree of noise protection, especially against capacitive-induced noise due to the proximity of signal wires to high-current power conductors.

5.8.3 **Current loop transmission**

The signal attenuation effect of conductor resistances can be minimized if varying voltage signals are transmitted as varying current signals. This requires a voltage-to-current converter of the form shown in Figure 5.21, which is commonly known as a 4–20 mA current loop interface. Two voltage-controlled current sources are used, one providing a constant 4 mA output which is used as the power supply current and the other providing a variable 0–16 mA output which is proportional to the input voltage level. The net output current therefore varies between 4 and 20 mA. This is a very common means of connecting remote instruments to a central control room.

5.8.4 **Voltage-to-frequency conversion**

Even better immunity to noise can be obtained in signal transmission if the signal is transmitted in a digital format. This is done by putting the input analog voltage signal into a voltage-to-frequency converter circuit which converts voltage variations into frequency variations. Such frequency variations can be readily transmitted in a digital

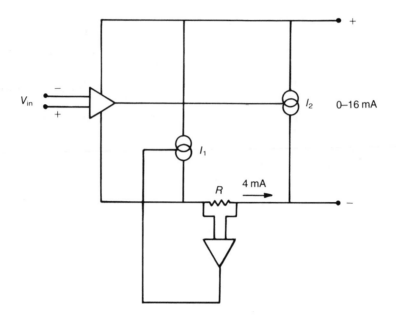

Figure 5.21 4–20 mA current loop transmitter

format. After transmission, reconversion back into an analog voltage signal is generally required using a frequency-to-voltage converter.

5.8.5 **Fiber optic transmission**

Noise corruption of signals is almost eliminated by the use of fiber optic transmission cables, but there is a cost penalty associated with this because of the higher cost of a fiber optic system compared with that of metal conductors. Apart from noise reduction, signal attenuation along a fiber optic link is much less than along an equivalent length of metal conductor. A more detailed study of fiber optic systems can be found in Chapter 9.

Signals are normally transmitted along a fiber optic cable in digital format, although analog transmission is sometimes used. If there is a requirement to transmit more than one signal, it is more economic to multiplex the signals on to a single cable rather than transmit the signals separately on multiple cables. **Multiplexing** involves switching the analog signals in turn, in a synchronized sequential manner, into an analog-to-digital converter which outputs on to the transmission line. At the other end of the transmission line, a digital-to-analog converter transforms the digital signal back into analog form and it is then switched in turn on to separate analog signal lines.

References and further reading

Blinchikoff, H. J. (1976) *Filtering in the Time and Frequency Domains*, Wiley: New York.

Hilburn, J. L. and Johnson, D. E. (1973) *Manual of Active Filter Design*, McGraw-Hill: New York.

Huelsman, L. P. (1970) *Active Filters: Lumped, Distributed, Integrated, Digital and Parametric*, McGraw-Hill: New York.

Lynn, P. A. (1973) *The Analysis and Processing of Signals*, Macmillan: London.

Skilling, H. H. (1967) *Electrical Engineering Circuits*, Wiley: New York.

Williams, E. (1963) *Electric Filter Circuits*, Pitman: London.

6 Bridge circuits

Bridge circuits are used very commonly as a variable conversion element in measurement systems and produce an output in the form of a voltage level which changes as the measured physical quantity changes in value. They provide an accurate method of measuring resistance, inductance and capacitance values, and enable very small changes in these quantities about a nominal value to be detected. They are of immense importance in measurement system technology because so many transducers measuring physical quantities have an output which is expressed as a change in resistance, inductance or capacitance. The displacement-measuring strain gauge, which has a varying resistance output, is but one example of this class of transducer. Excitation of the bridge is normally by a d.c. voltage for resistance measurement and by an a.c. voltage for inductance or capacitance measurement. Both null and deflection types of bridge exist, and, in a like manner to instruments in general, null types are mainly employed for calibration purposes and deflection types are used within closed-loop automatic control schemes.

6.1 Null-type d.c. bridge (Wheatstone bridge)

A null-type bridge with d.c. excitation, commonly known as a Wheatstone bridge, has the form shown in Figure 6.1. The four arms of the bridge consist of the unknown resistance R_u, two equal-value resistors R_2 and R_3 and a variable resistor R_v (usually a decade resistance box). A d.c. voltage V_i is applied across the points AC and the resistance R_v is varied until the voltage measured across points BD is zero. This null point is usually measured with a high-sensitivity galvanometer.

To analyze the Wheatstone bridge, define the current flowing in each arm to be I_1 ... I_4 as shown in Figure 6.1. Normally, if a high-impedance voltage measuring instrument is used, the current I_m drawn by the measuring instrument will be very small and can be approximated to zero. If this assumption is made, then, for $I_m = 0$:

$$I_1 = I_3 \quad \text{and} \quad I_2 = I_4$$

Looking at path ADC, we have a voltage V_i applied across a resistance $R_u + R_3$ and by Ohm's law:

$$I_1 = \frac{V_i}{R_u + R_3}$$

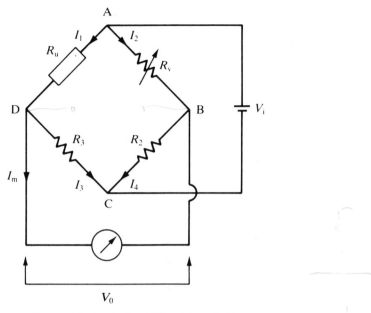

Figure 6.1 Analysis of Wheatstone bridge

Similarly for path ABC:

$$I_2 = \frac{V_i}{R_v + R_2}$$

Now we can calculate the voltage drop across AD and AB:

$$V_{AD} = I_1 R_u = \frac{V_i R_u}{R_u + R_3} \qquad \left(\frac{R_u}{R_u + R_3}\right) V_i$$

$$V_{AB} = I_2 R_v = \frac{V_i R_v}{R_v + R_2}$$

By the principle of superposition, $V_0 = V_{BD} = V_{BA} + V_{AD} = -V_{AB} + V_{AD}$. Thus:

$$V_0 = -\frac{V_i R_v}{R_v + R_2} + \frac{V_i R_u}{R_u + R_3} \qquad (6.1)$$

At the null point $V_0 = 0$, so:

$$\frac{R_u}{R_u + R_3} = \frac{R_v}{R_v + R_2}$$

Inverting both sides:

$$\frac{R_u + R_3}{R_u} = \frac{R_v + R_2}{R_v}$$

i.e.

$$\frac{R_3}{R_u} = \frac{R_2}{R_v}$$

or

$$R_u = \frac{R_3 R_v}{R_2} \tag{6.2}$$

Thus, if $R_2 = R_3$, then $R_u = R_v$.

As R_v is an accurately known value because it is derived from a variable decade resistance box, this means that R_u is also accurately known.

6.2 Deflection-type d.c. bridge

A deflection-type bridge with d.c. excitation is shown in Figure 6.2. This differs from the Wheatstone bridge mainly in that the variable resistance R_v is replaced by a fixed resistance R_1 of the same value as the nominal value of the unknown resistance R_u. As the resistance R_u changes, so the output voltage V_0 varies, and this relationship between V_0 and R_u must be calculated.

This relationship is simplified if we again assume that a high-impedance voltage measuring instrument is used and the current drawn by it, I_m, can be approximated to zero. (The case when this assumption does not hold is covered later in this section.) The analysis is then exactly the same as for the preceding example of the Wheatstone bridge, except that R_v is replaced by R_1. Thus, from Equation (6.1), we have:

$$V_0 = V_i \left(\frac{R_u}{R_u + R_3} - \frac{R_1}{R_1 + R_2} \right) \tag{6.3}$$

When R_u is at its nominal value, i.e. for $R_u = R_1$, it is clear that $V_0 = 0$ (since $R_2 = R_3$). For other values of R_u, V_0 has negative and positive values which vary in a non-linear way with R_u.

Example 6.1
A certain type of pressure transducer, designed to measure pressures in the range 0–10 bar, consists of a diaphragm with a strain gauge cemented to it to detect diaphragm deflections. The strain gauge has a nominal resistance of 120 Ω and

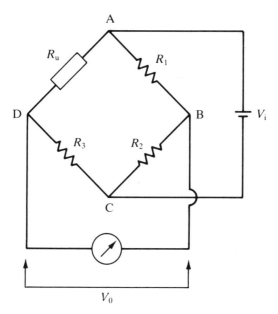

Figure 6.2 Deflection-type d.c. bridge

forms one arm of a Wheatstone bridge circuit, with the other three arms each having a resistance of 120 Ω. The bridge output is measured by an instrument whose input impedance can be assumed infinite. If, in order to limit heating effects, the maximum permissible gauge current is 30 mA, calculate the maximum permissible bridge excitation voltage. If the sensitivity of the strain gauge is 338 mΩ/bar and the maximum bridge excitation voltage is used, calculate the bridge output voltage when measuring a pressure of 10 bar.

Solution
This is the type of bridge circuit shown in Figure 6.2 in which the components have the following values:

$$R_1 = R_2 = R_3 = 120 \, \Omega$$

Defining I_1 to be the current flowing in path ADC of the bridge, we can write:

$$V_i = I_1(R_u + R_3)$$

At balance, $R_u = 120 \, \Omega$ and the maximum value allowable for I_1 is 0.03 A. Hence:

$$V_i = 0.03(120 + 120) = 7.2 \, V$$

Thus the maximum bridge excitation voltage allowable is 7.2 volts.
For a pressure of 10 bar applied, the resistance change is 3.38 Ω, i.e. R_u is then equal to 123.38 Ω. Applying Equation (6.3), we can write:

$$V_0 = V_i\left(\frac{R_u}{R_u + R_3} - \frac{R_1}{R_1 + R_2}\right) = 7.2\left(\frac{123.38}{243.38} - \frac{120}{240}\right) = 50\,\text{mV}$$

Thus, if the maximum permissible bridge excitation voltage is used, the output voltage is 50 mV when a pressure of 10 bar is measured.

The non-linear relationship between the output reading and the measured quantity exhibited by Equation (6.3) is inconvenient and does not conform with our normal requirement for a linear input–output relationship. The method of coping with this non-linearity varies according to the form of primary transducer involved in the measurement system.

One special case is where the change in the unknown resistance R_u is typically small compared with the nominal value of R_u.

If we calculate the new voltage V_0' when the resistance R_u in Equation (6.3) changes by an amount δR_u, we have:

$$V_0' = V_i\left(\frac{R_u + \delta R_u}{R_u + \delta R_u + R_3} - \frac{R_1}{R_1 + R_2}\right) \qquad (6.4)$$

The change of voltage output is therefore given by:

$$\delta V_0 = V_0' - V_0 = \frac{V_i\,\delta R_u}{R_u + \delta R_u + R_3}$$

If $\delta R_u \ll R_u$, then the following linear relationship is obtained:

$$\frac{\delta V_0}{\delta R_u} = \frac{V_i}{R_u + R_3} \qquad (6.5)$$

This expression describes the measurement sensitivity of the bridge. Such an approximation to make the relationship linear is valid for transducers such as strain gauges where the typical changes of resistance with strain are very small compared with the nominal gauge resistance.

However, many instruments which are inherently linear themselves at least over a limited measurement range, such as resistance thermometers, exhibit large changes in output as the input quantity changes, and the approximation of Equation (6.5) cannot be applied. In such cases, specific action must be taken to improve linearity in the relationship between the bridge output voltage and the measured quantity. One common solution to this problem is to make the values of the resistances R_2 and R_3 at least ten times those of R_1 and R_u (nominal). The effect of this is best observed by looking at a numerical example.

Consider a platinum resistance thermometer with a range of 0–50 °C, whose resistance at 0 °C is 500 Ω and which varies with temperature at the rate of 4 Ω/°C. Over this range of measurement, the output characteristic of the thermometer itself is nearly perfectly linear. (The subject of resistance thermometers is discussed further in Chapter 12.)

Taking first the case where $R_1 = R_2 = R_3 = 500\,\Omega$ and $V_i = 10\,\text{V}$, and applying Equation (6.3):

at 0°C

$$V_0 = 0$$

at 25 °C

$$R_u = 600\,\Omega \qquad V_0 = 10\left(\frac{600}{1100} - \frac{500}{1000}\right) = 0.455\,\text{V}$$

at 50 °C

$$R_u = 700\,\Omega \qquad V_0 = 10\left(\frac{700}{1200} - \frac{500}{1000}\right) = 0.833\,\text{V}$$

This relationship between V_0 and R_u is plotted as curve A in Figure 6.3 and the non-linearity is apparent. Inspection of the manner in which the output voltage V_0 above changes for equal steps of temperature change also clearly demonstrates the non-linearity:

for the temperature change from 0 to 25 °C, the change in V_0 is $(0.455 - 0) = 0.455\,\text{V}$

for the temperature change from 25 to 50 °C, the change in V_0 is $(0.833 - 0.455) = 0.378\,\text{V}$

If the relationship was linear, the change in V_0 for the 25–50 °C temperature step would also be 0.455 V, giving a value for V_0 of 0.910 V at 50 °C.

Now take the case where $R_1 = 500\,\Omega$ but $R_2 = R_3 = 5000\,\Omega$ and let $V_i = 26.1\,\text{V}$:

at 0°C

$$V_0 = 0$$

at 25 °C

$$R_u = 600\,\Omega \qquad V_0 = 26.1\left(\frac{600}{5600} - \frac{500}{5500}\right) = 0.424\,\text{V}$$

at 50 °C

$$R_u = 700\ \Omega \quad V_0 = 26.1\left(\frac{700}{5700} - \frac{500}{5500}\right) = 0.833\ \text{V}$$

This relationship is shown as curve B in Figure 6.3 and a considerable improvement in linearity is achieved. This is more apparent if the differences in values for V_0 over the two temperature steps are inspected:

from 0 to 25 °C, the change in V_0 is 0.424 V
from 25 to 50 °C, the change in V_0 is 0.409 V

The changes in V_0 over the two temperature steps are much closer to being equal than before, demonstrating the improvement in linearity.

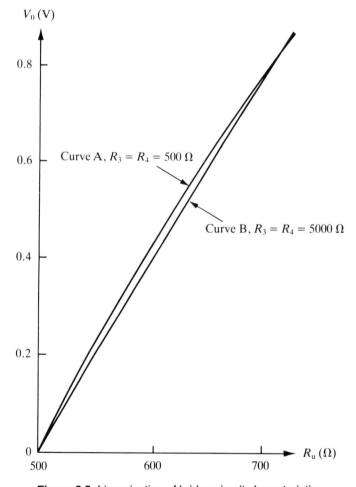

Figure 6.3 Linearization of bridge circuit characteristic

In increasing the values of R_2 and R_3, it was also necessary to increase the excitation voltage from 10 V to 26.1 V to obtain the same output levels. In practical applications, V_i would normally be set at the maximum level consistent with the limitation of the effect of circuit heating in order to maximize the measurement sensitivity ($\delta V_0/\delta R_u$ relationship). It would therefore not be possible to increase V_i further if R_2 and R_3 were increased, and the general effect of such an increase in R_2 and R_3 is thus a decrease in the sensitivity of the measurement system.

The importance of this inherent non-linearity in the bridge output relationship is greatly diminished if the primary transducer and bridge circuit are incorporated as elements within an intelligent instrument. In that case, digital computation is applied to produce an output in terms of the measured quantity which automatically compensates for the non-linearity in the bridge circuit.

6.2.1 **The case where current drawn by the measuring instrument is not negligible**

For various reasons, it is not always possible to meet the condition that the impedance of the instrument measuring the bridge output voltage is sufficiently large for the current drawn by it to be negligible. Wherever the measurement current is not negligible, an alternative relationship between the bridge input and output must be derived which takes the current drawn by the measuring instrument into account.

Thévenin's theorem is again a useful tool for this purpose. Replacing the voltage source V_i in Figure 6.4(a) by a zero internal resistance produces the circuit shown in Figure 6.4(b), or the equivalent representation shown in Figure 6.4(c). It is apparent from Figure 6.4(c) that the equivalent circuit resistance consists of a pair of parallel resistors R_u and R_3 in series with the parallel resistor pair R_1 and R_2. Thus, R_{DB} is given by:

$$R_{DB} = \frac{R_1 R_2}{R_1 + R_2} + \frac{R_u R_3}{R_u + R_3} \tag{6.6}$$

The equivalent circuit derived via Thévenin's theorem with the resistance R_m of the measuring instrument connected across the output is shown in Figure 6.4(d).

The open-circuit voltage across DB, E_0, is the output voltage calculated earlier (Equation (6.3)) for the case of $R_m = 0$:

$$E_0 = V_i \left(\frac{R_u}{R_u + R_3} - \frac{R_1}{R_1 + R_2} \right) \tag{6.7}$$

If the current flowing is I_m when the measuring instrument of resistance R_m is connected across DB, then, by Ohm's law, I_m is given by:

$$I_m = \frac{E_0}{R_{DB} + R_m} \tag{6.8}$$

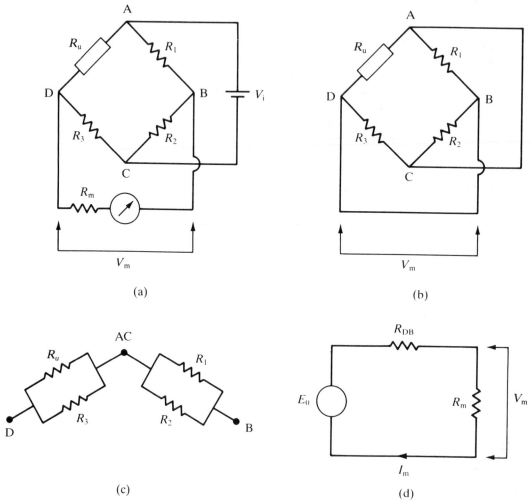

Figure 6.4 (a) A bridge circuit; (b) equivalent circuit by Thévenin's theorem; (c) alternative representation; (d) equivalent circuit via Thévenin's theorem

If V_m is the voltage measured across R_m, then again by Ohm's law:

$$V_m = I_m R_m = \frac{E_0 R_m}{R_{DB} + R_m} \qquad (6.9)$$

Substituting for E_0 and R_{DB} in Equation (6.9), using the relationships developed in Equations (6.6) and (6.7), we obtain:

$$V_m = \frac{V_i[R_u/(R_u + R_3) - R_1/(R_1 + R_2)]R_m}{R_1 R_2/(R_1 + R_2) + R_u R_3/(R_u + R_3) + R_m}$$

Simplifying:

$$V_m = \frac{V_i R_m (R_u R_2 - R_1 R_3)}{R_1 R_2 (R_u + R_3) + R_u R_3 (R_1 + R_2) + R_m (R_1 + R_2)(R_u + R_3)} \qquad (6.10)$$

Example 6.2
A bridge circuit, as shown in Figure 6.5, is used to measure the value of the unknown resistance R_u of a strain gauge of nominal value 500 Ω. The output voltage measured across points DB in the bridge is measured by a voltmeter. Calculate the measurement sensitivity in volts per ohm change in R_u if
(a) the resistance R_m of the measuring instrument is neglected, and
(b) account is taken of the value of R_m.

Solution
For $R_u = 500\ \Omega$, $V_m = 0$. To determine sensitivity, calculate V_m for $R_u = 501\ \Omega$.
(a) Applying Equation (6.3):

$$V_m = V_i \left(\frac{R_u}{R_u + R_3} - \frac{R_1}{R_1 + R_2} \right)$$

Substituting in values:

$$V_m = 10 \left(\frac{501}{1001} - \frac{500}{1000} \right) = 5.00\ \text{mV}$$

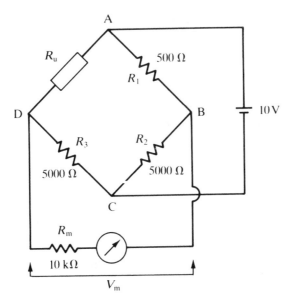

Figure 6.5 Bridge circuit

Thus, if the resistance of the measuring circuit is neglected, the measurement sensitivity is 5.00 mV per ohm change in R_u.

(b) applying Equation (6.10) and substituting in values:

$$V_m = \frac{10 \times 10^4 \times 500(501 - 500)}{500^2(1001) + 500 \times 501(1000) + 10^4 \times 1000 \times 1001} = 4.76\,\text{mV}$$

Thus, if proper account is taken of the 10 kΩ value of the resistance of R_m, the true measurement sensitivity is shown to be 4.76 mV per ohm change in R_u.

6.3 Error analysis

In the application of bridge circuits, the contribution of component-value tolerances to the limits of accuracy of the total measurement system must be clearly understood. The analysis below applies to a null-type (Wheatstone) bridge, but similar principles can be applied for a deflection-type bridge.

The maximum measurement error is determined by first finding the value of R_u in Equation (6.2) with each parameter in the equation set at that limit of its tolerance which produces the maximum value of R_u. Similarly, the minimum possible value of R_u is calculated, and the required error band is then the span between these maximum and minimum values.

Example 6.3
In the Wheatstone bridge circuit of Figure 6.1, R_v is a decade resistance box of accuracy ±0.2% and $R_2 = R_3 = 500\,\Omega \pm 0.1\%$. If the value of R_v at the null position is 520.4 Ω, determine the error band for R_u expressed as a percentage of its nominal value.

Solution
Applying Equation (6.2) with $R_v = 520.4\,\Omega + 0.2\% = 521.44\,\Omega$,
$R_3 = 5000\,\Omega + 0.1\% = 5005\,\Omega$, $R_2 = 5000\,\Omega - 0.1\% = 4995\,\Omega$, we get:

$$R_v = \frac{521.44 \times 5005}{4995} = 522.48\,\Omega\ (= +0.4\%)$$

Applying Equation (6.2) with $R_v = 520.4\,\Omega - 0.2\% = 519.36\,\Omega$,
$R_3 = 5000\,\Omega - 0.1\% = 4995\,\Omega$, $R_2 = 5000\,\Omega + 0.1\% = 5005\,\Omega$, we get:

$$R_v = \frac{519.36 \times 4995}{5005} = 518.32\,\Omega\ (= -0.4\%)$$

Thus the error band for R_u is ±0.4%.

The cumulative effect of errors in individual bridge circuit components is clearly seen. Although the maximum error in any one component is $\pm 0.2\%$, the possible error in the measured value of R_u is $\pm 0.4\%$. Such a magnitude of error is often not acceptable, and special measures are taken to overcome the introduction of error by component-value tolerances. One such practical measure is the introduction of apex balancing. This is one of many methods of bridge balancing which all produce a similar result.

6.3.1 Apex balancing

One form of apex balancing consists of placing an additional variable resistor R_5 at the junction C between the resistances R_2 and R_3 and applying the excitation voltage V_i to the wiper of this variable resistance, as shown in Figure 6.6.

For calibration purposes, R_u and R_v are replaced by two equal resistances whose values are accurately known, and R_5 is varied until the output voltage V_0 is zero. At this point, if the portions of resistance on either side of the wiper on R_5 are R_6 and R_7 (such that $R_5 = R_6 + R_7$), we can write:

$$R_3 + R_6 = R_2 + R_7$$

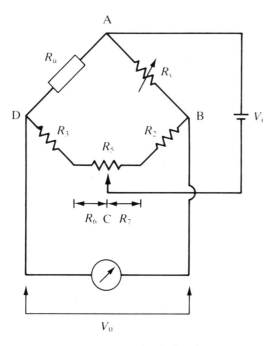

Figure 6.6 Apex balancing

We have thus eliminated any source of error due to the tolerance in the values of R_2 and R_3, and the error in the measured value of R_u depends only on the accuracy of one component, the decade resistance box R_v.

Example 6.4

A potentiometer R_5 is put into the apex of the bridge shown in Figure 6.6 to balance the circuit. The bridge components have the following values: $R_u = 500 \, \Omega$, $R_v = 500 \, \Omega$, $R_2 = 515 \, \Omega$, $R_3 = 480 \, \Omega$, $R_5 = 100 \, \Omega$. Determine the required value of the resistances R_6 and R_7 of the parts of the potentiometer track either side of the slider in order to balance the bridge and compensate for the unequal values of R_2 and R_3.

Solution

For balance, $R_2 + R_7 = R_3 + R_6$. Hence, $515 + R7 = 480 + R6$. Also, because R_6 and R_7 are the two parts of the potentiometer track R_5 whose resistance is 100 Ω:

$$R_6 + R_7 = 100$$

Thus $515 + R_7 = 480 + (100 - R_7)$, i.e. $2R_7 = 580 - 515 = 65$. Thus, $R_7 = 32.5$. Hence, $R_6 = 100 - 32.5 = 67.5 \, \Omega$.

6.4 Alternating current bridges

Bridges with a.c. excitation are used to measure unknown impedances. As for d.c. bridges, both null and deflection types exist, with null types being generally reserved for calibration duties.

6.4.1 Null-type impedance bridge

A typical null-type impedance bridge is shown in Figure 6.7. The null point can be conveniently detected by monitoring the output with a pair of headphones connected via an operational amplifier across the points BD. This is a much cheaper method of null detection than the application of an expensive galvanometer which is required for a d.c. Wheatstone bridge.

Referring to Figure 6.7, at the null point:

$$I_1 R_1 = I_2 R_2$$

$$I_1 Z_u = I_2 Z_v$$

Thus:

$$Z_u = \frac{Z_v R_1}{R_2} \tag{6.11}$$

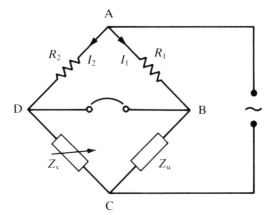

Figure 6.7 Null-type impedance bridge

If Z_u is capacitive, i.e. $Z_u = 1/j\omega C_u$, then Z_v must consist of a variable capacitance box, which is readily available.

If Z_u is inductive, then $Z_u = R_u + j\omega L_u$.

Notice that the expression for Z_u now has a resistive term in it because it is impossible to realize a pure inductor. An inductor coil always has a resistive component, though this is made as small as possible by designing the coil to have a high Q factor (Q factor is the ratio of inductance to resistance).

Therefore, Z_v must consist of a variable resistance box and a variable inductance box. However, the latter are not readily available because it is difficult and hence expensive to manufacture a set of fixed value inductors to make up a variable inductance box. For this reason, an alternative kind of null-type bridge circuit, known as the Maxwell bridge, is commonly used to measure unknown inductances.

6.4.2 **Maxwell bridge**

A Maxwell bridge is shown in Figure 6.8. The requirement for a variable inductance box is avoided by introducing instead a second variable resistance. The circuit requires one standard fixed value capacitor, two variable resistance boxes and one standard fixed value resistor, all of which are components which are readily available and inexpensive.

Referring to Figure 6.8, we have at the null-output point:

$$I_1 Z_{AD} = I_2 Z_{AB}$$

$$I_1 Z_{DC} = I_2 Z_{BC}$$

Thus:

$$\frac{Z_{BC}}{Z_{AB}} = \frac{Z_{DC}}{Z_{AD}}$$

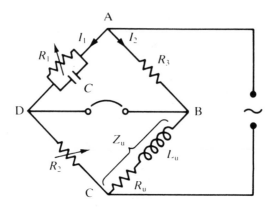

Figure 6.8 Maxwell bridge

or:

$$Z_{BC} = \frac{Z_{DC}Z_{AB}}{Z_{AD}}$$

(6.12)

The quantities in Equation (6.12) have the following values:

$$\frac{1}{Z_{AD}} = \frac{1}{R_1} + j\omega C \qquad \text{or} \qquad Z_{AD} = \frac{R_1}{1 + j\omega C R_1}$$

$$Z_{AB} = R_3 \quad Z_{BC} = R_u + j\omega L_u \quad Z_{DC} = R_2$$

Substituting the values into Equation (6.12):

$$R_u + j\omega L_u = \frac{R_2 R_3 (1 + j\omega C R_1)}{R_1}$$

Taking real and imaginary parts:

$$R_u = \frac{R_2 R_3}{R_1} \quad L_u = R_2 R_3 C$$

(6.13)

This expression (6.13) can be used to calculate the quality factor (Q value) of the coil:

$$Q = \frac{\omega L_u}{R_u} = \frac{\omega R_2 R_3 C R_1}{R_2 R_3} = \omega C R_1$$

If a constant frequency ω is used:

$$Q \propto R_1$$

Thus, the Maxwell bridge can be used to measure the Q value of a coil directly using this relationship.

Example 6.5

In the Maxwell bridge shown in Figure 6.8, let the fixed-value bridge components have the following values: $R_3 = 5\,\Omega$, $C = 1$ mF. Calculate the value of the unknown impedance (L_u, R_u) if $R_1 = 159\,\Omega$ and $R_2 = 10\,\Omega$ at balance.

Solution

Substituting values into the relations developed in Equation (6.13) above:

$$R_u = \frac{R_2 R_3}{R_1} = \frac{10 \times 5}{159} = 0.3145\,\Omega$$

$$L_u = R_2 R_3 C = \frac{10 \times 5}{10^3} = 50\,\text{mH}$$

Example 6.6

Calculate the Q factor for the unknown impedance in Example 6.5 above at a supply frequency of 50 Hz.

Solution

$$Q = \frac{\omega L_u}{R_u} = \frac{2\pi 50(0.05)}{0.3145} = 49.9$$

6.4.3 Deflection-type a.c. bridge

A common deflection type of a.c. bridge circuit is shown in Figure 6.9.
   For capacitance measurement:

$$Z_u = 1/j\omega C_u \quad Z_1 = 1/j\omega C_1$$

For inductance measurement (making the simplification that the resistive component of the inductor is small and approximates to zero):

$$Z_u = j\omega L_u \quad Z_1 = j\omega L_1$$

Analysis of the circuit to find the relationship between V_0 and Z_u is greatly simplified if one assumes that I_m is very small.

For $I_m = 0$, currents in the two branches of the bridge, as defined in Figure 6.9, are given by:

$$I_1 = \frac{V_s}{Z_1 + Z_u} \qquad I_2 = \frac{V_s}{R_2 + R_3}$$

Also:

$$V_{AD} = I_1 Z_u \quad \text{and} \quad V_{AB} = I_2 R_3$$

Then:

$$V_0 = V_{BD} = V_{AD} - V_{AB} = V_s \left(\frac{Z_u}{Z_1 + Z_u} - \frac{R_3}{R_2 + R_3} \right)$$

Thus for capacitances:

$$V_0 = V_s \left(\frac{1/C_u}{1/C_1 + 1/C_u} - \frac{R_3}{R_2 + R_3} \right) = V_s \left(\frac{C_1}{C_1 + C_u} - \frac{R_3}{R_2 + R_3} \right) \qquad (6.14)$$

and for inductances:

$$V_0 = V_s \left(\frac{L_u}{L_1 + L_u} - \frac{R_3}{R_2 + R_3} \right) \qquad (6.15)$$

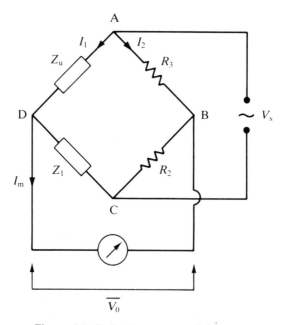

Figure 6.9 Deflection-type a.c. bridge

This latter relationship (6.15) is in practice only approximate since inductive impedances are never pure inductances as assumed but always contain a finite resistance (i.e. $Z_u = j\omega L_u + R$). However, the approximation is valid in many circumstances.

Example 6.7

A deflection bridge as shown in Figure 6.9 is used to measure an unknown capacitance, C_u. The components in the bridge have the following values: $V_s = 20\ V_{rms}$, $C_1 = 100\ \mu F$, $R_2 = 60\ \Omega$, $R_3 = 40\ \Omega$. If $C_u = 100\ \mu F$, calculate the output voltage V_0.

Solution
From Equation (6.14),

$$V_0 = V_s\left(\frac{C_1}{C_1 + C_u} - \frac{R_3}{R_2 + R_3}\right) = 20(0.5 - 0.4) = 2\ V_{rms}$$

Example 6.8

An unknown inductance L_u is measured using a deflection type of bridge as shown in Figure 6.9. The components in the bridge have the following values: $V_s = 10\ V_{rms}$, $L_1 = 20\ mH$, $R_2 = 100\ \Omega$, $R_3 = 100\ \Omega$. If the output voltage V_0 is $1\ V_{rms}$, calculate the value of L_u.

Solution
From Equation (6.15),

$$\frac{L_u}{L_1 + L_u} = \frac{V_0}{V_s} + \frac{R_3}{R_2 + R_3} = 0.1 + 0.5 = 0.6$$

Thus $L_u = 0.6(L_1 + L_u)$, so:

$$0.4L_u = 0.6L_1$$

$$L_u = \frac{0.6L_1}{0.4} = 30\ mH$$

6.5 Exercises

6.1 (a) A Wheatstone bridge circuit is shown in Figure 6.10. Derive an expression for the output voltage V_0 in terms of R_1, R_2, R_3, R_4 and V_s (assuming that the impedance of the voltage measuring instrument is infinite).
   (b) If the elements have the following values, $R_1 = 110\ \Omega$, $R_2 = 100\ \Omega$, $R_3 = 1000\ \Omega$, $R_4 = 1000\ \Omega$ and $V_s = 10\ V$, calculate the output voltage V_0.

6.2 In the Wheatstone bridge shown in Figure 6.10, the resistive components have the following nominal values: $R_1 = 3\ k\Omega$, $R_2 = 6\ k\Omega$, $R_3 = 8\ k\Omega$ and $R_4 = 4\ k\Omega$. The

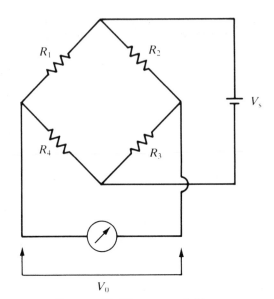

Figure 6.10 Wheatstone bridge

actual value of each resistance is related to the nominal value according to:
$R_{actual} = R_{nominal} + \delta R$ where δR has the following values: $\delta R_1 = 30\ \Omega$,
$\delta R_2 = -20\ \Omega$, $\delta R_3 = 40\ \Omega$ and $\delta R_4 = -50\ \Omega$. Calculate the open-circuit bridge output voltage if the bridge supply voltage V_s is 50 V.

6.3 (a) Figure 6.2 shows a d.c. bridge circuit designed to measure the value of an unknown resistance R_u. The impedance of the instrument measuring the output voltage V_0 is such that it draws negligible current. Derive an expression for the output voltage V_0 in terms of the input voltage V_i and the four resistances in the bridge, R_u, R_1, R_2 and R_3.

(b) Suppose that the unknown resistance R_u is a resistance thermometer whose resistance at 100 °C is 500 Ω and which varies with temperature at the rate of 0.5 Ω/°C for small temperature changes around 100 °C. Calculate the sensitivity of the total measurement system for small changes in temperature around 100 °C, given the following resistance and voltage values measured at 15 °C by instruments calibrated at 15 °C: $R_1 = 500\ \Omega$, $R_2 = R_3 = 5000\ \Omega$ and $V_i = 10$ V.

(c) If the resistance thermometer is measuring a fluid whose true temperature is 104 °C, calculate the error in the indicated temperature if the ambient temperature around the bridge circuit is 20 °C instead of the calibration temperature of 15 °C, given the following additional information:

voltage measuring instrument zero drift coefficient $= +1.3$ mV/°C
voltage measuring instrument sensitivity drift coefficient $= 0$
resistances R_1, R_2 and R_3 have a positive temperature coefficient of $+0.2\%$ of nominal value per °C
voltage source V_i is unaffected by temperature changes.

6.4   Four strain gauges of resistance 120 Ω each are arranged into a Wheatstone bridge configuration such that each of the four arms in the bridge has one strain gauge in it. The maximum permissible current in each strain gauge is 100 mA. What is the maximum bridge supply voltage allowable, and what power is dissipated in each strain gauge with that supply voltage?

6.5   (a) Suppose that the variables shown in Figure 6.2 have the following values: $R_1 = 100\,\Omega$, $R_2 = 100\,\Omega$, $R_3 = 100\,\Omega$, $V_i = 12$ V. R_u is a resistance thermometer with a resistance of 100 Ω at 100 °C and a temperature coefficient of $+0.3\,\Omega/°C$ over the temperature range from 50 °C to 150 °C (i.e. the resistance increases as the temperature goes up). Draw a graph of bridge output voltage V_0 for ten-degree steps in temperature between 100 °C and 150 °C (calculating V_0 according to Equation (6.3)).

(b) Draw a graph of V_0 for similar temperature values if $R_2 = R_3 = 1000\,\Omega$ and all other components have the same values as given in part (a) above. Notice that the line through the data points is straighter than that drawn in part (a) but the output voltage is much less at each temperature point.

6.6   The unknown resistance R_u in a d.c. bridge circuit, connected as shown in Figure 6.4(a), is a resistance thermometer. The thermometer has a resistance of 350 Ω at 50 °C and its temperature coefficient is $+1\,\Omega/°C$ (the resistance increases as the temperature rises). The components of the system have the following values: $R_1 = 350\,\Omega$, $R_2 = R_3 = 2$ kΩ, $R_m = 20$ kΩ, $V_i = 5$ V. What is the output voltage reading when the temperature is 100 °C?
(*Hint:* Use Equation (6.10).)

6.7   In the d.c. bridge circuit shown in Figure 6.11, the resistive components have the

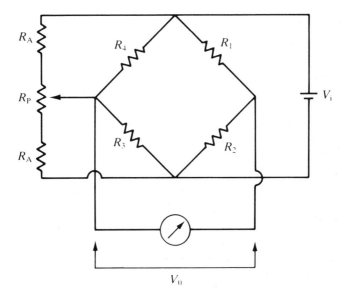

Figure 6.11 Direct current bridge

following values: $R_1 = R_2 = 120\,\Omega$, $R_3 = 117\,\Omega$, $R_4 = 123\,\Omega$, $R_A = R_P = 1000\,\Omega$.

(a) What are the resistance values of the parts of the potentiometer track either side of the slider when the potentiometer is adjusted to balance the bridge?

(b) What then is the effective resistance of each of the two left-hand arms of the bridge when the bridge is balanced?

6.8 A Maxwell bridge, designed to measure the unknown impedance (R_u, L_u) of a coil, is shown in Figure 6.8.

(a) Derive an expression for R_u and L_u under balance conditions.

(b) If the fixed bridge component values are $R_3 = 100\,\Omega$ and $C = 20\,\mu\text{F}$, calculate the value of the unknown impedance if $R_1 = 3183\,\Omega$ and $R_2 = 50\,\Omega$ at balance.

(c) Calculate the Q factor for the coil if the supply frequency is 50 Hz.

6.9 A deflection bridge as shown in Figure 6.9 is used to measure an unknown inductance L_u. The components in the bridge have the following values: $V_s = 30\,V_{rms}$, $L_1 = 80\,\text{mH}$, $R_2 = 70\,\Omega$, $R_3 = 30\,\Omega$. If $L_u = 50\,\text{mH}$, calculate the output voltage V_0.

6.10 An unknown capacitance C_u is measured using a deflection bridge as shown in Figure 6.9. The components of the bridge have the following values: $V_s = 10\,V_{rms}$, $C_1 = 50\,\mu\text{F}$, $R_2 = 80\,\Omega$, $R_3 = 20\,\Omega$. If the output voltage is $3\,V_{rms}$, calculate the value of C_u.

6.11 A Hays bridge is often used for measuring the inductance of high-Q coils and has the configuration shown in Figure 6.12.

(a) Obtain the bridge balance conditions.

(b) Show that if the Q-value of an unknown inductor is high, the expression for the inductance value when the bridge is balanced is independent of frequency.

(c) If the Q-value is high, calculate the value of the inductor if the bridge component values at balance are as follows:

$$R_2 = R_3 = 1000\,\Omega \quad C = 0.02\,\mu\text{F}$$

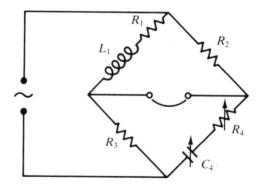

Figure 6.12 Hays bridge

References and further reading

Jones, C. (1962) *An Introduction to Advanced Electrical Engineering,* English
Universities Press: London.
Smith, R. J. (1976) *Circuits, Devices and Systems,* Wiley: New York.

7 Measurement of electrical signals

7.1 Introduction

A large number of measuring instruments and transducers have an output which is either in the form of a resistance, capacitance or inductance change or alternatively a modulation of some other electrical parameter such as the voltage level, current, power level, frequency or phase of a signal. Study of the various techniques available for measuring these quantities is therefore important.

For some purposes, it is only necessary to have a visual indication of the signal level when monitoring the output of a measuring instrument, whilst for other purposes there is a requirement to record the output signal continuously in a form that can be saved for future study and use.

The instruments used to give a visual indication of the level of electrical signals are meters, of both analog and digital varieties, and the cathode ray oscilloscope. These are discussed in sections 7.2 and 7.3 respectively. The later sections in this chapter then discuss measurement of the various other electrical parameters which may be required in a particular measurement system.

7.2 Meters

Meters exist in both digital and analog forms. The digital forms consist of various versions of the digital voltmeter (normally abbreviated to DVM), which are differentiated according to the method used to convert an analog voltage signal into a digital output reading. Analog meters can be divided into six subclassifications, according to their principles of operation. These six types are moving coil, moving iron, electrodynamic, induction, clamp on and electrostatic.

Analog meters are electromechanical devices driving a pointer against a scale. They are prone to measurement errors from a number of sources, with quoted accuracy figures of between ±0.1% and ±3%. Inaccurate scale marking during manufacture, bearing friction, bent pointers and ambient temperature variations all limit measurement accuracy. Further human errors are introduced through parallax error (not reading the scale from directly above) and mistakes in interpolating between scale markings.

Digital meters give a reading in the form of a digital display. There are no problems of parallax and every observer sees the same value. In fact, digital meters are technically superior in every respect to analog meters and have quoted accuracy figures of between $\pm 0.005\%$ (measuring d.c. voltages) and $\pm 2\%$. Unfortunately, this is only achieved at a considerably higher manufacturing cost than that involved in producing analog meters.

Additional advantages of digital voltmeters are their very high input impedance (10 MΩ compared with 1–20 kΩ for analog meters), the ability to measure signals of frequency up to 1 MHz and the common inclusion of features such as automatic ranging, which prevents overload and reverse polarity connection, etc.

7.2.1 Voltage-to-time conversion digital voltmeter

This is known as a ramp type of DVM. When an unknown voltage signal is applied to the input terminals of the instrument, a negative-slope ramp waveform is generated internally and compared with the input signal. When the two are equal, a pulse is generated which opens a gate, and at a later point in time a second pulse closes the gate when the negative ramp voltage reaches zero. The length of time between the gate opening and closing is monitored by an electronic counter, which produces a digital display according to the level of the input voltage signal. The best accuracy of this type is about $\pm 0.05\%$.

7.2.2 Potentiometric digital voltmeter

This type of DVM uses the servo principle, where the error between the unknown input voltage level and a reference voltage is applied to a servo-driven potentiometer which adjusts the reference voltage until it balances the unknown voltage. The output reading is produced by a mechanical drum-type digital display driven by the potentiometer. This is a relatively cheap form of DVM.

7.2.3 Dual-slope integration digital voltmeter

In this type of DVM, the unknown voltage is applied to an integrator for a fixed time T_1, following which a reference voltage of opposite sign is applied to the integrator, which discharges down to a zero output in an interval T_2 measured by a counter. The output–time relationship for the integrator is shown in Figure 7.1, from which the unknown voltage V_i can be calculated geometrically from the triangle as:

$$V_i = V_{ref}(T_1/T_2)$$

This is another relatively simple and cheap form of DVM.

7.2.4 Voltage-to-frequency conversion digital voltmeter

In this form, the unknown voltage signal is fed via a range switch and an amplifier to a converter circuit whose output is in the form of a train of voltage pulses at a frequency

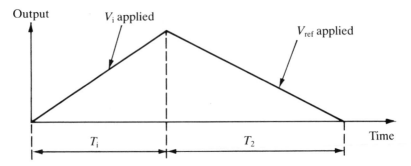

Figure 7.1 Output–time relationship for integrator in a dual-slope digital voltmeter (DVM)

proportional to the magnitude of the input signal. The main advantage of this type of DVM is its ability to reject a.c. noise.

7.2.5 Moving-coil meters

A moving-coil meter is an analog instrument which responds only to direct current inputs and is shown schematically in Figure 7.2. It consists of a rectangular coil wound round a soft iron core which is suspended in the field of a permanent magnet. The signal being measured is applied to the coil and this produces a radial magnetic field. The interaction between this induced field and the field produced by the permanent magnet causes a torque, which results in rotation of the coil. The amount of rotation of the coil is measured by attaching a pointer to it which moves past a graduated scale.

The theoretical torque produced is given by:

$$T = BIhwN \tag{7.1}$$

where B is the flux density of the radial field, I is the current flowing in the coil, h is the height of the coil, w is the width of the coil and N is the number of turns in the coil.

If the iron core is cylindrical and the air gap between the coil and pole faces of the permanent magnet is uniform, then the flux density B is constant, and Equation (7.1) can be rewritten as:

$$T = KI \tag{7.2}$$

i.e. the torque is proportional to the coil current and the instrument scale is linear.

In certain situations, a non-linear scale such as a logarithmic one is required, which is achieved by using either a specially shaped core or specially shaped magnet pole-faces.

Whilst Figure 7.2 shows the traditional moving-coil instrument with a long U-shaped permanent magnet, many newer instruments employ much shorter magnets made

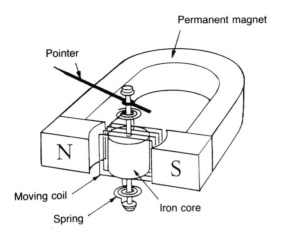

Permanent magnet

Pointer

N

S

Moving coil

Iron core

Spring

Figure 7.2 Moving-coil meter

from recently developed magnetic materials such as Alnico and Alcomax. These materials produce a substantially greater flux density, which, besides allowing the magnet to be smaller, has additional advantages in allowing reductions to be made in the size of the coil and in increasing the usable range of deflection of the coil to about 120 degrees.

Example 7.1

Calculate the reading which would be observed on a moving-coil ammeter when it is measuring the current in the circuit shown in Figure 7.3.

Solution

A moving coil meter measures **mean** current:

$$I_{mean} = \frac{1}{2\pi}\left(\int_0^{\pi} \frac{5\omega t}{\pi}\, d\omega t + \int_{\pi}^{2\pi} 5\sin(\omega t)\, d\omega t \right)$$

$$= \frac{1}{2\pi}\left(\left[\frac{5(\omega t)^2}{2\pi} \right]_0^{\pi} + 5\left[-\cos(\omega t) \right]_{\pi}^{2\pi} \right)$$

$$= \frac{1}{2\pi}\left(\frac{5\pi^2}{2\pi} - 0 - 5 - 5 \right)$$

$$= \frac{1}{2\pi}\left(\frac{5\pi}{2} - 10 \right) = \frac{5}{2\pi}\left(\frac{\pi}{2} - 2 \right)$$

$$= -0.342 \text{ amps}$$

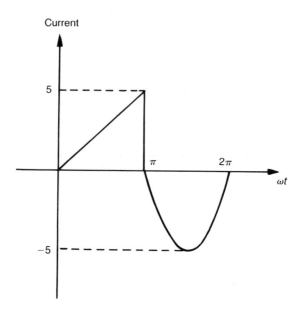

Figure 7.3 Circuit for Example 7.1

7.2.6 **Moving-iron meters**

This type of analog meter is suitable for measuring both direct current and alternating current signals up to frequencies of 125 Hz. It is the cheapest form of meter available and consequently the one most commonly found in voltage and current measurement situations. The signal to be measured is applied to a stationary coil, and the associated field produced is often amplified by the presence of an iron structure associated with the fixed coil. The moving element in the instrument consists of an iron vane which is suspended within the field of the fixed coil. When the fixed coil is excited, the iron vane turns in a direction which increases the flux through it.

The majority of moving-iron instruments are either of the attraction type or of the repulsion type. A few instruments belong to a third combination type. The attraction type, where the iron vane is drawn into the field of the coil as the current is increased, is shown schematically in Figure 7.4(a). The alternative repulsion type is sketched in Figure 7.4(b).

For an excitation current I, the torque produced which causes the vane to turn is given by:

$$T = \frac{I^2 \, dM}{2 \, d\theta}$$

where M is the mutual inductance and θ is the angular deflection.

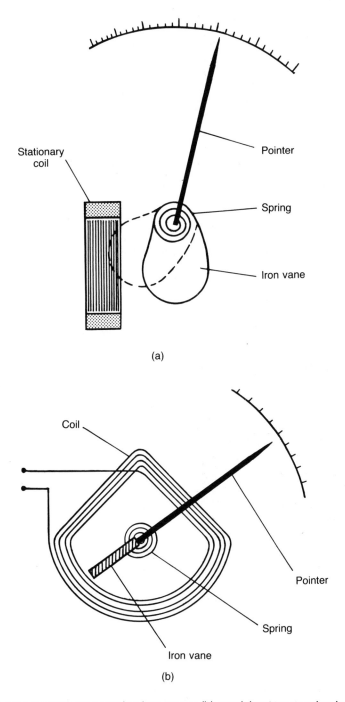

Figure 7.4 (a) Attraction-type moving-iron meter; (b) repulsion-type moving-iron meter

Rotation is opposed by a spring which produces a backwards torque given by:

$$T_s = K\theta$$

At equilibrium, $T = T_s$, and θ is therefore given by:

$$\theta = \frac{I^2 \, dM}{2K \, d\theta}$$

The instrument thus has a square-law response where the deflection is proportional to the square of the signal being measured, i.e. the output reading is a root-mean-square (r.m.s.) quantity.

Example 7.2
Calculate the reading which would be observed on a moving-iron ammeter when it is measuring the current in the circuit shown in Figure 7.3.

Solution
A moving-iron meter measures r.m.s. current:

$$I^2_{rms} = \frac{1}{2\pi} \left(\int_0^\pi \frac{25(\omega t)^2}{\pi^2} \, d\omega t + \int_\pi^{2\pi} 25 \sin^2(\omega t) \, d\omega t \right)$$

$$= \frac{1}{2\pi} \left(\int_0^\pi \frac{25(\omega t)^2}{\pi^2} \, d\omega t + \int_\pi^{2\pi} \frac{25(1 - \cos 2\omega t)}{2} \, d\omega t \right)$$

$$= \frac{25}{2\pi} \left(\left[\frac{(\omega t)^3}{3\pi^2} \right]_0^\pi + \left[\frac{\omega t}{2} - \frac{\sin 2\omega t}{4} \right]_\pi^{2\pi} \right)$$

$$= \frac{25}{2\pi} \left(\frac{\pi}{3} + \frac{2\pi}{2} - \frac{\pi}{2} \right)$$

$$= \frac{25}{2\pi} \left(\frac{\pi}{3} + \frac{\pi}{2} \right) = \frac{25}{2} \left[\frac{1}{3} + \frac{1}{2} \right] = 10.416$$

$$I_{rms} = (I^2_{rms})^{1/2} = 3.23 \text{ amps}$$

7.2.7 Electrodynamic meters

Electrodynamic meters (or dynamometers) are analog instruments which can measure both d.c. and a.c. quantities up to a frequency of 2 kHz. The typical construction of an electrodynamic meter is illustrated in Figure 7.5. It consists of a circular moving coil

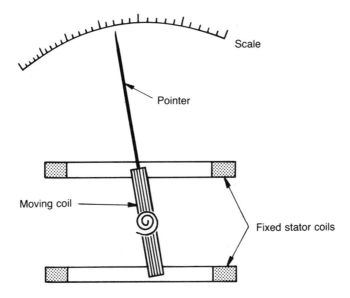

Figure 7.5 Electrodynamic meter

mounted in the magnetic field produced by two separately wound circular stator coils which are connected in series. The torque is dependent upon the mutual inductance between the coils and is given by:

$$T = I_1 I_2 \frac{\mathrm{d}M}{\mathrm{d}\theta} \tag{7.3}$$

where I_1 and I_2 are the currents flowing in the fixed and moving coils, M is the mutual inductance and θ represents the angular displacement between the coils.

When the meter is used as an ammeter, the measured current is applied to both coils. The torque is thus proportional to the square of the current. If the measured current is a.c., the meter is unable to follow the alternating torque values and instead it displays the mean value of the square of the current. By suitable drawing of the scale, the position of the pointer shows the square root of this value, i.e. the r.m.s. current.

Electrodynamic meters are typically expensive but have the advantage of being quite accurate. Voltage, current and power can all be measured if the fixed and moving coils are connected appropriately.

Example 7.3
A dynamometer ammeter is connected in series with a 500 Ω resistor, a rectifying device and a 240 V_{rms} alternating sinusoidal power supply. The rectifier behaves as a resistance of 200 Ω to current flowing in one direction and as a resistance of 2 kΩ to current in the opposite direction. Calculate the reading on the meter.

Solution

$V_{peak} = V_{rms}(2)^{1/2} = 339.4$ V

For $0 < \omega t < \pi$, $R = 700$ Ω, and for $\pi < \omega t < 2\pi$, $R = 2500$ Ω. Thus:

$$I^2_{rms} = \frac{1}{2\pi} \left(\int_0^\pi \frac{(339.4 \sin \omega t)^2}{700^2} \, d\omega t + \int_\pi^{2\pi} \frac{(339.4 \sin \omega t)^2}{2500^2} \, d\omega t \right)$$

$$= \frac{339.4^2}{2\pi(10^4)} \left(\int_0^\pi \frac{\sin^2 \omega t}{49} \, d\omega t + \int_\pi^{2\pi} \frac{\sin^2 \omega t}{625} \, d\omega t \right)$$

$$= \frac{339.4^2}{4\pi(10^4)} \left(\int_0^\pi \frac{(1 - \cos 2\omega t)}{49} \, d\omega t + \int_\pi^{2\pi} \frac{(1 - \cos 2\omega t)}{625} \, d\omega t \right)$$

$$= \frac{339.4^2}{4\pi(10^4)} \left(\left[\frac{\omega t}{49} - \frac{\sin 2\omega t}{98} \right]_0^\pi + \left[\frac{\omega t}{625} - \frac{\sin 2\omega t}{1250} \right]_\pi^{2\pi} \right)$$

$$I^2_{rms} = \frac{339.4^2}{4\pi(10^4)} \left(\frac{\pi}{49} + \frac{\pi}{625} \right) = 0.0634$$

$$I_{rms} = (0.0634)^{1/2} = 0.25 \text{ amps}$$

7.2.8 Induction meters

Induction meters are analog instruments which respond only to alternating current signals. The meter consists of two electromagnets arranged such that a phase difference θ exists between the fluxes ϕ_1 and ϕ_2 associated with the two coils, as sketched in Figure 7.6. These two fluxes produce an induced e.m.f. and circulating concentric currents in the moving element of the instrument which consists of a conducting disk. The conducting disk tries to follow the rotating flux under the influence of a torque given by:

$$T = K\phi_1 \phi_2 \sin \theta$$

For current and voltage measurement, the signal to be measured is applied to both electromagnet coils, which results in the meter having a square-law output relationship of the form $T \propto I_2$ or $T \propto V_2$. The phase difference is obtained by shunting one coil with a non-inductive resistance, and rotation of the coil is opposed by a restraining spring.

When the instrument is used as an induction-type energy meter, the disk is unrestrained and free to rotate continuously. Circuit current is applied to one coil and circuit voltage to the other, thereby making the output reading a power measurement.

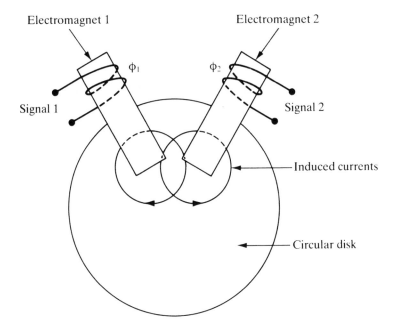

Figure 7.6 Induction meter

7.2.9 **Clamp-on meters**

These are used for measuring circuit currents and voltages in a non-invasive manner which avoids having to break the circuit being measured. The meter clamps on to a current-carrying conductor and the output reading is obtained by transformer action. The principle of operation is illustrated in Figure 7.7, where it can be seen that the clamp-on jaws of the instrument act as a transformer core and the current-carrying conductor acts as a primary winding. Current induced in the secondary winding is rectified and applied to a moving-coil meter. Although it is a very convenient instrument to use, the clamp-on meter has low sensitivity and the minimum current measurable is usually about 1 amp.

7.2.10 **Electrostatic meters**

Electrostatic meters use the principle of attraction between two oppositely charged circular plates, as illustrated in Figure 7.8. The force produced is proportional to the square of the potential difference between the plates and is insufficient to provide an adequate deflection at voltages less than about 50 V. Electrostatic meters are therefore normally only used for high-voltage applications, where voltages up to 500 kV can be measured directly. This type of instrument is very accurate, and it has the particular advantage of a very high input impedance, thereby drawing an extremely low current from the measured system.

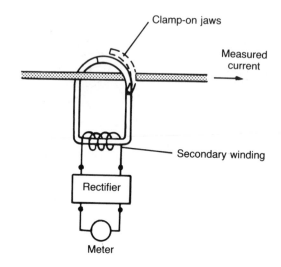

Figure 7.7 Clamp-on meter

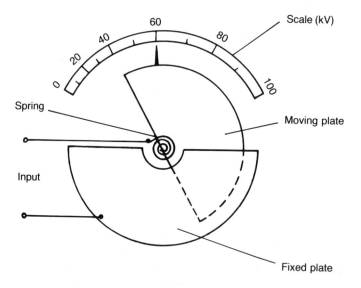

Figure 7.8 Electrostatic meter

7.3 **Cathode ray oscilloscope**

The cathode ray oscilloscope is probably the most versatile and useful instrument available for signal measurement. The more expensive models can measure signals at frequencies up to 500 MHz and even the cheapest models can measure signals up to 20 MHz. One particularly strong merit of the oscilloscope is its high input impedance, typically 1 MΩ, which means that the instrument has a negligible loading effect in most measurement situations. Some disadvantages of oscilloscopes include their fragility (being built around a cathode ray tube) and their moderately high cost. Also, whilst the accuracy of the best instruments is about ±1%, the accuracy of the cheapest ones is only ±10%. The most important aspects in the specification of an oscilloscope are its bandwidth, its rise time and its accuracy.

The bandwidth is defined as the range of frequencies over which the oscilloscope amplifier gain is within 3 dB* of its peak value, as illustrated in Figure 7.9. The −3 dB point is where the gain is 0.707 times its maximum value. In most oscilloscopes, the amplifier is direct coupled, which means that it amplifies d.c. voltages by the same factor as low-frequency a.c. ones. For such instruments, the minimum frequency measurable is zero and the bandwidth can be interpreted as the maximum frequency where the sensitivity (deflection per volt) is within 3 dB of the peak value. The oscilloscope must be chosen such that the maximum frequency to be measured is well within the bandwidth. The −3 dB specification means that an oscilloscope with a specified accuracy of ±2% and bandwidth of 100 MHz will only have an accuracy of ±5% when measuring 30 MHz signals and this accuracy will decrease still further at higher frequencies. Thus when applied to signal amplitude measurement, the oscilloscope is only usable at frequencies up to about 0.3 times its specified bandwidth.

The rise time is the transit time between the 10% and 90% levels of the response when a step input is applied to the oscilloscope. Oscilloscopes are normally designed such that:

$$\text{bandwidth} \times \text{rise time} = 0.35$$

Thus, for a bandwidth of 100 MHz, rise time = 0.35/100 000 000 = 3.5 ns.

An oscilloscope is a relatively complicated instrument which is constructed from a number of subsystems, and it is necessary to consider each of these in turn in order to understand how the complete instrument functions.

7.3.1 **Cathode ray tube**

The cathode ray tube is the fundamental part of an oscilloscope and is shown in Figure 7.10. The cathode consists of a barium- and strontium-oxide-coated, thin, heated filament from which a stream of electrons is emitted. The stream of electrons is

---

*The decibel, commonly written dB, is used to express the ratio between two quantities. For two voltage levels V_1 and V_2, the difference between the two levels is expressed in decibels as $20\log_{10}(V_1/V_2)$. It follows from this that $20\log_{10}(0.707/1) = -3$ dB.

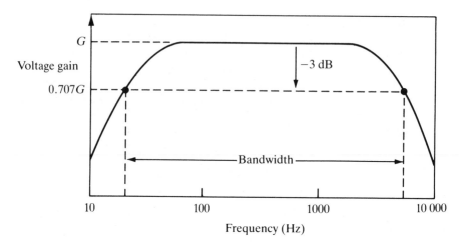

Figure 7.9 Bandwidth

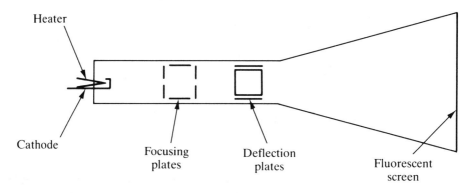

Figure 7.10 Cathode ray tube

focused on to a well-defined spot on a fluorescent screen by an electrostatic focusing system which consists of a series of metal disks and cylinders charged at various potentials. Adjustment of this focusing mechanism is provided by controls on the front panel of an oscilloscope. An 'Intensity' control varies the cathode heater current, and therefore the rate of emission of electrons, and thus adjusts the intensity of the display on the screen. These and other typical controls are shown in the illustration of the front panel of a simple oscilloscope given in Figure 7.11.

Application of potentials to two sets of deflector plates mounted at right angles to one another within the tube provide for deflection of the stream of electrons, such that the spot where the electrons are focused on the screen is moved. The two sets of deflector plates are normally known as the horizontal and vertical deflection plates,

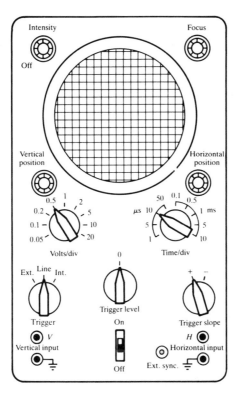

Figure 7.11 Controls of a simple oscilloscope

according to the respective motion caused to the spot on the screen. The magnitude of any signal applied to the deflector plates can be calculated by measuring the deflection of the spot against a crossed-wires graticule etched on the screen.

In the oscilloscope's most common mode of usage measuring time-varying signals, the unknown signal is applied, via an amplifier, to the y-axis (vertical) deflector plates and a time base to the x-axis (horizontal) deflector plates. In this mode of operation, the display on the oscilloscope screen is in the form of a graph with the magnitude of the unknown signal on the vertical axis and time on the horizontal axis.

7.3.2 Channels

The basic subsystem of an electron source, focusing system and deflector plates is often duplicated one or more times within the cathode ray tube to provide a capability of displaying two or more signals together on the screen. Each basic subsystem is known as a channel, and therefore the commonest oscilloscope configuration having two channels can display two separate signals simultaneously.

7.3.3 **Single-ended inputs**

Most simple oscilloscopes have this type of input, which only allows signal voltages to be measured relative to ground. This type can be recognized by the presence of only one input terminal plus a ground terminal per oscilloscope channel.

7.3.4 **Differential inputs**

This type of input is provided on more expensive oscilloscopes and allows the potentials at two non-grounded points in a circuit to be compared. Two input terminals plus a ground terminal are provided for each channel. This arrangement can also be used in single-ended mode to measure a signal relative to ground by using just one of the input terminals plus ground.

7.3.5 **Time base circuit**

The purpose of a time base is to apply a voltage to the horizontal deflector plates such that the horizontal position of the spot is proportional to time. This voltage, in the form of a ramp known as a sweep waveform, must be applied repetitively, such that the motion of the spot across the screen appears as a straight line when a d.c. level is applied to the input channel. Furthermore, this time base voltage must be synchronized with the input signal in the general case of a time-varying signal, such that a steady picture is obtained on the oscilloscope screen. The length of time taken for the spot to traverse the screen is controlled by a 'Time/div' switch, which sets the length of time taken by the spot to travel between two marked divisions on the screen, thereby allowing signals at a wide range of frequencies to be measured.

Each cycle of the sweep waveform is initiated by a pulse from a pulse generator. The input to the pulse generator is a sinusoidal signal known as a triggering signal, with a pulse being generated every time the triggering signal crosses a preselected slope and voltage-level condition. This condition is defined by the 'Trigger level' and 'Trigger slope' switches. The former selects the voltage level on the trigger signal, commonly zero, at which a pulse is generated, whilst the latter selects whether pulsing occurs on a positive- or negative-going part of the triggering waveform.

Synchronization of the sweep waveform with the measured signal is most easily achieved by deriving the trigger signal from the measured signal, a procedure which is known as 'internal triggering'. Alternatively, 'external triggering' can be applied if the frequencies of the triggering signal and measured signals are related by an integer constant such that the display is stationary. External triggering is necessary when the amplitude of the measured signal is too small to drive the pulse generator, and it is also used in applications where there is a requirement to measure the phase difference between two sinusoidal signals of the same frequency. It is very convenient to use the 50 Hz line voltage for external triggering when measuring signals at mains frequency, and this is often given the name 'line triggering'.

7.3.6 **Vertical sensitivity control**

In order to make the range of signal magnitudes measurable by an oscilloscope as wide as possible, a series of attenuators and pre-amplifiers are provided at the input. These condition the measured signal to the optimum magnitude for input to the main amplifier and vertical deflection plates. Selection of the appropriate input amplifier/attenuator is made by setting a 'Volts/div' control associated with each oscilloscope channel. This defines the magnitude of the input signal which will cause a deflection of one division on the screen.

7.3.7 **Display position control**

The position on the screen at which a signal is displayed can be controlled in two ways. The horizontal position can be adjusted by a 'Horizontal position' knob on the oscilloscope front panel and similarly a 'Vertical position' knob controls the vertical position. These controls adjust the position of the display by biasing the measured signal with d.c. voltage levels.

7.4 **Voltage measurement**

Atoms are composed of electrically charged particles, called electrons and protons, which occur in equal numbers in the equilibrium state. If a deficit in either type of particle exists, the atom is said to have either a positive or negative charge, which is measured in units called **coulombs** (C). Such a charge is a source of potential energy and it has the capacity to do work. If the charge is present at some position within an electrical circuit, it will move to some lower energy level if released, and expend energy in so doing. This net positive or negative charge is called a **potential difference** or **voltage** and is measured in units called **volts** (V).

The particular concern of this chapter is not with means of generating voltages, but merely with ways of measuring them. The measuring instruments available include electromechanical (analog) voltmeters, electronic (analog) voltmeters, the cathode ray oscilloscope and digital voltmeters. The first of these, electromechanical voltmeters, are a passive type of instrument, but all the others are active instruments which include signal amplification stages of some form.

The electromechanical group consists of the moving-iron voltmeter, the moving-coil voltmeter, the electrostatic voltmeter, the dynamometer voltmeter and the thermo-couple meter. Most of these have already been discussed in detail, and therefore the coverage of them in the following paragraphs is brief. Electronic instruments with an analog form of output are a relatively recent development, and they have some important advantages. First, they have a high input impedance which avoids the circuit loading problems associated with many applications of electromechanical instruments. Secondly, they have an amplification capability which enables them to measure small signal levels accurately. Direct current electronic voltmeters exist as both direct-coupled and chopper-amplifier types. In the case of a.c. versions, three different types exist, known as average responding, peak responding and r.m.s. responding.

Digital voltmeters have been developed to satisfy a need for higher measurement accuracies and a faster speed of response to voltage changes. As a voltage-indicating instrument, the binary nature of the output reading is readily applied to a display in the form of discrete numerals. Where human operators are required to monitor system voltage levels, this form of output makes an important contribution to measurement reliability and accuracy, since the problem of meter parallax error is eliminated and the possibility of gross error through misreading the meter output is greatly reduced. The availability of a direct output in digital form has also become very valuable to the rapidly expanding range of computer control applications. Digital voltmeters differ mainly in the principle used to effect the analog-to-digital conversion between the measured analog voltage and the output digital reading.

A further class of instruments exists for measuring the very high voltages associated with the electricity supply grid system and certain other systems. The applications for these are so specialized that it is not appropriate to discuss them within this text and the reader is referred elsewhere if such information is required (Buckingham and Price 1966).

7.4.1 **Electromechanical voltmeters**

The very earliest instruments used for measuring voltage relied on the principle of interaction between charged bodies. This idea was exploited by Kelvin in the nineteenth century in the development of the electrostatic voltmeter, which measures the forces of attraction between two oppositely charged plates. Such forces are relatively small, and consequently this instrument is now only used for measuring d.c. voltages greater than 100 volts in magnitude.

The most popular d.c. analog voltmeter used today is the moving-coil type, because of its sensitivity, accuracy and linear scale. It operates at low current levels of 1 milliamp or so, and hence the basic instrument is only suitable for measuring voltages in the millivolt range and up to around 2 volts. For higher voltages, the measuring range of the instrument is increased by placing a resistance in series with the coil, such that a known proportion of the applied voltage is measured by the meter.

Alternative instruments for measuring d.c. voltages are the dynamometer voltmeter and the moving-iron meter. In their basic form, these instruments are most suitable for measuring voltages in the range of 0–30 volts. Higher voltages are measured by placing a resistance in series with the instrument, as in the case of the moving-coil type.

For measuring a.c. voltages, the usual choice is between the dynamometer voltmeter and the moving-iron meter. Both of these have an r.m.s. form of output. Of the two, the dynamometer voltmeter is more accurate but also more expensive. In both cases, range extension is obtained by using the instruments with a series resistance. This is beneficial to a.c. voltage measurements, as it compensates for the effect of coil inductance by reducing the total resistance/inductance ratio, and hence measurement accuracy is improved. For measurements above about 300 volts, it becomes impractical to include a suitable resistance within the case of the instrument because of heat-dissipation problems, and instead an external resistance is used.

For the measurement of high a.c. voltages, an alternative to using one of these instruments with very large resistances is to transform the voltage to a lower value before applying the measuring instrument. This procedure is the recommended industrial practice for measuring voltages above 750 volts.

One major limitation in applying analog meters to a.c. voltage measurement is that the maximum frequency measurable directly is low, 2 kHz for the dynamometer voltmeter and only 100 Hz in the case of the moving-iron instrument. A partial solution to this limitation is to rectify the voltage signal and then apply it to a moving-coil meter, as shown in Figure 7.12. This extends the upper measurable frequency limit to 20 kHz. However, the inclusion of the bridge rectifier makes the measurement system temperature sensitive, and non-linearities significantly affect measurement accuracy for voltages which are small relative to the full-scale value.

An alternative solution to the upper frequency limitation is provided by the **thermocouple meter**, shown in Figure 7.13. In this, the a.c. voltage signal heats a small element, and the resulting temperature rise is measured by a thermocouple. The d.c. voltage generated in the thermocouple is applied to a moving-coil meter. The output meter reading is an r.m.s. quantity which varies in a non-linear fashion with the magnitude of the measured voltage. Very high frequencies up to 50 MHz can be measured by this method.

Before ending this section on electromechanical meters, mention must also be made of the **analog multi-meter**, whose working arrangement is shown in Figure 7.14. This combines a d.c. moving-coil meter with a bridge rectifier to extend its measuring capability to a.c. quantities. A set of rotary switches allows the selection of various series and shunt resistors, which make the instrument capable of measuring both voltage and current over a number of ranges. An internal power source is also often

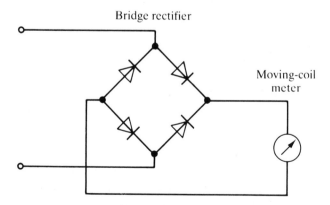

Figure 7.12 Measurement of high-frequency voltage signals

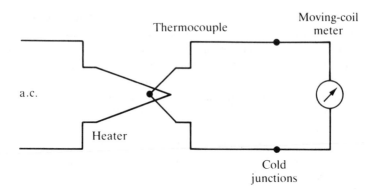

Figure 7.13 The thermocouple meter

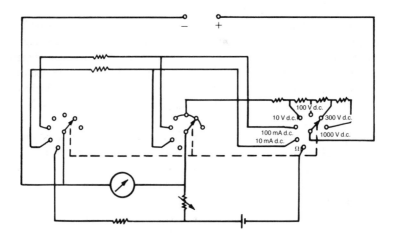

Figure 7.14 The analog multi-meter

provided to allow it to measure resistances as well. Whilst this instrument is very useful for giving an indication of voltage levels, the compromises in its design which enable it to measure so many different quantities necessarily mean that its accuracy is not as good as instruments which are purpose designed to measure just one quantity over a single measuring range.

7.4.2 Electronic analog voltmeters

The standard electronic voltmeter for d.c. measurements consists of a simple direct-coupled amplifier and a moving-coil meter, as shown in Figure 7.15(a). For the measurement of very low-level voltages of a few microvolts, a more sophisticated

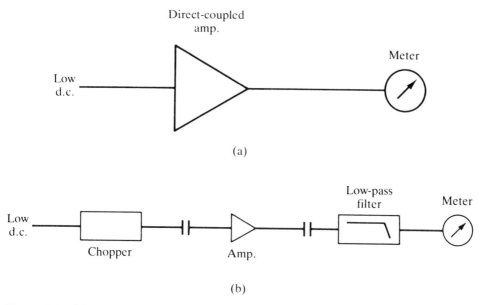

Figure 7.15 (a) Simple d.c. electronic voltmeter; (b) d.c. electronic voltmeter with chopper amplifier

circuit, known as a chopper amplifier, is used. This is shown in Figure 7.15(b). In this, the d.c. input is chopped at a low frequency of around 250 Hz, passed through a blocking capacitor, amplified, passed through another blocking capacitor to remove drift, demodulated, filtered and applied to a moving-coil meter.

Three versions of electronic voltmeter exist for measuring a.c. signals. The **average-responding type** is essentially a direct-coupled d.c. electronic voltmeter with an additional rectifying stage at the input. The output is a measure of the average value of the measured voltage waveform. The second form, known as a **peak-responding type**, has a half-wave rectifier at the input followed by a capacitor. The final part of the circuit consists of an amplifier and moving-coil meter. The capacitor is charged to the peak value of the input signal, and therefore the amplified signal applied to the moving-coil meter gives a reading of the peak voltage in the input waveform. Finally, a third type is available, known as an **r.m.s.-responding type**, which gives an output reading in terms of the r.m.s. value of the input waveform. This type is essentially a thermocouple meter in which an amplification stage has been inserted at the input.

7.4.3 Cathode ray oscilloscope

The principles of operation of the cathode ray oscilloscope have already been discussed in section 7.3. The instrument is widely used for voltage measurement, especially as an item of test equipment for circuit fault finding. It is not a particularly accurate instrument, and is best used where only an approximate measurement is required. Its great value lies in its versatility and applicability to measuring a very wide range of a.c. and d.c. voltages. As a test instrument, it is often required to measure voltages whose frequency and magnitude are totally unknown. The set of rotary switches which alter its range and time base so easily, and the circuitry which protects it from damage when high voltages are applied to it on the wrong range, make it ideally suitable for such applications.

7.4.4 Digital voltmeter (DVM)

The major part of the digital voltmeter is the circuitry that converts the analog voltage being measured into a digital quantity. As the instrument only measures d.c. quantities in its basic mode, another necessary component within it is one which performs a.c.–d.c. conversion and thereby gives it the capacity to measure a.c. signals. After conversion, the voltage value is displayed by means of indicating tubes or a set of solid-state light-emitting diodes. Four-, five- or even six-figure output displays are commonly used, and although the instrument itself may not be inherently more accurate than some analog types, this form of display enables measurements to be recorded with much greater precision than that obtainable by reading an analog meter scale.

Various techniques are used to effect the analog-to-digital conversion, as discussed in Chapter 10. As a general rule, the more expensive and complicated conversion methods achieve a faster conversion speed. The simplest form of DVM is the ramp type. Its main drawbacks are non-linearities in the shape of the ramp waveform used and lack of noise rejection, and these problems limit its accuracy to ±0.05%. It is relatively cheap, however. Another cheap form of DVM is the potentiometric type, and this gives excellent performance for its price. Better noise-rejection capabilities and correspondingly better measurement accuracy (up to ±0.005%) are obtained with integrating-type DVMs. Unfortunately, these are quite expensive.

The digital multi-meter is an extension of the DVM which can measure a.c. and d.c. voltages over a number of ranges through inclusion within it of a set of switchable amplifiers and attenuators. It is widely used in circuit test applications as an alternative to the analog multi-meter, and includes protection circuits which prevent damage if high voltages are applied to the wrong range.

7.5 Current measurement

If atoms with a net positive or negative charge are released from some point in an electrical circuit, the electrically charged particles will flow away from the point of unequal charge distribution until an equilibrium state is reached. The source of charged particles is known as a potential difference, and their rate of flow is known as a current. When current flows in opposite directions along two parallel conductors a force is exerted, and this phenomenon is used to define the unit of **amperes** (A) by which current is measured. When two currents, of magnitude 1 ampere each, flow in opposite directions along two infinitely long parallel wires positioned 1 metre apart in vacuum, the force produced on each conductor is 2×10^{-7} per metre length of conductor.

Any of the current-operated electromechanical voltmeters discussed in section 7.2 can be used to measure current by placing them in series with the current-carrying circuit, and the same frequency limits apply for the measured signal. As in the case of voltage measurement, this upper frequency limit can be raised by rectifying the a.c. current prior to measurement or by using a thermocouple meter. To minimize the effect on the measured system, any current measuring instrument should have a small resistance. This is opposite to the case of voltage measurement where the instrument is required to have a high resistance for minimal circuit loading.

Moving-coil meters are suitable for direct measurement of d.c. currents in the milliamp range up to 1 ampere and the range for dynamometer ammeters is up to a few amps. Moving-iron meters can measure d.c. currents up to several hundred amps directly. Similar measurement ranges apply when moving-iron and dynamometer-type instruments are used to measure a.c. currents.

To measure currents higher than allowed by the basic design of electromechanical instruments, it is necessary to insert a shunt resistance into the circuit and measure the voltage drop across it. One difficulty which results from this technique is the large power dissipation in the shunt when high-magnitude currents are measured. In the case of a.c. current measurement, care must also be taken to match the resistance and reactance of the shunt to that of the measuring instrument so that frequency and waveform distortion in the measured signal are avoided.

Current transformers provide an alternative method of measuring high-magnitude currents, which avoids the difficulty of designing a suitable shunt. Different versions of these exist for transforming both d.c. and a.c. currents. A d.c. current transformer is shown in Figure 7.16. The central d.c. conductor in the instrument is threaded through two magnetic cores which carry two high-impedance windings connected in series opposition. It can be shown (Baldwin 1973) that the current flowing in the windings when excited with an a.c. voltage is proportional to the d.c.

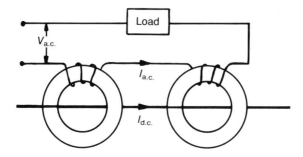

Figure 7.16 Direct current transformer

current in the central conductor. This output current is commonly rectified and then measured by a moving-coil instrument.

An a.c. current transformer typically has a primary winding consisting of only a few copper turns wound on a rectangular or ring-shaped core. The secondary winding on the other hand would normally have several hundred turns according to the current step-down ratio required. The output of the secondary winding is measured by any suitable current measuring instrument. The design of current transformers is substantially different from that of voltage transformers. The rigidity of their mechanical construction has to be sufficient to withstand the large forces arising from short-circuit currents, and special attention has to be paid to the insulation between their windings for similar reasons. A low-loss core material is used and flux densities are kept as small as possible to reduce losses. In the case of very high currents, the primary winding often consists of a single copper bar which behaves as a single-turn winding. The clamp-on meter, described in section 7.2, is a good example of this.

Apart from electromechanical meters, all the other instruments for measuring voltage discussed in section 7.2 can be applied to current measurement by using them to measure the voltage drop across a known resistance placed in series with the current-carrying circuit. The digital voltmeter and electronic meters are widely applied for measuring currents accurately by this method, and the cathode ray oscilloscope is frequently used to obtain approximate measurements in circuit-test applications. Finally, mention must also be made of the use of digital and analog multi-meters for current measurement, particularly in circuit-test applications. These instruments include a set of switchable dropping resistors and so can measure currents over a wide range. Protective circuitry within such instruments prevents damage when high currents are applied on the wrong input range.

7.6 **Resistance measurement**

The standard methods available for measuring resistance, which is measured in units of **ohms** (Ω), include the voltmeter–ammeter method, the resistance-substitution method, the ohmmeter and the d.c. bridge circuit. The digital voltmeter can also be used for measuring resistance if an accurate current source is included within it which passes current through the resistance. This can give a measurement accuracy of $\pm0.1\%$. Apart from the ohmmeter, these instruments are normally only used to measure medium-value resistances in the range of $1\,\Omega$ to $1\,M\Omega$. Special instruments are available for obtaining high-accuracy resistance measurements outside this range (see Baldwin 1973).

7.6.1 **Voltmeter–ammeter method**

The voltmeter–ammeter method consists of applying a measured d.c. voltage across the unknown resistance and measuring the current flowing. Two alternatives exist for connecting the two meters, as shown in Figure 7.17. In Figure 7.17(a), the ammeter

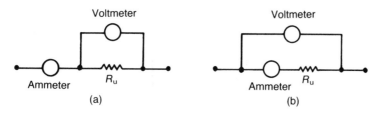

Figure 7.17 Voltmeter–ammeter method of measuring resistance

measures the current flowing in both the voltmeter and the resistance. The error due to this is minimized when the measured resistance is small relative to the voltmeter resistance. In the alternative form of connection, Figure 7.17(b), the voltmeter measures the voltage drop across the unknown resistance and the ammeter. Here, the measurement error is minimized when the unknown resistance is large with respect to the ammeter resistance. Thus method (a) is best for the measurement of small resistances and method (b) for large ones.

Having thus measured the voltage and current, the value of the resistance is then calculated very simply by Ohm's law. This is a suitable method wherever the measurement accuracy of ±1% which it gives is adequate.

7.6.2 **Resistance-substitution method**

In the voltmeter–ammeter method above, either the voltmeter is measuring the voltage across the ammeter as well as across the resistance, or the ammeter is measuring the current flow through the voltmeter as well as through the resistance. The measurement error caused by this is avoided in the resistance-substitution technique. In this method, an unknown resistance in a circuit is temporarily replaced by a variable resistance. The variable resistance is adjusted until the measured circuit voltage and current are the same as existed with the unknown resistance in place. The variable resistance at this point is equal in value to the unknown resistance.

7.6.3 **Ohmmeter**

The ohmmeter is a simple instrument in which a battery applies a known voltage across a combination of the unknown resistance and a known resistance in series, as shown in Figure 7.18. Measurement of the voltage, V_m, across the known resistance, R, allows the unknown resistance, R_u, to be calculated from:

$$R_u = \frac{R(V_b - V_m)}{V_m}$$

where V_b is the battery voltage.

Ohmmeters are used to measure resistances over a wide range from a few milliohms up to 50 MΩ. The best measurement accuracy is only ±2% and ohmmeters

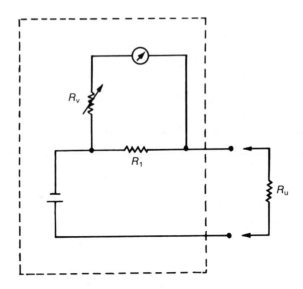

Figure 7.18 The ohmmeter

are therefore more suitable for use as test equipment rather than in applications where high accuracy is required. Most of the available versions contain a switchable set of standard resistances, so that measurements of reasonable accuracy over a number of ranges can be made.

Most **digital and analog multi-meters** contain circuitry of the same form as in an ohmmeter, and hence can be similarly used to obtain approximate measurements of resistance.

7.6.4 **Direct current bridge circuit**

Direct current bridge circuits, as discussed in Chapter 6, provide the most commonly used method of measuring medium-value resistances. The best measurement accuracy is provided by the null-output-type Wheatstone bridge and figures of ±0.02% are achievable with commercially available instruments. Deflection-type bridge circuits are simpler to use in practice than the null-output type, but their measurement accuracy is inferior and the non-linear output relationship is an additional difficulty. One advantage they have is in enabling resistance measurements to be used as inputs to automatic control schemes.

7.7 **Power measurement**

Measurement of the power consumption in an electrical circuit is needed for many reasons. Circuit components have to be designed to have an adequate power-handling capability, and appropriate means of heat dissipation have to be provided. Such measures are necessary to protect equipment from damage and also to avoid fire hazards due to equipment overheating. Measurement of both the useful power consumption and the power losses in a circuit also allows power utilization efficiency to be optimized.

Power measurement is also important to the electricity supply authorities, to ensure that the power consumption by customers is properly costed and charged for. It is also required to ensure that the total system load is within the capacity of the available generating plant.

Power is measured in units of **watts** (W). The energy required to move 1 coulomb of charge between two energy levels with a potential difference equal to 1 volt is called a **joule** (J), and the rate at which this is done is a measure of power expressed in joules per second or in watts.

In the case of d.c. circuits, the power consumption is the simple product of the measured voltage and current. In the case of a.c. circuits, where the voltage and current are varying cyclically but often with a phase difference, it is possible to define a quantity called the instantaneous power which is the product of the voltage level and current level at any instant of time. The instantaneous power varies in a cyclic manner, however, and is of little practical value. Power consumption in a.c. circuits is therefore normally quantified by the average power in watts, which is calculated according to:

$$P_{avg} = V_{rms} I_{rms} \cos(\phi)$$

where ϕ is the phase difference between the current and voltage waveforms.

It is appropriate in some circumstances to measure the reactive power as well, in units of 'volt amps reactive' (V A$_r$). This is given by:

$$P_{reactive} = V_{rms} I_{rms} \sin(\phi)$$

These two quantities are related to the instantaneous power by the vector diagram shown in Figure 7.19.

The two common instruments used for measuring power are the dynamometer wattmeter and the electronic wattmeter. Certain other instruments are also used for measuring high power levels in special applications, such as within the electricity supply industry, as described in Wolf (1973).

7.7.1 **Dynamometer wattmeter**

The dynamometer wattmeter is the standard instrument used for most power measurements at frequencies below 400 Hz. Products of voltages up to 300 volts and currents up to 20 amps can be measured. The instrument is the same standard type of dynamometer as described in section 7.2 for measuring voltage, except that the coils

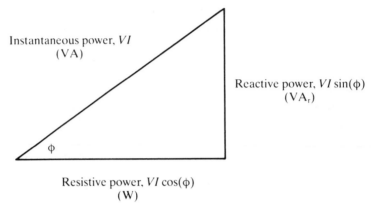

Resistive power, $VI \cos(\phi)$
(W)

Figure 7.19 Vector diagram showing relationship between resistive and reactive power components

are connected differently. When the instrument is used for measuring power dissipation, the stationary coil is connected in series with the measured circuit and carries the circuit current. The rotating coil has a large resistance placed in series with it and is connected across the measured circuit, thus measuring the voltage drop. Theoretically, the instrument measures the product of instantaneous voltage and current and therefore the output pointer reading could be expected to be a measure of instantaneous power, which would make it oscillate at a high frequency. In practice, system inertia prevents such oscillation if the measured circuit has a frequency greater than a few hertz and the pointer settles to a reading proportional to the average circuit power.

The measurement accuracy of commercial instruments varies between ±0.1% and ±0.5% of full scale, according to the design and cost of the instrument. A major source of error in dynamometer wattmeters is power dissipation in the coils. The error due to this becomes very large when the power level in the measured circuit is low. Consequently, the dynamometer wattmeter is not suitable for measuring power levels below 5 watts.

Apart from measuring power dissipation, the dynamometer can also measure both the reactive power and the power factor, cos (ϕ), in a circuit. Modification to the coils and their mode of connection is necessary for this as described in Wolf (1973).

7.7.2 The electronic wattmeter

The electronic wattmeter is capable of measuring power levels between 0.1 W and 100 kW in circuits operating at frequencies up to 100 kHz. Below 40 kHz, its accuracy is ±3%, but at higher frequencies accuracy can be as poor as ±10%. The instrument is used wherever its accuracy is superior to that of the dynamometer, i.e. at power levels below 5 W and at all frequencies above 400 Hz.

The basic operation of such instruments relies on a circuit containing two amplifiers.

One amplifier measures the current through the load and the other measures the voltage across the load. The amplifier outputs are multiplied together by a circuit which uses the logarithmic input relationship of a transistor. The multiplied output is proportional to the instantaneous power level, and is converted to a reading of the average power consumption by the inertial properties of the moving-coil meter to which it is applied. The moving-coil meter ceases to average the instantaneous power properly below a certain frequency, and consequently the electronic wattmeter is not used below frequencies of 10 Hz.

7.7.3 **Power measurement in three-phase circuits**

For power measurement in three-phase circuits, two wattmeters are required, as shown in Figure 7.20. The total average power is then given by the readings on the two wattmeters. To verify this, the analysis is different according to whether the circuit is star or delta connected.

For a star-connected circuit, referring to Figure 7.20(a), the net current at the neutral point is zero, i.e.

$$I_1 + I_2 + I_3 = 0 \qquad (7.4)$$

The instantaneous power P is given by:

$$P = V_1 I_1 + V_2 I_2 + V_3 I_3 \qquad (7.5)$$

From Equation (7.4), $I_3 = -(I_1 + I_2)$. Hence, substituting for I_3 in (7.5):

$$P = V_1 I_1 + V_2 I_2 - V_3(I_1 + I_2)$$

$$= I_1(V_1 - V_3) + I_2(V_2 - V_3) \qquad (7.6)$$

In Equation (7.6), I_1 and I_2 are the currents flowing through wattmeters 1 and 2 respectively. Also, $(V_1 - V_3)$ and $(V_2 - V_3)$ are the voltages across wattmeters 1 and 2 respectively. Thus, Equation (7.6) can be rewritten as:

$$P = p_1 + p_2 \qquad (7.7)$$

where p_1 and p_2 are the instantaneous powers measured by W_1 and W_2 respectively.

Because of the inertial properties of wattmeters, the readings p_1 and p_2 are in fact the average powers. Thus the average power P is given by Equation (7.7).

For a delta-connected network, referring to Figure 7.20(b):

$$V_1 + V_2 + V_3 = 0 \qquad (7.8)$$

The instantaneous power P is given by:

$$P = V_1 I_1 + V_2 I_2 + V_3 I_3 \qquad (7.9)$$

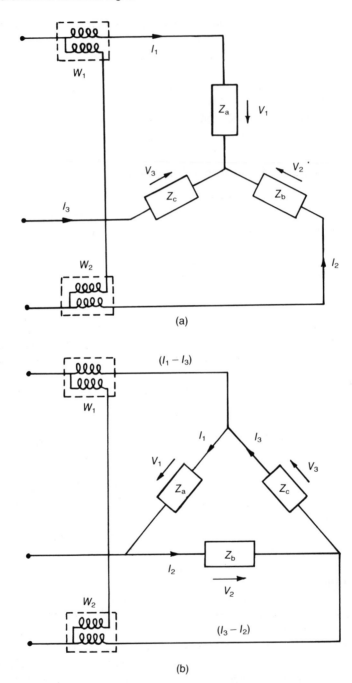

Figure 7.20 Two-wattmeter method for measuring power in three-phase circuits: (a) star-connected circuit; (b) delta-connected circuit

From Equation (7.8), $V_3 = -(V_1 + V_2)$. Substituting for V_3 in Equation (7.9):

$$P = V_1 I_1 + V_2 I_2 - I_3(V_1 + V_2)$$

$$= V_1(I_1 - I_3) + V_2(I_2 - I_3) \tag{7.10}$$

In Equation (7.10), V_1 and V_2 are the voltages across wattmeters 1 and 2 respectively. Also, $(I_1 - I_3)$ and $(I_2 - I_3)$ are the currents through wattmeters 1 and 2 respectively. Hence, Equation (7.10) can be rewritten as:

$$P = p_1 + p_2 \tag{7.11}$$

where p_1 and p_2 are the instantaneous powers measured by W_1 and W_2 respectively.

Because of the inertial properties of wattmeters, the readings p_1 and p_2 are in fact the average powers. Thus the average power P is given by Equation (7.11).

7.8 Frequency measurement

Frequency measurement is required as part of those instruments which convert the measured physical quantity into a frequency change, such as the variable reluctance velocity transducer, stroboscopes, the vibrating-wire force sensor, resonant-wire pressure measuring instruments and the quartz thermometer. The output relationship of some forms of a.c. bridge circuit used for measuring inductance and capacitance also requires the excitation frequency to be known.

The instrument that is now most widely used for measuring frequency, which is measured in units of **hertz** (Hz), is the digital counter–timer. The oscilloscope is also commonly used for obtaining approximate measurements of frequency, especially in circuit-test and fault-diagnosis applications. Within the audio frequency range, the Wien bridge is a further instrument which is often used. Certain other frequency measuring instruments also exist which are used in some special applications. These include the dynamometer frequency meter, the zero-beat meter, the grid-dip meter, Campbell's circuit and Maxwell's commutator bridge, and these are discussed in detail elsewhere (Wolf 1973, Baldwin 1973).

7.8.1 Digital counter–timers

Digital counter–timers are the most accurate and flexible instruments available for measuring frequency. An accuracy up to 1 part in 10^8 can be obtained, and all frequencies between d.c. and a few gigahertz can be measured.

The essential component within a counter–timer instrument is an oscillator which provides a very accurately known and stable reference frequency, which is typically either 100 kHz or 1 MHz. This is often maintained in a temperature-regulated environment within the instrument to guarantee its accuracy. The oscillator output is transformed by a pulse-shaper circuit into a train of pulses and applied to an electronic gate, as shown in Figure 7.21. Successive pulses at the reference frequency alternately open and close the gate. The input signal of unknown frequency is similarly transformed into a train of pulses and applied to the gate. The number of these pulses which get through the gate during the time that it is open during each gate cycle is proportional to the frequency of the unknown signal.

The accuracy of measurement obviously depends upon how far the unknown frequency is above the reference frequency. As it stands therefore, the instrument can only accurately measure frequencies which are substantially above 1 MHz. To enable it to measure much lower frequencies, a series of decade frequency dividers are provided within it. These increase the time between the reference frequency pulses by factors of ten, and a typical instrument can have gate pulses separated in time by between 1 μs and 1 s.

Improvement in the accuracy of low-frequency measurement can be obtained by modifying the gating arrangements such that the signal of unknown frequency is made to control the opening and closing of the gate. The number of pulses at the reference frequency which pass through the gate during the open period is then a measure of the frequency of the unknown signal.

7.8.2 **Phase-locked loop**

A phase-locked loop is a circuit consisting of a phase-sensitive detector, a low-pass filter and a voltage-controlled oscillator (VCO), connected in a closed-loop system as shown in Figure 7.22. In a VCO, the oscillation frequency is proportional to the applied voltage.

Operation of a phase-locked loop is as follows. The phase-sensitive detector compares the phase of the input signal with the phase of the VCO output. Any phase difference generates an error signal, which is amplified and fed back to the VCO. This adjusts the frequency of the VCO until the error signal goes to zero, and thus the VCO becomes locked to the frequency of the input signal and has a d.c. output proportional to the input signal frequency.

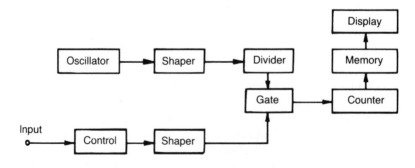

Figure 7.21 Digital counter–timer system

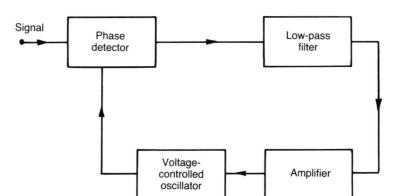

Figure 7.22 Phase-locked loop

7.8.3 **Cathode ray oscilloscope**

The cathode ray oscilloscope can be used in two ways to measure frequency. First, the internal time base can be adjusted until the distance between two successive cycles of the measured signal can be read against the calibrated graticule on the screen. The accuracy of measurement by this method is limited, but can be optimized by measuring between points in the cycle where the slope of the waveform is steep, generally where it is crossing through from the negative to the positive part of the cycle. Calculation of the unknown frequency from this measured time interval is relatively simple. For example, suppose that the distance between two cycles is 2.5 divisions when the internal time base is set at 10 ms per division. The cycle time is therefore 25 ms and hence the frequency is 1000/25, i.e. 40 Hz. The accuracy of measurement is dependent upon how accurately the distance between two cycles is read, and it is very difficult to reduce the error level below ±5% of the reading.

The alternative way of using an oscilloscope to measure frequency is to generate Lissajous patterns. These are produced by applying a known reference frequency sine wave to the y input (vertical deflection plates) of the oscilloscope and the unknown frequency sinusoidal signal to the x input (horizontal deflection plates). A pattern is produced on the screen according to the frequency ratio between the two signals, and if the numerator and denominator in the ratio of the two signals both represent an integral number of cycles, the pattern is stationary. Examples of these patterns are shown in Figure 7.23, which also shows that the phase difference between the waveforms has an effect on the shape. Frequency measurement proceeds by adjusting the reference frequency until a steady pattern is obtained on the screen and then calculating the unknown frequency according to the frequency ratio which the pattern obtained represents.

Frequency
ratio
y/x

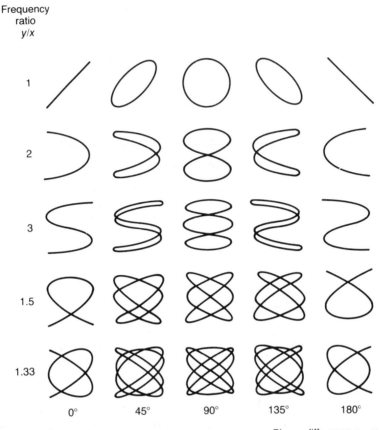

Phase difference x − y

Figure 7.23 Lissajous patterns

7.8.4 **The Wien bridge**

The Wien bridge, shown in Figure 7.24, is a special form of a.c. bridge circuit which can be used to measure frequencies in the audio range. An alternative use of the instrument is as a source of audio frequency signals of accurately known frequency.

A simple set of headphones is often used to detect the null-output balance condition. Other suitable instruments for this purpose are the oscilloscope and the electronic voltmeter. At balance, the unknown frequency is calculated according to:

$$f = \frac{1}{2\pi R_3 C_3}$$

The instrument is very accurate at audio frequencies, but at higher frequencies errors due to losses in the capacitors and stray capacitance effects become significant.

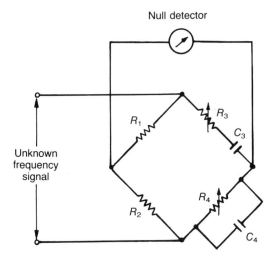

Figure 7.24 The Wien bridge

7.9 **Phase measurement**

The most accurate instrument for measuring the phase difference between two signals is the electronic counter–timer. However, two other methods also exist which are less accurate but very useful in some circumstances. These are plotting the signals on an X–Y plotter and using a dual-beam oscilloscope.

7.9.1 **Electronic counter–timer**

In principle, the phase difference between two sinusoidal signals can be determined by measuring the time which elapses between the two signals crossing the time axis. However, in practice, this is inaccurate because the zero crossings are susceptible to noise contamination. The normal solution to this problem is to amplify/attenuate the two signals so that they have the same amplitude and then measure the time which elapses between the two signals crossing some non-zero threshold value.

The basis of phase measurement by this method is a digital counter–timer with a quartz-controlled oscillator providing a frequency standard which is typically 10 MHz. The crossing points of the two signals through the reference threshold voltage level are applied to a gate which starts and then stops pulses from the oscillator to an electronic counter, as shown in Figure 7.25. The elapsed time, and hence phase difference, between the two input signals is then measured in terms of the counter display.

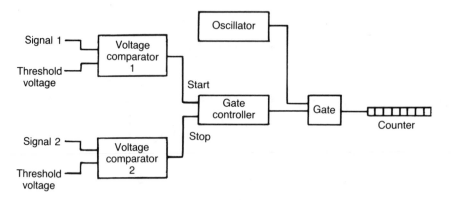

Figure 7.25 Phase measurement with digital counter–timer

7.9.2 *X–Y* plotter

This is a useful technique for phase measurement but is limited to low frequencies because of the very limited bandwidth of the *X–Y* plotter. If two input signals of equal magnitude are applied to the *X* and *Y* inputs of a plotter, the plot obtained is an ellipse, as shown in Figure 7.26. If the *X* and *Y* inputs are given by:

$$V_X = V \sin \omega t$$

$$V_Y = V \sin (\omega t + \phi)$$

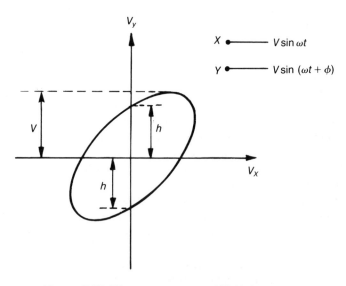

Figure 7.26 Phase measurement *X–Y* plotter

At $t = 0$, $V_X = 0$ and $V_Y = V \sin \phi$. From Figure 7.26, for $V_X = 0$, $V_Y = \pm h$:

$$\sin \phi = \pm h/V \qquad (7.12)$$

Solution of Equation (7.12) gives four possible values for ϕ but the ambiguity about which quadrant ϕ is in can usually be solved by observing the two signals plotted against time on a dual-beam oscilloscope.

7.9.3 Oscilloscope

Approximate measurement of the phase difference between signals can be made using a dual-beam oscilloscope. The two signals are applied to the two oscilloscope inputs and a suitable time base chosen such that the time between the crossing points of the two signals can be measured. The phase difference of both low- and high-frequency signals can be measured by this method, the upper frequency limit measurable being dictated by the bandwidth of the oscilloscope (which is normally very high).

7.10 Capacitance measurement

The only methods available for accurately measuring capacitance, which is measured in units of **farads** (F), are the various forms of a.c. bridge described in Chapter 6. Various types of capacitance bridge are available commercially, and these would normally be the recommended way of measuring capacitance. In some circumstances, such instruments are not immediately available, and if an approximate measurement of capacitance is acceptable, one of the following two methods can be considered.

The first of these, shown in Figure 7.27, consists of connecting the unknown capacitor in series with a known resistance in a circuit excited at a known frequency.

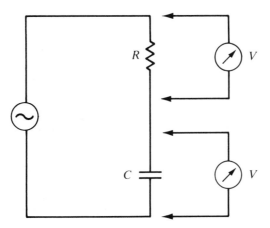

Figure 7.27 Approximate method of measuring capacitance

An a.c. voltmeter is used to measure the voltage drop across both the resistor and the capacitor. The capacitance value is then given by:

$$C = \frac{V_r}{2\pi f R V_c}$$

where V_r and V_c are the voltages measured across the resistance and capacitance respectively, f is the excitation frequency and R is the known resistance.

An alternative method of measurement is to measure the time constant of the capacitor connected in an RC circuit.

7.11 Inductance measurement

Inductance is measured in units of **henrys** (H). As in the case of capacitance, it can only be measured accurately by an a.c. bridge circuit. Various commercial inductance bridges are available for this purpose. Such instruments are not always immediately available, however, and the following method can be applied in such circumstances to give an approximate measurement of inductance.

This approximate method consists of connecting the unknown inductance in series with a variable resistance, in a circuit excited with a sinusoidal voltage, as shown in Figure 7.28. The variable resistance is adjusted until the voltage measured across the resistance is equal to that measured across the inductance. The two impedances are then equal, and the value of the inductance L can be calculated from:

$$L = \frac{(R^2 - r^2)^{1/2}}{2\pi f}$$

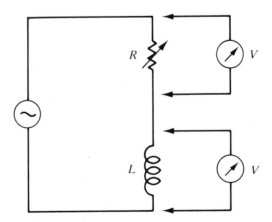

Figure 7.28 Approximate method of measuring inductance

where R is the value of the variable resistance, r is the value of the inductor resistance and f is the excitation frequency.

7.12 Exercises

7.1 Discuss the dual-ramp technique of analog-to-digital conversion used in some types of digital voltmeter.

7.2 The waveform of a cyclic current is shown in Figure 7.29. The current is measured using (a) a moving-coil ammeter and (b) a moving-iron ammeter. Calculate the readings on the two instruments.

(*Hint:* The current is given by $I = 10 \sin \omega t$ for $0 < \omega t < \pi$, and by $I = 10(1 - \omega t/\pi)$ for $\pi < \omega t < 2\pi$.)

7.3 (a) A moving-iron ammeter and a simple moving-coil ammeter are placed in series with a load in a half-wave rectifier circuit. The reading on the moving-iron instrument is 5 amps. What is the reading on the other instrument?

(b) Calculate the readings on the two instruments if the other half-wave is also rectified.

(Assume sinusoidal waveforms throughout.)

7.4 (a) Describe the construction and operating principles of a repulsion-type moving-iron meter.

(b) A compound voltage consisting of a d.c. level component of 5 V, together with a superimposed half-wave-rectified sinusoidal waveform of peak value 4 V, occurs at a certain point in an electrical circuit. What would be the reading on (i) a moving-coil meter and (ii) a moving-iron meter when used to measure this voltage waveform?

7.5 If the waveform shown in Figure 7.30 was fed to
(a) a true r.m.s. meter,

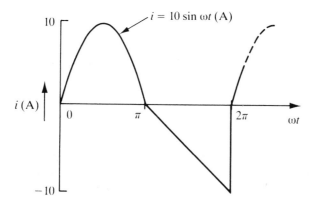

Figure 7.29 Waveform for Exercise 7.2

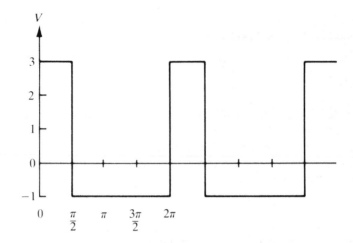

Figure 7.30 Waveform for Exercise 7.5

    (b) an average-measuring/r.m.s.-indicating meter,

    (c) a peak-measuring/r.m.s.-indicating meter, what would the reading be on each instrument?

7.6   (a) Describe the construction and operating principles of a dynamometer (electrodynamic meter).

    (b) A dynamometer ammeter, a d.c. moving-coil ammeter and a 200 Ω resistor are connected in series with a rectifying device across a 240 V_{rms} alternating sinusoidal supply. The rectifier can be considered to behave as a resistance of 100 Ω to current in one direction and of 1 kΩ to that in the opposite direction. Calculate: (i) the readings on the two ammeters and (ii) the power dissipated in the 200 Ω resistor.

References and further reading

Baldwin, C.T. (1973) *Fundamentals of Electrical Measurements*, Prentice Hall: Hemel Hempstead.

Buckingham, H. and Price, E. M. (1966) *Principles of Electrical Measurements*, English Universities Press: London.

Edwards, D. F. A. (1971) *Electronic Measurement Techniques*, Butterworths: London.

Harris, F. K. (1952) *Electrical Measurements*, Wiley: New York.

Smith, R. J. (1976) *Circuits, Devices and Systems*, Wiley: New York.

Wolf, S. (1973) *Guide to Electronics Measurements and Laboratory Practice*, Prentice Hall: Englewood Cliffs.

8 Data recording and presentation

8.1 Introduction

The earlier chapters in this book have been essentially concerned with describing ways of producing high-quality, error-free data at the output of a measurement system. Having got the data, the next consideration is how to present it in a form where it can be readily used and analyzed. This generally means writing it down or displaying it in some other way on paper. However, as an intermediate step, there is frequently a requirement to record the data in some other, non-paper form, from where it can be transferred to paper later.

This chapter therefore starts with a discussion of the various means of recording data, and includes chart recorders, ultraviolet recorders, fiber optic recorders, magnetic tape recorders and computer data logging. Following this the relative merits of presenting data on paper in tabular and graphical form are compared. This leads on to a discussion of mathematical regression techniques for fitting the best lines through data points on a graph. Confidence tests to assess the correctness of the line fitted are also described. Finally, correlation tests are described which determine the degree of association between two sets of data when they are both subject to random fluctuations.

8.2 Recording of data

Chart recorders are probably the most widely used method of recording data and carry out this task simply, cheaply and reliably. However, their dynamic response is poor and they cannot record signals with frequencies greater than about 30 Hz. Ultraviolet recorders can cope with much higher frequencies up to about 13 kHz, but they are very delicate and easily damaged instruments which are also quite expensive. Fiber optic recorders are even more expensive than ultraviolet recorders but have a higher bandwidth still. These different types of recorder, together with magnetic tape and computer data logging, are discussed below.

8.2.1 Chart recorders

Chart recorders provide a relatively simple and cheap way of making permanent records of electrical signals. The two main kinds of chart recorder are the galvanometric type and the potentiometric type, with the former accounting for about 80% of the market.

Galvanometric recorders

These work on the same principle as a moving-coil meter except that the pointer draws an ink trace on paper, as illustrated in Figure 8.1, instead of merely moving against a scale. The measured signal is applied to the coil, and the angular deflection of this coil and its attached pointer is proportional to the magnitude of the signal applied. The pointer normally carries a pen, and, by using a motor running at constant speed to drive the chart paper with which the pointer is in contact, a time history of the measured signal is obtained.

Inspection of Figure 8.2 shows that the displacement y of the pen across the chart recorder is given by $y = R \sin \theta$. This sine relationship between the input signal and the displacement y is non-linear, and results in an error of 0.7% for deflections of ± 10 degrees. A more serious problem resulting from the pen moving in an arc is that it is difficult to relate the magnitude of deflection to the time axis. One way of overcoming this is to print a grid on the chart paper in the form of circular arcs, as illustrated in Figure 8.3. Unfortunately, measurement errors often occur in reading this type of chart, as interpolation for points drawn between the curved grid lines is difficult. An alternative solution is to use heat-sensitive chart paper directed over a knife edge, and to replace the pen by a heated stylus, as illustrated in Figure 8.4. The input–output

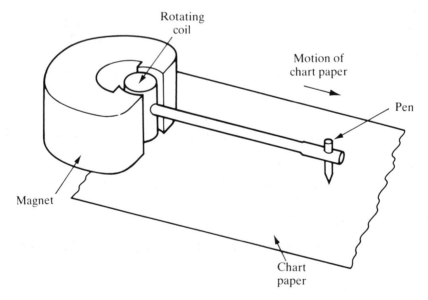

Figure 8.1 Simple galvanometric recorder

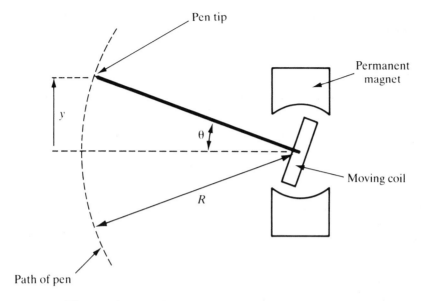

Figure 8.2 y versus θ relationship for standard chart recorder

relationship is still non-linear, with the deflection y being proportional to $\tan \theta$ as shown in Figure 8.5, and the reading error for excursions of ±10 degrees is still 0.7%. However, the rectilinearly scaled chart paper now required, as shown in Figure 8.6, allows much easier interpolation between grid lines.

Galvanometric recorders have a typical accuracy of ±2% and a resolution of 1%. Measurement with galvanometric chart recorders is limited to signals below about 30 Hz in frequency, the reason for which is readily understood if their dynamic behaviour is examined.

Dynamic behaviour of galvanometric recorders
Neglecting friction, the torque equation with the recorder in steady state can be expressed as:

torque due to current in coil = torque due to spring

Following a step input, we can write:

torque due to current in coil = torque due to spring + accelerating torque

or:

$$K_i I = K_s \theta + J\ddot{\theta} \qquad (8.1)$$

Direction of
paper travel

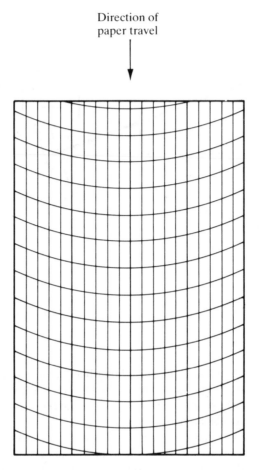

Figure 8.3 Curvilinear paper for standard galvanometric recorder

where I is the coil current, θ is the angular displacement, J is the moment of inertia and K_i and K_s are constants.

Consider now what happens if a recorder with resistance R_r is connected to a transducer with resistance R_t and output voltage V_t, as shown in Figure 8.7. The current flowing in steady state is given by:

$$I = V_t/(R_t + R_r)$$

When the transducer voltage V_t is first applied to the recorder coil, the coil will accelerate and, because the coil is moving in a magnetic field, a backwards voltage will be induced in it given by:

$$V_i = -K_i \dot{\theta}$$

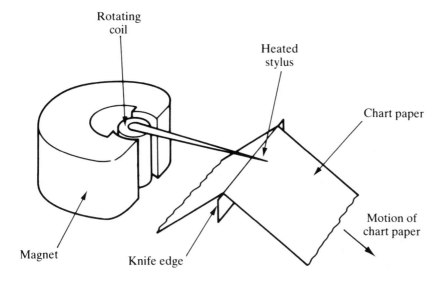

Figure 8.4 Knife-edge galvanometric recorder

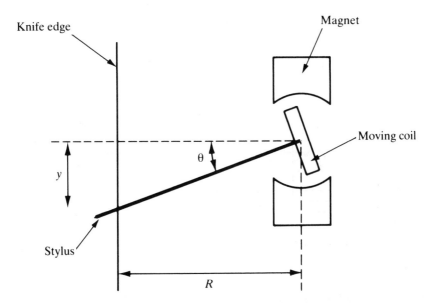

Figure 8.5 y versus θ relationship for knife-edge recorder

Direction of
paper travel

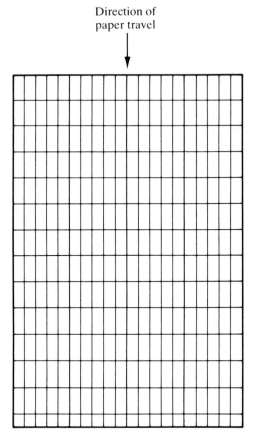

Figure 8.6 Rectilinear chart paper for the knife-edge galvanometric recorder

Transducer   Chart recorder

Figure 8.7 Connection of transducer to chart recorder

Hence the coil current is now given by:

$$I = \frac{V_t - K_i \dot{\theta}}{R_t + R_r}$$

Now substituting for I in the system Equation (8.1):

$$K_i\left(\frac{V_t - K_i\dot{\theta}}{R_t + R_r}\right) = K_s\theta + J\ddot{\theta}$$

or rearranging:

$$\ddot{\theta} + \frac{K_i^2\dot{\theta}}{J(R_t + R_r)} + \frac{K_s\theta}{J} = \frac{K_iV_t}{J(R_t + R_r)} \qquad (8.2)$$

This is the standard equation of a second-order dynamic system, with natural frequency ω and damping factor β given by:

$$\omega_n = \frac{K_s}{J} \qquad \beta = \frac{K_i^2}{2K_sJ(R_t + R_r)}$$

In steady state, $\ddot{\theta} = \dot{\theta} = 0$ and Equation (8.2) reduces to:

$$\frac{\theta}{V_t} = \frac{K_i}{K_s(R_t + R_r)} \qquad (8.3)$$

which is an expression describing the measurement sensitivity of the system.

The typical frequency response of a chart recorder for different damping factors is shown in Figure 8.8. This shows that the best bandwidth is obtained at a damping factor of 0.7, and this is therefore the target when the instrument is designed. The damping factor depends not only on the coil and spring constants (K_i and K_s) but also on the total circuit resistance ($R_t + R_r$). Adding a series or parallel resistance between the transducer and recorder as illustrated in Figure 8.9 respectively reduces or increases the damping factor. However, consideration of the sensitivity expression (8.3) shows that any reduction in the damping factor takes place at the expense of a reduction in measurement sensitivity. Other methods to alter the damping factor are therefore usually necessary and these techniques include decreasing the spring constant and system moment of inertia. In practice, there is a limit as to how far the dynamic performance of a chart recorder can be improved and other instruments such as ultraviolet recorders have to be used for higher-frequency signals.

Potentiometric recorders

Potentiometric recorders have much better specifications than galvanometric recorders, with an accuracy of $\pm 0.1\%$ of full scale and a measurement resolution of 0.2% f.s. being achievable. Such instruments employ a servo system, as shown in Figure 8.10, in which the pen is driven by a servomotor, and a potentiometer on the pen feeds back a signal proportional to pen position. This position signal is compared with

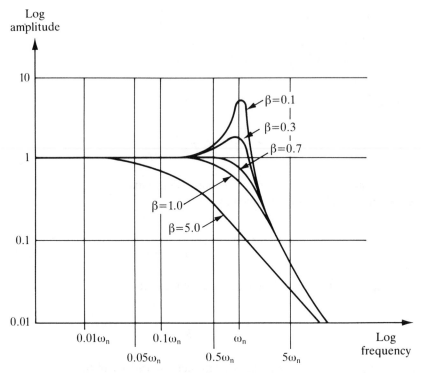

Figure 8.8 Frequency response of galvanometric chart recorder

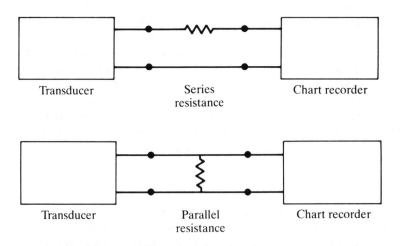

Figure 8.9 Addition of series and parallel resistances between transducer and chart recorder

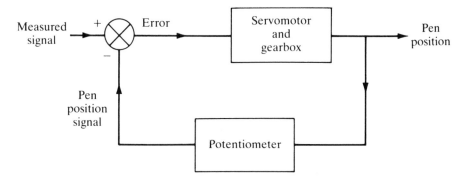

Figure 8.10 Servo system of potentiometer recorder

the measured signal, and the difference is applied as an error signal which drives the motor. However, a consequence of this electromechanical balancing mechanism is to give the instrument a slow response time in the range of 0.2–2.0 seconds. This means that potentiometric recorders are only suitable for measuring d.c. and slowly time-varying signals.

8.2.2 **Ultraviolet recorders**

The earlier discussion about galvanometric recorders concluded that restrictions on how far the system moment of inertia and spring constants can be reduced limited the maximum bandwidth to about 100 Hz. Ultraviolet recorders work on very similar principles to standard galvanometric chart recorders, but achieve a very significant reduction in system inertia and spring constants by mounting a narrow mirror rather than a pen system on the moving coil. This mirror reflects a beam of ultraviolet light on to ultraviolet-sensitive paper. It is usual to find several of these mirror–galvanometer systems mounted in parallel within one instrument to provide a multi-channel recording capability, as illustrated in Figure 8.11. This arrangement enables signals at frequencies up to 13 kHz to be recorded with a typical accuracy of ±2% f.s. Whilst it is possible to obtain satisfactory permanent signal recordings by this method, special precautions are necessary to protect the ultraviolet-sensitive paper from light before use and to spray a fixing lacquer on it after recording. Such instruments must also be handled with extreme care, because the mirror galvano-meters and their delicate mounting systems are easily damaged by relatively small shocks.

8.2.3 **Fiber optic recorders**

The fiber optic recorder is a very recent development which uses a fiber optic system to direct light on to light-sensitive paper. Fiber optic recorders are similar to oscilloscopes in construction, in so far as they have an electron gun and focusing

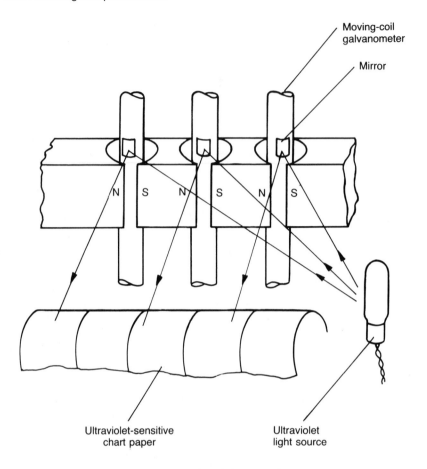

Moving-coil
galvanometer

Mirror

N S   N S   N S

Ultraviolet-sensitive
chart paper

Ultraviolet
light source

Figure 8.11 Ultraviolet recorder

system which directs a stream of electrons on to one point on a fluorescent screen. The screen is usually a long thin one instead of the square type found in an oscilloscope and only one set of deflection plates is provided. The signal to be recorded is applied to the deflection plates and the movement of the focused spot of electrons on the screen is proportional to the signal amplitude. A narrow strip of fiber optics in contact with the fluorescent screen transmits the motion of the spot to photosensitive paper held in close proximity to the other end of the fiber optic strip. By driving the photosensitive paper at a constant speed past the fiber optic strip, a time history of the measured signal is obtained. Such recorders are much more expensive than ultraviolet recorders but have an even higher bandwidth.

Whilst the construction above is the more common in fiber optic recorders, a second type also exists which uses a conventional square screen instead of a long thin one. This has a square faceplate attached to the screen housing a square array of fiber optics. The other side of the fiber optic system is in contact with chart paper. The effect of this is to provide a hard copy of the typical form of display obtainable on a cathode ray oscilloscope.

8.2.4 **Magnetic tape recorders**

Magnetic tape recorders can record analog signals up to 80 kHz in frequency. As the speed of the tape transport can be switched between several values, signals can be recorded at high speed and replayed at a lower speed. Such time scaling of the recorded information allows a hard copy of the signal behaviour to be obtained from instruments such as ultraviolet and galvanometric recorders whose bandwidth is insufficient to allow direct signal recording. A 200 Hz signal cannot be recorded directly on a chart recorder, but if it is recorded on a magnetic tape recorder running at high speed and then replayed at a speed ten times lower, its frequency will be time scaled to 20 Hz which can be recorded on a chart recorder. Instrumentation tape recorders typically have between four and ten channels, allowing many signals to be recorded simultaneously.

The two basic types of analog tape recording technique are direct recording and frequency-modulated recording. Direct recording offers the best data bandwidth but the accuracy of signal amplitude recording is quite poor, and this seriously limits the usefulness of this technique in most applications. The frequency-modulated technique offers much better amplitude recording, with an accuracy of $\pm5\%$ at signal frequencies of 80 kHz. In consequence, this technique is very much more common than direct recording.

8.2.5 **Computer data logging**

The only technique available for recording signals at frequencies higher than 80 kHz involves using a digital computer. As the signals to be recorded are in analog form, a prerequisite for this is an analog-to-digital converter board within the computer to sample the analog signals and convert them to the digital form in which the computer operates. The relevant aspects of the computer hardware involved will be covered in Chapter 10. Careful choice of the sampling interval is necessary so that an accurate digital record of the signal is obtained as explained in Chapter 5, so that problems of aliasing, etc., are not encountered. Some prior analog signal conditioning may also be required in some circumstances when a computer is used for data logging, again as mentioned in Chapter 5.

8.3 **Presentation of data**

The two formats available for presenting data on paper are tabular and graphical ones and the relative merits of these are compared below. In some circumstances, it is clearly best to use only one or other of these two alternatives alone. However, in many data collection exercises, part of the measurements and calculations is expressed in tabular form and part graphically, so making best use of the merits of each technique.

8.3.1 **Tabular presentation of data**

A tabular presentation allows data values to be recorded in a precise way which maintains exactly the accuracy to which the data values were measured. In other words, the data values are written down exactly as measured. Besides recording the raw data values as measured, tables often also contain further values calculated from the raw data.

An example of a tabular data presentation is given in Table 8.1. This records the results of an experiment to determine the strain induced in a bar of material under a range of stresses. Data was obtained by applying a sequence of forces to the end of the bar and using an extensometer to measure the change in length. Values of the stress and strain in the bar are calculated from these measurements and are also included in the table. The final row, which is of crucial importance in any tabular presentation, is the estimate of possible error in each calculated result.

A table of measurements and calculations should conform to several rules as illustrated in Table 8.1:

1. The table should have a title which explains what data is being presented within the table.
2. Each column of figures in the table should refer to the measurements or calculations associated with one quantity only.
3. Each column of figures should be headed by a title which identifies the data values contained in the column.
4. The units in which quantities in each column are measured should be stated at the top of the column.
5. All headings and columns should be separated by bold horizontal (and sometimes vertical) lines.
6. The errors associated with each data value quoted in the table should be given. The form shown in Table 8.1 is a suitable way to do this when the error level is the same for all data values in a particular column. However, if error levels vary, then it is preferable to write the error boundaries alongside each entry in the table.

8.3.2 **Graphical presentation of data**

The presentation of data in graphical form involves some compromise in the accuracy to which the data is recorded, as the exact values of measurements are lost. However, graphical presentation has two important advantages over tabular presentation:

1. Graphs provide a pictorial representation of results which is more readily comprehended than a set of tabular results.

Table 8.1 Sample tabular presentation of data

Table of measured applied forces and extensometer readings and calculations of stress
and strain

Force applied (kN)	Extensometer reading (divisions)	Stress (N/m²)	Strain
0	0	0	0
2	4.0	15.5	19.8×10^{-5}
4	5.8	31.0	28.6×10^{-5}
6	7.4	46.5	36.6×10^{-5}
8	9.0	62.0	44.4×10^{-5}
10	10.6	77.5	52.4×10^{-5}
12	12.2	93.0	60.2×10^{-5}
14	13.7	108.5	67.6×10^{-5}
Possible error in measurements (%)	±0.2	±0.2	±1.5	±1.0 × 10⁻⁵

2. Graphs are particularly useful for expressing the quantitative significance of results and showing whether a linear relationship exists between two variables.

Figure 8.12 shows a graph drawn from the stress and strain values given in Table 8.1. Construction of the graph involves first of all marking the points corresponding to the stress and strain values. The next step is to draw some line through these data points which best represents the relationship between the two variables. This line will normally be either a straight one or a smooth curve. The data points will not usually lie exactly on this line but instead will lie on either side of it. The magnitude of the excursions of the data points from the line drawn will depend on the magnitude of the random measurement errors associated with the data.

As for tables, certain rules exist for the proper representation of data in graphical form:

1. The graph should have a title or caption which explains what data is being presented in the graph.
2. Both axes of the graph should be labelled to express clearly what variable is associated with each axis and to define the units in which the variables are expressed.
3. The number of points marked along each axis should be kept reasonably small – about five divisions is often a suitable number.
4. No attempt should be made to draw the graph outside the boundaries corresponding to the maximum and minimum data values measured; in Figure 8.12, for

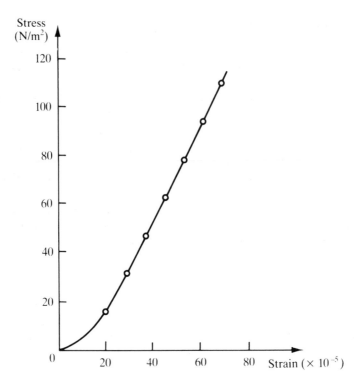

Figure 8.12 Sample graphical presentation data: graph of stress versus strain

example, the graph stops at a point corresponding to the highest measured stress value of 108.5.

8.3.3 Fitting curves to data points on a graph

The procedure of drawing a straight line or smooth curve as appropriate which passes close to all data points on a graph, rather than joining the data points by a jagged line which passes through each data point, is justified on account of the random errors which are known to affect measurements. Any line between the data points is mathematically acceptable as a graphical representation of the data if the maximum deviation of any data point from the line is within the boundaries of the identified level of possible measurement errors. However, within the range of possible lines which could be drawn, only one will be optimum. This optimum line is where the sum of negative errors in data points on one side of the line is balanced by the sum of positive errors in data points on the other side of the line. The nature of the data points is often such that a perfectly acceptable approximation to the optimum can be obtained by drawing a line through the data points by eye. In other cases, however, it is necessary to fit a line by regression techniques.

8.3.4 **Regression techniques**

Regression techniques consist of finding a mathematical relationship between measurements of two variables y and x, such that the value of one variable y can be predicted from a measurement of the other variable x. It is assumed for this that random errors only affect the y-values and that the values of x are exact. Regression procedures are simplest if a straight-line relationship exists between the variables, which can then be estimated by linear least-squares regression.

In many cases, inspection of the raw data points plotted on a graph shows that a straight-line relationship between the points does not exist. However, knowledge of the physical laws governing the data can often suggest a suitable alternative form of relationship between the two sets of measurements.

In some cases, the measured variables can be transformed such that a linear relationship is obtained. For example, suppose that two variables y and x are related according to $y = ax^c$. A linear relationship from this can be derived as:

$$\log(y) = \log(a) + c \log(x)$$

Thus if a graph is constructed of $\log(y)$ plotted against $\log(x)$, the parameters of a straight-line relationship can be estimated by linear least-squares regression.

Apart from linear relationships, physical laws often suggest a relationship where one variable y is related to another variable x by a power series of the form:

$$y = a_0 + a_1 x + a_2 x^2 + \ldots + a_p x^p$$

Estimation of the parameters $a_0 \ldots a_p$ is very difficult if p has a large value. Fortunately, a relationship where p only has a small value can be fitted to most data sets. Quadratic least-squares regression is used to estimate parameters where p has a value of two, and for larger values of p, polynomial least-squares regression is used.

Where the appropriate form of relationship between variables in measurement data sets is not obvious either from visual inspection or from consideration of physical laws, a method which is effectively a trial-and-error one has to be applied. This consists of estimating the parameters of successively higher-order relationships between y and x until a curve is found which fits the data sufficiently closely. What level of closeness is acceptable is considered in the later section on confidence tests.

Linear least-squares regression

If a linear relationship between y and x exists for a set of n measurements $y_1 \ldots y_n$, $x_1 \ldots x_n$, then this relationship can be expressed as $y = a + bx$, where the coefficients a and b are constants. The purpose of least-squares regression is to select the optimum values for a and b such that the line gives the best fit to the measurement data.

The deviation of each point (x_i, y_i) from the line can be expressed as d_i, where $d_i = y_i - (a + bx_i)$.

The best-fit line is obtained when the sum of the squared deviations, S, is a minimum, i.e. when:

$$S = \sum_{i=1}^{n} (d_i^2) = \sum_{i=1}^{n} (y_i - a - bx_i)^2$$

is a minimum. The minimum can be found by setting the partial derivatives $\partial S/\partial a$ and $\partial S/\partial b$ to zero and solving the resulting two simultaneous (normal) equations:

$$\partial S/\partial a = \Sigma\, 2(y_i - a - bx_i)(-1) = 0$$

$$\partial S/\partial b = \Sigma\, 2(y_i - a - bx_i)(-x_i) = 0$$

Using x_m and y_m to represent the mean values of x and y, the least-squares estimates $\hat{a}$ and $\hat{b}$ of the coefficients a and b can be written as:

$$\hat{b} = \frac{\Sigma(x_i - x_m)(y_i - y_m)}{\Sigma(x_i - x_m)^2} \tag{8.4}$$

$$\hat{a} = y_m - \hat{b}x_m \tag{8.5}$$

Example 8.1

In an experiment to determine the variation of the specific heat of a substance with temperature, the following table of results was obtained:

Temp. (°C)	40	45	50	55	60	65	70	75	80	85	90	95	100
Specific heat	1.38	1.43	1.46	1.49	1.56	1.57	1.59	1.64	1.69	1.72	1.78	1.83	1.85

Fit a straight line to this data set using least-squares regression and estimate the specific heat at a temperature of 72 °C.

Solution

Let y represent the specific heat and x represent the temperature. Then a suitable straight line is given by $y = a + bx$. We can now proceed to calculate estimates for the coefficients a and b using Equations (8.4) and (8.5) above.

The first step is to calculate the mean values of x and y. These are found to be $x_m = 70$ and $y_m = 1.6146$. Next, we need to tabulate $(x_i - x_m)$ and $(y_i - y_m)$ for each pair of data values:

i	$(x_i - x_m)$	$(y_i - y_m)$
1	−30	−0.2346
2	−25	−0.1846
.	.	.
.	.	.
.	.	.
13	+30	+0.2354

Now, applying Equations (8.4) and (8.5), we obtain:

$$\hat{b} = 0.007\ 817\ 7 \quad \text{and} \quad \hat{a} = 1.066\ 923$$

i.e. $y = 1.066\ 923 + 0.007\ 817\ 7x$. Using this equation, at a temperature (x) of 72 °C, the specific heat (y) is 1.63. Note that in this solution, we have only specified the answer to an accuracy of three figures, which is the same accuracy as the measurements. Any greater number of figures in the answer would be meaningless.

Least-squares regression is often appropriate for situations where a straight-line relationship is not immediately obvious, for example where $y \propto x^2$ or $y \propto \exp(x)$.

Example 8.2
From theoretical considerations, it is known that the voltage (V) across a charged capacitor decays with time (t) according to the relationship:

$$V = K \exp(-t/\tau)$$

Estimate values for K and τ if the following values of V and t are measured:

V	4.3	2.8	1.8	1.2	0.8	0.5	0.3
t	0	1	2	3	4	5	6

Solution
If $V = K \exp(-t/\tau)$ then:

$$\log_e(V) = \log_e(K) - t/\tau$$

Let $y = \log(V)$, $a = \log(K)$, $b = -1/\tau$, $x = t$. Then $y = a + bx$ which is the equation of a straight line. The best estimates of coefficients a and b are given by (using Equations (8.4) and (8.5)):

$$b = \frac{\Sigma(x_i - x_m)(y_i - y_m)}{\Sigma(x_i - x_m)^2} \qquad a = y_m - bx_m$$

From the given data, $x_m = 3$ and $y_m = 0.1626$.

Now calculate the deviations of the data from the means, and hence the quantities required to calculate a and b, as in Table 8.2.

Table 8.2

V (y_i)	log V (y_i)	t (x_i)	x_i − x_m	y_i − y_m	(x_i − x_m)(y_i − y_m)	(x_i − x_m)²
4.3	1.46	0	−3	1.296	−3.89	9
2.8	1.03	1	−2	0.867	−1.73	4
1.8	0.588	2	−1	0.425	−0.425	1
1.2	0.182	3	0	0.0197	0	0
0.8	−0.223	4	1	−0.386	−0.386	1
0.5	−0.693	5	2	−0.856	−1.71	4
0.3	−1.20	6	3	−1.367	−4.10	9

$\Sigma(x_i - x_m)(y_i - y_m) = -12.24$, $\Sigma(x_i - x_m)^2 = 28$
Hence, $b = -12.24/28 = -0.437$ and $a = 1.47$
$K = \exp(a) = \exp(1.47) = 4.369$
$\tau = -1/b = -1/(-0.437) = 2.287$.

Quadratic least-squares regression

Quadratic least-squares regression is used to estimate the parameters of a relationship $y = a + bx + cx^2$ between two sets of measurements $y_1 \ldots y_n$ and $x_1 \ldots x_n$.

The deviation of each point (x_i, y_i) from the line can be expressed as d_i, where $d_i = y_i - (a + bx_i + cx_i)^2$.

The best-fit line is obtained when the sum of the squared deviations, S, is a minimum, i.e. when:

$$S = \sum_{i=1}^{n} (d_i^2) = \sum_{i=1}^{n} (y_i - a - bx_i - cx_i^2)^2$$

is a minimum. The minimum can be found by setting the partial derivatives $\partial S/\partial a$, $\partial S/\partial b$ and $\partial S/\partial c$ to zero and solving the resulting simultaneous equations, as for the linear least-squares regression case above. Standard computer programs to estimate the parameters a, b and c by numerical methods are widely available and therefore a detailed solution is not presented here.

Polynomial least-squares regression

Polynomial least-squares regression is used to estimate the parameters of the pth-order relationship $y = a_0 + a_1 x + a_2 x^2 + \ldots + a_p x_p$ between two sets of measurements $y_1 \ldots y_n$ and $x_1 \ldots x_n$.

The deviation of each point (x_i, y_i) from the line can be expressed as d_i, where $d_i = y_i - (a_0 + a_1 x_i + a_2 x_i^2 + \ldots + a_p x_i^p)$.

The best-fit line is obtained when the sum of the squared deviations given by

$$S = \sum_{i=1}^{n} (d_i^2)$$

is a minimum. The minimum can be found as before by setting the p partial derivatives $\partial S/\partial a_0 \ldots \partial S/\partial a_p$ to zero and solving the resulting simultaneous equations. Again, as for the quadratic least-squares regression case, standard computer programs to estimate the parameters $a_0 \ldots a_p$ by numerical methods are widely available and therefore a detailed solution is not presented here.

8.3.5 Confidence tests in curve fitting

Having applied least-squares regression to estimate the parameters of a chosen relationship, some form of follow-up procedure is clearly required to assess how well the estimated relationship fits the data points. One fundamental requirement in curve fitting is that the maximum deviation d_i of any data point (y_i, x_i) from the fitted curve is less than the calculated maximum measurement error level. For some data sets it is impossible to find a relationship between the data points which satisfies this requirement. This normally occurs when both variables in a measurement data set are subject to random variation, such as where the two sets of data values are measurements of human height and weight. Correlation analysis is applied in such cases to determine the degree of association between the variables.

Assuming that a curve can be fitted to the data points in a measurement set without violating the above fundamental requirement that only one, not both, of the measured variables is subject to random errors, a further simple curve-fitting confidence test is to calculate the sum of squared deviations S for the chosen y/x relationship and to compare it with the value of S calculated for the next higher-order regression line which could be fitted to the data. Thus if a straight-line relationship is chosen, the value of S calculated should be of a similar magnitude to that obtained by fitting a quadratic relationship. If the value of S were substantially lower for a quadratic relationship, this would indicate that a quadratic relationship was a better fit to the data than a straight-line one and further tests would be needed to examine whether a cubic or higher-order relationship was a better fit still.

Other, more sophisticated confidence tests exist such as the F-ratio test. However, these are outside the scope of this book.

8.3.6 Correlation tests

Where both variables in a measurement data set are subject to random fluctuations, correlation analysis is applied to determine the degree of association between the variables. For example, in the case already quoted of a data set containing measurements of human height and weight, we certainly expect some relationship between the variables of height and weight because a tall person is heavier *on average* than a short person. Correlation tests determine the strength of the relationship (or interdependence) between the measured variables, which is expressed in the form of a correlation coefficient.

For two sets of measurements $x_1 \ldots x_n$ and $y_1 \ldots y_n$ with means x_m and y_m, the correlation coefficient Φ is given by:

$$\Phi = \frac{\Sigma(x_i - x_m)(y_i - y_m)}{\{[\Sigma(x_i - x_m)^2][\Sigma(y_i - y_m)^2]\}^{1/2}}$$

The value of $|\Phi|$ always lies between 0 and 1, with 0 representing the case where the variables are completely independent of one another and 1 the case where they are totally related to one another.

For $0 < |\Phi| < 1$, linear least-squares regression can be applied to find relationships between the variables, which allow x to be predicted from a measurement of y, and y to be predicted from a measurement of x. This involves finding two separate regression lines of the form:

$$y = a + bx \quad \text{and} \quad x = c + dy$$

These two lines are not normally coincident as shown in Figure 8.13. Both lines pass through the centroid of the data points but their slopes are different.

As $|\Phi| \rightarrow 1$, the lines tend to coincidence, representing the case where the two variables are totally dependent upon one another. As $|\Phi| \rightarrow 0$, the lines tend to

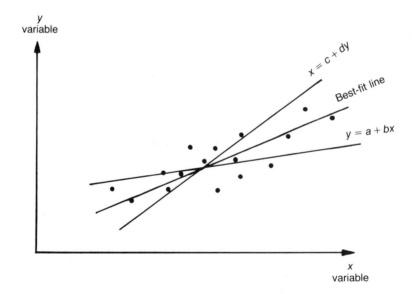

Figure 8.13 Relationship between two variables with random fluctuations

orthogonal ones parallel to the x- and y-axes. In this case, the two sets of variables are totally independent. The best estimate of x given any measurement of y is x_m and the best estimate of y given any measurement of x is y_m.

For the general case, the best fit to the data is that line which bisects the angle between the lines on Figure 8.13.

8.4 **Exercises**

8.1 (a) Explain the derivation of the expression

$$\ddot{\theta} + \frac{K_f^2\dot{\theta}}{JR} + \frac{K_s\theta}{J} = \frac{K_lV_t}{JR}$$

describing the dynamic response of a chart recorder following a step change in the electrical voltage output of a transducer connected to its input. Explain also what all the terms in the expression stand for. (Assume that the impedances of both the transducer and the recorder have a resistive component only and that there is negligible friction in the system.)

(b) Derive expressions for the measuring system natural frequency ω_n, the damping factor β and the steady-state sensitivity.

(c) Explain simple ways of increasing and decreasing the damping factor and describe the corresponding effect on measurement sensitivity.

(d) What damping factor gives the best system bandwidth?

(e) What aspects of the design of a chart recorder would you modify in order to improve the system bandwidth? What is the maximum bandwidth typically attainable in chart recorders, and if such a maximum bandwidth instrument is available, what is the highest-frequency signal that would generally be regarded as suitable for measuring if the accuracy of the signal amplitude measurement is important?

8.2 Theoretical considerations show that quantities x and y are related in a linear fashion such that:

$$y = ax + b$$

Show that the best estimates of the constants a and b are given by:

$$a = \frac{\overline{xy} - \overline{x}\,\overline{y}}{\overline{x^2} - (\overline{x})^2}$$

$$b = \overline{y} - a\overline{x}$$

Explain carefully the meaning of all the terms in the above two equations.

8.3 The characteristics of a chromel–constantan thermocouple are known to be approximately linear over the temperature range 300–800 °C. The output e.m.f. was measured practically at a range of temperatures and the following table of results obtained. Using least-squares regression, calculate the coefficients a and b for the relationship $T = aE + b$ which best describes the temperature–e.m.f. characteristic.

Temp. (°C)	300	325	350	375	400	425	450	475	500	525	550
e.m.f. (mV)	21.0	23.2	25.0	26.9	28.6	31.3	32.8	35.0	37.2	38.5	40.7
Temp. (°C)		575	600	625	650	675	700	725	750	775	800
e.m.f. (mV)		43.0	45.2	47.6	49.5	51.1	53.0	55.5	57.2	59.0	61.0

8.4 Measurements of the current (I) flowing through a resistor and the corresponding voltage drop (V) are shown below:

I	1	2	3	4	5
V	10.8	20.4	30.7	40.5	50.0

The instruments used to measure voltage and current were accurate in all respects except that they each had a zero error which the observer failed to take account of or to correct at the time of measurement. Determine the value of the resistor from the data measured.

8.5 A measured quantity y is known from theoretical considerations to depend on a variable x according to the relationship:

$$y = a + bx^2$$

For the following set of measurements of x and y, use linear least-squares regression to determine the estimates of the parameters a and b which fit the data best.

x	0	1	2	3	4	5
y	0.9	9.2	33.4	72.5	130.1	200.8

8.6 The mean time to failure (MTTF) of an integrated circuit is known to obey a law of the form:

$$MTTF = C \exp(T_0/T)$$

where T is the operating temperature and C and T_0 are constants. The following values of MTTF at various temperatures were obtained from accelerated-life tests.

MTTF (h)	54	105	206	411	941	2145
Temp. (K)	600	580	560	540	520	500

(a) Estimate the values of C and T_0.

(*Hint:* $\log_e(\text{MTTF}) = \log_e(C) + T_0/T$. This equation is now a straight-line relationship between $\log(\text{MTTF})$ and $1/T$, where $\log(C)$ and T_0 are constants.)

(b) For an MTTF of 10 years, calculate the maximum allowable temperature.

References and further reading

Chatfield, C. (1983) *Statistics for Technology*, Chapman and Hall: London.

Topping, J. (1960) *Errors of Observation and their Treatment*, Chapman and Hall: London.

9 Fiber optic sensors and transmission systems

Fiber optics is the technology of using light to transmit information. Light has a number of advantages over electricity as a medium for transmitting information: it is immune to corruption by neighbouring electromagnetic fields, the attenuation over a given transmitted distance is much less and it is also intrinsically safe.

Fiber optic technology is used in two major ways. First, the fiber optic cable itself can be used directly as a sensor in which the variable being measured causes some measurable change in the characteristics of the light transmitted by the cable. Secondly, fiber optic cables can be used as a transmission medium for data.

This latter use of fiber optic cables for transmitting information can be further divided into three separate areas of application. First, relatively short fiber optic cables are used as part of various instruments to transmit light from conventional sensors to a more convenient location for processing, often in situations where space is very short at the point of measurement. Secondly, longer fiber optic cables are used to connect remote instruments to controllers in instrumentation networks. Thirdly, longer links are still used for data transmission systems in telephone and computer networks. These three application classes have different requirements and tend to use different types of fiber optic cable.

9.1 Principles of fiber optics

The central part of a fiber optic system is a light-transmitting cable containing at least one but more often a bundle of glass or plastic fibers. This is terminated at each end by a transducer, as shown in Figure 9.1. At the input end, the transducer converts the signal from the electrical form in which most signals originate into light. At the output

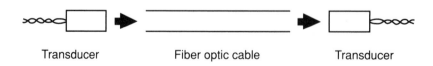

Transducer   Fiber optic cable   Transducer

Figure 9.1 Cables and transducers

end, the transducer converts the transmitted light back into an electrical form suitable for use by data recording, manipulation and display systems. These two transducers are often known as the transmitter and receiver respectively.

The arrangement described above is somewhat modified when the cable is used directly as a sensor. In that case, the light injected into the cable comes directly from a light source and does not originate as an electrical signal. However, the same mechanisms described later have to be used to get the light into the cable.

Further consideration of each of the above fiber optic system elements is given below.

9.1.1 Fiber optic cable

Fiber optic cable consists of an inner cylindrical core surrounded by an outer cylindrical cladding, as shown in Figure 9.2. The refractive index of the inner material is greater than that of the outer material, and the relationship between the two refractive indices affects the transmission characteristics of light along the cable. The amount of attenuation of light as it travels along the cable varies with the wavelength of the light transmitted. This characteristic is very non-linear and a graph of attenuation against wavelength shows a number of peaks and troughs. The position of these peaks and troughs varies according to the material used for the fibers. It should be noted that fiber manufacturers rarely mention these non-linear attenuation characteristics and quote the value of attenuation which occurs at the most favourable wavelength.

Two forms of cable exist, known as monomode and multimode. Monomode cables have a small-diameter core, typically 6 μm, whereas multimode cables have a much larger core, typically between 50 μm and 200 μm in diameter. Both glass and plastic in different combinations are used in various forms of cable. One option is to use different types of glass fiber for both the core and the cladding. A second, and cheaper, option is to have a glass fiber core and a plastic cladding. This has the additional advantage of being less brittle than the all-glass version. Finally, all-plastic cables also exist, where two types of plastic fiber with different refractive indices are

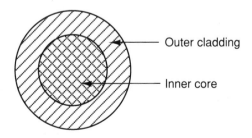

Outer cladding

Inner core

Figure 9.2 Cross-section through fiber optic cable

used. This is the cheapest form of all but it has the disadvantage of having high attenuation characteristics, making it unsuitable for the transmission of light over medium to large distances.

Protection is normally given to the cable by enclosing it in the same types of insulating and armouring materials that are used for copper cables. This protects the cable against various hostile operating environments and also against mechanical damage. When suitably protected, fiber optic cables can even withstand being engulfed in flames.

9.1.2 **Fiber optic transmitter**

The light-emitting diode (LED) is commonly used as the transducer which converts an electrical signal into light and transmits it into the cable. The LED is particularly suitable for this task as it has an approximately linear relationship between the input current and the light output. The type of LED chosen must closely match the attenuation characteristics of the light path through the cable and the spectral response of the receiving transducer.

An important characteristic of the transmitter is the proportion of its power which is coupled into the fiber optic cable: this is more important than its absolute output power. This proportion is maximized by making purpose-designed LED transmitters which have a spherical lens incorporated into the chip during manufacture. This produces an approximately parallel beam of light into the cable with a typical diameter of 400 μm.

The proportion of light entering the fiber optic cable is also governed by the quality of the end face of the cable and the way it is bonded to the transmitter. A good end face can be produced by either polishing or cleaving. Polishing involves grinding the fiber end down with progressively finer polishing compounds until a surface of the required quality is obtained. Attachment to the transmitter is then normally achieved by gluing. This is a time-consuming process but uses cheap materials. Cleaving makes use of special kits which nick the fiber, break it very cleanly by applying mechanical force and then attach it to the transmitter by crimping. This is a much faster method but cleaving kits are quite expensive. Both methods produce good results.

The proportion of light transmitted into the cable is also dependent on the proper alignment of the transmitter with the centre of the cable. The effect of misalignment depends on the relative diameters of the cable. Figure 9.3 shows the effect on the proportion of power transmitted into the cable for the cases of cable diameter greater than beam diameter, cable diameter equal to beam diameter and cable diameter less than beam diameter. This shows that some degree of misalignment can be tolerated except where the beam and cable diameters are equal. The cost of producing exact alignment of the transmitter and cable is very high, as it requires the LED to be exactly aligned in its housing, the fiber to be exactly aligned in its connector and the housing to be exactly aligned with the connector. Therefore, great cost savings can be achieved wherever some misalignment can be tolerated in the specification for the cable.

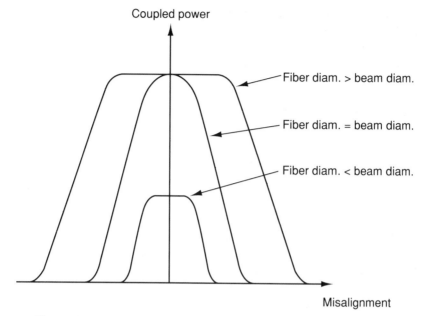

Figure 9.3 Effect of transmitter alignment on light power transmitted

9.1.3 **Fiber optic receiver**

The device used to convert the signal back from light to electrical form is usually either a PIN diode or a phototransistor. Phototransistors have good sensitivity but only a low bandwidth. On the other hand, PIN diodes have a much higher bandwidth but a lower sensitivity. If both high bandwidth and high sensitivity are required, then special avalanche photodiodes are used, but at a severe cost penalty. The same considerations about losses at the interface between the cable and receiver apply as for the transmitter, and both polishing and cleaving are used to prepare the fiber ends.

The output voltages from the receiver are very small and amplification is always necessary. The system is very prone to noise corruption at this point. However, the development of receivers which incorporate an amplifier are finding great success in reducing the scale of this noise problem.

9.2 **Transmission characteristics**

Monomode cables have very simple transmission characteristics because the core has a very small diameter and light can only travel in a straight line down it. On the other hand, multimode cables have quite complicated transmission characteristics because of the relatively large diameter of the core.

Whilst the transmitter is designed to maximize the amount of light which enters the cable in a direction parallel to its length, some light will inevitably enter multimode cables at other angles. Light which enters a multimode cable at any angle other than normal to the end face will be refracted in the core. It will then travel in a straight line until it meets the boundary between the core and cladding materials. At this boundary, some of the light will be reflected back into the core and some will be refracted in the cladding.

For materials of refractive indices n_1 and n_2 as shown in Figure 9.4, light entering from the external medium with refractive index n_0 at an angle α_0 will be refracted at an angle α_1 in the core and, when it meets the core–cladding boundary, part will be reflected at an angle β_1 back into the core and part will be refracted at an angle β_2 in the cladding. α_1 and α_0 are related by Snell's law according to:

$$n_0 \sin \alpha_0 = n_1 \sin \alpha_1 \tag{9.1}$$

Similarly, β_1 and β_2 are related by:

$$n_1 \sin \beta_1 = n_2 \sin \beta_2 \tag{9.2}$$

Light which enters the cladding is lost and contributes to the attenuation of the transmitted signal in the cable. However, examination of Equation (9.1) shows how this loss can be prevented. If $\beta_2 = 90$ degrees, then the refracted ray will travel along the boundary between the core and cladding and if $\beta_2 > 90$ degrees, all of the beam will be reflected back into the core. The case where $\beta_2 = 90$ degrees, corresponding to incident light at an angle α_c, is therefore the critical angle for total internal reflection to occur at the core–cladding boundary. The condition for this is that $\sin \beta_2 = 1$.

Setting $\sin \beta_2 = 1$ in Equation (9.1):

$$\frac{n_1 \sin \beta_1}{n_2} = 1$$

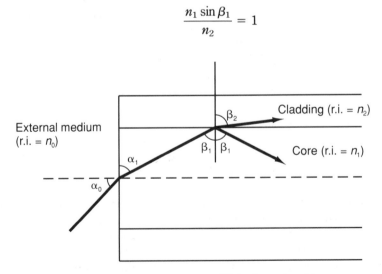

Figure 9.4 Transmission of light through cable

Thus:

$$\sin \beta_1 = n_2/n_1$$

Inspection of Figure 9.4 shows that $\cos \alpha_1 = \sin \beta_1$. Hence:

$$\sin \alpha_1 = (1 - \cos^2 \alpha_1)^{1/2} = (1 - \sin^2 \beta_1)^{1/2} = [1 - (n_2/n_1)^2]^{1/2}$$

From Equation (9.1):

$$\sin \alpha_c = \sin \alpha_0 = \frac{n_1}{n_0} \sin \alpha_1$$

Hence:

$$\sin \alpha_c = \frac{n_1}{n_0} [1 - (n_2/n_1)^2]^{1/2}$$

Therefore, provided that the angle of incidence of the light into the cable is greater than the critical angle given by $\theta = \sin^{-1} \alpha_c$, all of the light will be internally reflected at the core–cladding boundary. Further reflections will occur as the light passes down the fibers and it will thus travel in a zigzag fashion to the end of the cable.

Whilst attenuation has been minimized, there is a remaining problem in that the transmission time of the parts of the beam which travel in this zigzag manner will be greater than light which enters the fiber at 90 degrees to the face and so travels in a straight line to the other end. In practice, the incident light rays to the cable will be spread over the range given by $\sin^{-1} \alpha_c < \theta < 90$ degrees and so the transmission times of these separate parts of the beam will be distributed over a corresponding range. These differential delay characteristics of the light beam are known as modal dispersion. The practical effect is that a step change in light intensity at the input end of the cable will be received over a finite period of time at the output.

It is possible largely to overcome this latter problem in multimode cables by using cables made solely from glass fibers in which the refractive index changes gradually over the cross-section of the core rather than abruptly at the core–cladding interface as in the step index cable discussed so far. This special type of cable is known as graded index cable and it progressively bends light incident at less than 90 degrees to its end face rather than reflecting it off the core–cladding boundary. Although the parts of the beam away from the centre of the cable travel further, they also travel faster than the beam passing straight down the centre of the cable because the refractive index is lower away from the centre. Hence, all parts of the beam are subject to approximately the same propagation delay. In consequence, a step change in light intensity at the input produces approximately a step change of light intensity at the output. The alternative solution is to use a monomode cable. This propagates light in a single mode only, which means that time dispersion of the signal is almost eliminated.

9.3 **Fiber optic sensors**

The basis of operation of fiber optic sensors is the translation of the physical quantity measured into a change in one or more parameters of a light beam. The light parameters that can be modulated are one or more of the following:

1. Intensity
2. Phase
3. Polarization
4. Wavelength
5. Transmission time.

Fiber optic sensors usually incorporate either glass/plastic cables or all-plastic cables. All-glass types are rarely used because of their fragility. Plastic cables have particular advantages for sensor applications because they are cheap and have a relatively large diameter of 0.5–1.0 mm, making connection to the transmitter and receiver easy. However, plastic cables should not be used in certain hostile environments where they may be severely damaged. The cost of the fiber optic cable itself is insignificant for sensing applications, as the total cost of the sensor is dominated by the cost of the transmitter and receiver.

Fiber optic sensors characteristically enjoy long life. For example, the life expectancy of reflective fiber optic switches is quoted at 10 million operations. Their accuracy is also good, with for instance ±1% of full-scale reading being quoted as a typical inaccuracy level for a fiber optic pressure sensor. Further advantages are their simplicity, low cost, small size, high reliability and capability of working in many kinds of hostile environment. However, in spite of these obvious merits, industrial usage is currently quite low. This may well be an example of a vicious circle, where new users do not appear because there is no substantial body of existing users to consult and gain experience from.

Two major classes of fiber optic sensor exist, intrinsic sensors and extrinsic sensors. In **intrinsic sensors**, the fiber optic cable itself is the sensor, whereas in **extrinsic sensors**, the fiber optic cable is only used to guide light to/from a conventional sensor.

9.3.1 **Intrinsic sensors**

Intrinsic sensors can modulate the intensity, phase, polarization, wavelength or transit time of light. Sensors which modulate light intensity tend to use mainly multimode fibers, but only monomode cables are used to modulate other light parameters. A particularly useful feature of intrinsic fiber optic sensors is that they can, if required, provide distributed sensing over distances of up to 1 metre.

Light intensity is the simplest parameter to manipulate in intrinsic sensors because only a simple source and detector are required. The various forms of switches shown in Figure 9.5 are perhaps the simplest form of these, as the light path is simply blocked and unblocked as the switch changes state.

Modulation of the intensity of transmitted light takes place in various simple forms of proximity, displacement, pressure, pH and smoke sensors. Some of these are sketched in Figure 9.6. In proximity and displacement sensors (the latter are often given the special name *Fotonic* sensors), the amount of reflected light varies with the distance between the fiber ends and a boundary. In pressure sensors, the refractive index of the fiber, and hence the intensity of light transmitted, varies according to the

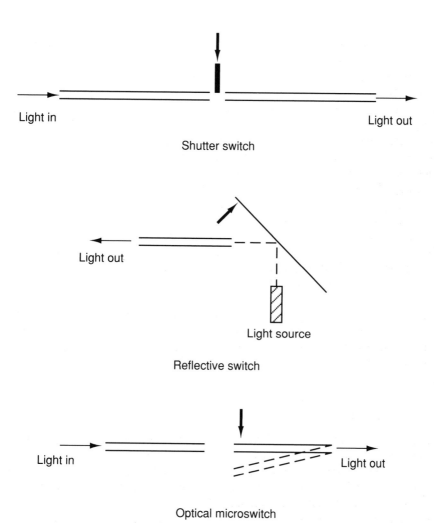

Figure 9.5 Fiber optic switches

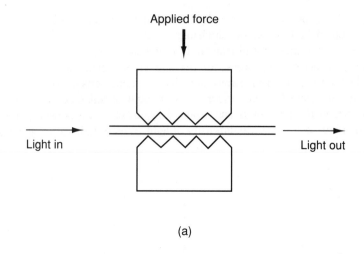

(a)

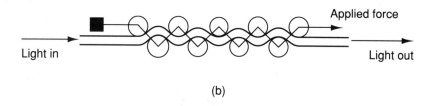

(b)

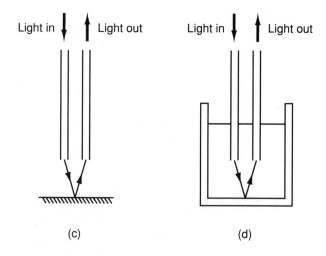

Figure 9.6 Intensity-modulating sensors: (a) simple pressure sensor; (b) roller-chain pressure sensor (microbend sensor); (c) proximity sensor; (d) pH sensor

mechanical deformation of the fibers caused by pressure. In the pH probe, the amount of light reflected back into the fibers depends on the pH-dependent colour of the chemical indicator in the solution around the probe tip. Finally, in a form of smoke detector, two fiber optic cables placed either side of a space detect any reduction in the intensity of light transmission between them caused by the presence of smoke.

A simple form of accelerometer can be made by placing a mass subject to the acceleration on a multimode fiber. The force exerted by the mass on the fiber causes a change in the intensity of light transmitted, thus allowing the acceleration to be determined. The typical accuracy quoted for this device is ±0.02 g in the measurement range of ±5 g and $\pm2\%$ in the measurement range up to 100 g.

A similar principle is used in probes which measure the internal diameter of tubes. The probe consists of eight strain-gauged cantilever beams which track changes in diameter, giving a measurement resolution of 20 μm.

A slightly more complicated method of effecting light intensity modulation is the variable shutter sensor shown in Figure 9.7. This consists of two fixed fibers with two collimating lenses and a variable shutter between. Movement of the shutter changes the intensity of light transmitted between the fibers. This is used to measure the displacement of various devices such as Bourdon tubes, diaphragms and bimetallic thermometers.

Yet another type of intrinsic sensor uses cable where the core and cladding have similar refractive indices but different temperature coefficients. This is used as a temperature sensor. Temperature rises cause the refractive indices to become even closer together and losses from the core to increase, thus reducing the quantity of light transmitted.

Refractive index variation is also used in a form of intrinsic sensor used for cryogenic leak detection. The fiber used for this has a cladding whose refractive index becomes greater than that of the core when it is cooled to cryogenic temperatures. The fiber optic cable is laid in the location where cryogenic leaks might occur. If any leaks do occur, light travelling in the core is transferred to the cladding, where it is attenuated. Cryogenic leakage is thus indicated by monitoring the light transmission characteristics of the fiber.

Yet another use of refractive index variation is found in devices which detect oil in water. These use a special form of cable where the cladding used is sensitive to oil. Any oil present diffuses into the cladding and changes the refractive index, thus

Figure 9.7 Variable shutter sensor

increasing light losses from the core. Unclad fibers are used in a similar way. In these, any oil present settles on the core and allows light to escape.

The **cross-talk sensor** measures several different variables by modulating the intensity of light transmitted. It consists of two parallel fibers which are close together and where one or more short lengths of adjacent cladding are removed from the fibers. When it is immersed in a transparent liquid, there are three different effects which each cause a variation in the intensity of light transmitted. Thus, the sensor can perform three separate functions. First, it can measure temperature according to the temperature-induced variation in the refractive index of the liquid. Secondly, it can act as a level detector according to the depth of the liquid which changes the transmission characteristics between the fibers. Thirdly, it can measure the refractive index of the liquid itself when used under controlled temperature conditions.

The refractive index of a liquid can be measured in an alternative way by using an arrangement where light travels across the liquid between two cable ends which are fairly close together. The angle of the cone of light emitted from the source cable, and hence the amount of light transmitted into the detector, is dependent on the refractive index of the liquid.

The use of materials where the fluorescence varies according to the value of the measurand can also be used as part of intensity-modulating intrinsic sensors. Fluorescence-modulating sensors can give very high sensitivity and are potentially very attractive in biomedical applications where requirements to measure very small quantities such as low oxygen and carbon monoxide concentrations, low blood pressure levels, etc., exist. Similarly, low concentrations of hormones, steroids, etc., may also be measured (Grattan 1989).

Further examples of intrinsic fiber optic sensors which modulate light intensity are described in Chapter 15 (level measurement) and Chapter 17 (measuring small displacements).

As mentioned previously, light phase, polarization, wavelength and transit time can be modulated as well as intensity in intrinsic sensors. Monomode cables are used almost exclusively in these types of intrinsic sensor.

Phase modulation normally requires a coherent (laser) light source. It can provide very high sensitivity in displacement measurement but cross-sensitivity to temperature and strain degrades its performance. Maintaining frequency stability of the light source and manufacturing difficulties in coupling the light source to the fiber are further problems. Various versions of this class of instrument exist to measure temperature, pressure, strain, magnetic fields and electric fields. Field-generated quantities such as electric current and voltage can also be measured. In each case, the measurand causes a phase change between a measuring and a reference light beam which is detected by an interferometer. Fuller details can be found in Harmer (1982) and Medlock (1986).

The principle of phase modulation has also been used in the fiber optic accelerometer (where a mass subject to acceleration rests on a fiber), and in fiber strain gauges (where two fibers are fixed on the upper and lower surfaces of a bar under strain). These are discussed in more detail in Harmer (1982). The fiber optic gyroscope described in Chapter 19 is a further example of a phase-modulating device.

Devices using polarization modulation require special forms of fiber which maintain polarization. Polarization changes can be effected by electrical fields, magnetic fields, temperature changes and mechanical strain. Each of these parameters can therefore be measured by polarization modulation.

Various devices which modulate the wavelength of light are used for special purposes, as described in Medlock (1986). However, the only common wavelength-modulating fiber optic device is the form of laser Doppler flowmeter which uses fiber optic cables, as described in Chapter 14.

Fiber optic devices using modulation of the transit time of light are uncommon because of the speed of light. Measurement of the transit time for light to travel from a source, be reflected off an object and travel back to a detector is only viable for extremely large distances. However, a few special arrangements have evolved which use transit time modulation, as described in Medlock (1986). These include instruments such as the optical resonator, which can measure both mechanical strain and temperature. Temperature-dependent wavelength variation also occurs in semi-conductor crystal beads (e.g. aluminium gallium arsenide). These are bonded to the end of a fiber optic cable and excited from an LED at the other end of the cable. Light from the LED is reflected back along the cable by the beads at a different wavelength. Measurement of the wavelength change allows temperatures in the range up to 200 °C to be measured accurately. A particular advantage of this sensor is its small size, typically of 0.5 mm diameter at the sensing tip. Finally, to complete the catalogue of transit time devices, the frequency modulation in a piezoelectric quartz crystal used for gas sensing can also be regarded as a form of time domain modulation.

9.3.2 Extrinsic sensors

Extrinsic fiber optic sensors use a fiber optic cable, normally a multimode one, to transmit modulated light from a conventional sensor. A major feature of extrinsic sensors, which makes them so useful in such a large number of applications, is their ability to reach places which are otherwise inaccessible. One example of this is the insertion of fiber optic cables into the jet engines of aircraft to measure temperature by transmitting radiation into a radiation pyrometer located remotely from the engine. Fiber optic cable can be used in the same way to measure the internal temperature of electrical transformers, where the extreme electromagnetic fields present make other measurement techniques impossible.

Extrinsic fiber optic sensors provide excellent protection of measurement signals against noise corruption. Unfortunately, the output of many forms of conventional sensor is not in a form which can be transmitted by a fiber optic cable. Conversion into a suitable form must therefore take place prior to transmission. For example, in the case of a platinum resistance thermometer (PRT), the temperature changes are translated into resistance changes. The PRT must therefore have an electrical power supply. The modulated voltage level at the output of the PRT can then be injected into the fiber optic cable via the usual type of transmitter. This complicates the measurement process and means that low-voltage power cables must be routed with the fiber optic cable to the transducer. One particular adverse effect of this is that the advantage of intrinsic safety is lost.

Recent research has been directed to these kinds of problems which beset some extrinsic sensors, and this has resulted in the development of power sources in the form of electronically generated pulses driven by a lithium battery (Grattan 1989). This avoids having to transmit electrical power to the sensor via cables and provides intrinsically safe operation.

Piezoelectric sensors lend themselves to use in extrinsic sensors because the modulated frequency of a quartz crystal can be readily transmitted into a fiber optic cable by fitting electrodes to the crystal which are connected to a low-power LED. Resonance of the crystal can be created either by electrical means or by optical means using the photothermal effect. The photothermal effect describes the principle where, if light is pulsed at the required oscillation frequency and directed at a quartz crystal, the localized heating and thermal stress caused in the crystal results in it oscillating at the pulse frequency. Piezoelectric extrinsic sensors can be used as part of various pressure, force and displacement sensors. At the other end of the cable, a phase-locked loop is typically used to measure the transmitted frequency.

One extremely accurate form of extrinsic sensor is a device known as the Accufibre temperature sensor. This is a form of radiation pyrometer which has a black-box cavity at the focal point of the lens system. A fiber optic cable is used to transmit radiation from the black-box cavity to a spectrometric device which computes the temperature. Fuller details are given in Chapter 12.

Fiber optic cables are also now commonly included in digital encoders, where the use of fibers to transmit light to and from the disks allows the light source and detectors to be located remotely. This allows the devices to be smaller, which is a great advantage in many applications where space is at a premium.

9.4 Fiber optic instrumentation networks

Little if any saving arises out of installing fiber optic links in instrumentation networks; indeed there may be a cost penalty. However, there are great advantages in terms of the links' immunity to contamination of the signals carried. Immunity to current surges caused by stray electromagnetic fields also affords protection to computers in the network, as such surges can cause software corruption.

As with fiber optic sensors, the cost of short fiber optic links in instrumentation networks is dominated by the cost of the terminating transducers. However, as the length of the link becomes greater, the cost of the fiber optic cable becomes more significant. The cheapness of plastic cables is attractive for instrumentation networks but they cannot generally be used over distances greater than about 30 m because signal attenuation is very high.

One disadvantage of fiber optics compared with electrical conductors in networks is that light connections at the ends of the cable are much more costly than electrical connections. Branches are impossible with glass and silica fibers, though a technique is available for creating branches in plastic cables.

Current research is looking at ways of distributing a range of discrete sensors measuring different variables along a fiber optic cable. Alternatively, sensors of the same type, which are located at various points along a cable, are being investigated as a means of providing distributed sensing of a single measured variable. For example, the use of a 2 km long cable to measure the temperature distribution along its entire length has been demonstrated, measuring temperature at 400 separate points to a resolution of 1 °C.

Various types of branching network and multiplexing schemes have been proposed, some of which have been implemented as described in Grattan (1989). Wavelength division multiplexing is particularly well suited to fiber optic applications and the technique is now becoming well established. A single fibre is capable of propagating a large number of different wavelengths without cross-interference, and multiplexing thus allows a large number of distributed sensors to be addressed. A single optical light source is often sufficient for this, particularly if the modulated parameter is not light intensity.

The subject of instrumentation networks, and the use of fiber optic cables within them, is discussed further in Chapter 11.

9.5 **Data networks using fiber optics**

Fiber optic cables have major advantages over copper for data transmission over large distances. Attenuation is much smaller and bandwidth is much larger. This means that fiber optic links can replace several copper cables. Also, the requirement for signal boosters along the line to compensate for attenuation is greatly reduced. Both of these factors mean that fiber optic links can lead to great cost savings.

In contrast to fiber optic sensors, the cost of fiber optic links in long-distance data networks is dominated by the cost of the cable, with the cost of the transmitter and receiver being relatively insignificant.

References and further reading

Grattan, K. T. V. (1989) 'New developments in sensor technology – fibre- and electro-optics', *Measurement and Control*, **22**(6), 165–75.

Harmer, A. L. (1982) 'Principles of optical fiber sensors and instrumentation', *Measurement and Control*, **15**(4), 143–51.

Medlock, R. S. (1986) 'Review of modulating techniques for fibre optic sensors', *Measurement and Control*, **19**(1), 4–17.

10 Intelligent instruments

The impact that the microprocessor has had on the domestic consumer is very evident, with its inclusion in washing machines, vehicle fuel control systems and home computers to name but a few applications. The advent of the microprocessor has had an equally significant, though perhaps not so obviously apparent, impact on the field of instrumentation. It is therefore fitting that a chapter of this book should be devoted to describing the principles of operation of microprocessors and explaining how they are included and programmed as a microcomputer within a measuring instrument, thereby producing what has come to be known as an intelligent instrument.

An intelligent instrument comprises all the usual elements of a measurement system as shown in Figure 1.3, and is only distinguished from dumb (non-intelligent) measurement systems by the inclusion of a microprocessor to fulfil the signal processing function. The effect of this computerization of the signal processing function is an improvement in the quality of the instrument output measurements and a general simplification of the signal processing task. Some examples of the signal processing which a microprocessor can readily perform include correction of the instrument output for bias caused by environmental variations (e.g. temperature changes), and conversion to produce a linear output from a transducer whose characteristic is fundamentally non-linear. A fuller discussion about the techniques of digital signal processing is to be found in section 5.6.

10.1 **Elements of a microcomputer**

The primary function of a digital computer is the manipulation of data. The three elements which are essential to the fulfilment of this task are the central processing unit, the memory and the input–output interface, as shown in Figure 10.1. These elements are collectively known as the computer hardware, and each element exists physically as one or more integrated circuit chips mounted on a printed circuit board. Where the central processing unit (CPU) consists of a single microprocessor, it is usual to regard the system as a microcomputer. The distinction between microcomputer, minicomputer and mainframe computer is a very arbitrary division made according to relative computer power. However, this classification has become somewhat meaningless, with present-day 'microcomputers' being more powerful than the mainframe computers of only a few years ago.

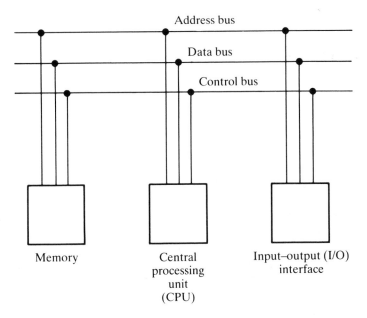

Figure 10.1 Elements of a microcomputer

The **central processing unit** (CPU) part of a computer can be regarded as the brain of the system. The CPU determines what computational operations are carried out and the sequence in which they are executed. During such operations, the CPU makes use of one or more special storage locations within itself known as **registers**. Another part of the CPU is the **arithmetic and logic unit** (ALU) which is where all arithmetic operations are evaluated. The CPU operates according to a sequential list of required operations defined by a computer program, known as the computer software. This program is held in the second of the three system components known as the computer memory.

The **computer memory** also serves several other functions besides this role of holding the computer program. One of these is to provide temporary storage locations which the CPU uses to store variables during execution of the computer program. A further common use of memory is to store data tables which are used for scaling and variable conversion purposes during program execution.

Memory can be visualized as a consecutive sequence of boxes in which various items are stored, as shown in Figure 10.2 for a typical memory size of 65 536 storage units. If this storage mechanism is to be useful, then it is essential that a means is provided for giving a unique label to each storage box. This is achieved by labelling the first box as 0, the next one as 1 and so on for the rest of the storage locations. These numbers are known as the **memory addresses**. Whilst they can be labelled by decimal numbers, it is more usual to use hexadecimal notation (see section 10.2).

Two main types of computer memory exist and there are important differences

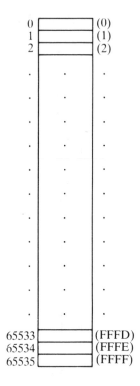

Figure 10.2 Schematic representation of computer memory (numbers in parentheses are memory addresses in hexadecimal notation)

between them. The two kinds are **random-access memory** (RAM) and **read-only memory** (ROM). The CPU can both read from and write to the former, but it can only read from the latter. The importance of ROM becomes apparent if the behaviour of each kind of memory is considered when the power supply is turned off. At power-off time, RAM loses its contents but ROM maintains them, and this is the value of ROM. Intelligent instruments normally use ROM for storage of the program and data tables and just have a small amount of RAM which is used by the CPU for temporary variable storage during program execution.

The third essential element of a computer system is the **input–output (I/O) interface**, which allows the computer to communicate with the outside world by reading in data values and outputting results after the appropriate computation has been executed. In the case of a microcomputer performing a signal processing function within an intelligent instrument, this means reading in the values obtained from one or more transducers and outputting a processed value for presentation at the instrument output. All such external peripherals are identified by a unique number as for memory addresses.

Communication between these three computer elements is provided by three electronic highways known as the **data bus**, the **address bus** and the **control bus**. At each data transfer operation executed by the CPU, two items of information must be conveyed along the electronic highway: the item of data being transferred and the address where it is being sent. Whilst both of these items of information could be conveyed along a single bus, it is more usual to use two buses which are called the data bus and the address bus. The timing of data transfer operations is important, particularly when transfers take place to peripherals such as disk drives and keyboards where the CPU often has to wait until the peripheral is free before it can initialize a data transfer. This timing information is carried by a third highway known as the control bus.

The latest trend made possible by advances in very large-scale integration (VLSI) technology is to incorporate all three functions of central processor unit, memory and I/O within a single chip (known as the computer on a chip). These chips are already well established in domestic appliances and vehicle fuel control systems, and their exploitation within intelligent instruments is likely to grow rapidly.

10.2 **Microcomputer operation**

As has already been mentioned, the fundamental role of a computer is the manipulation of data. Numbers are used both in quantifying items of data and in the form of codes which define the computational operations to be executed. All numbers which are used for these two purposes must be stored within the computer memory and also transported along the communication buses. A detailed consideration of the conventions used for representing numbers within the computer is therefore required.

10.2.1 **Number systems**

The decimal system is the best-known number system, but it is not very suitable for use by digital computers. It uses a base of ten, such that each digit in a number can have any one of ten values within the range 0–9. Items of electronic equipment such as the digital counter, which are often used as computer peripherals, have liquid crystal display elements which can each display any of the ten decimal digits, and therefore a four-element display can directly represent decimal numbers in the range 0–9999. The decimal system is therefore perfectly suitable for use with such output devices.

The fundamental unit of data storage within a digital computer is a memory element known as a **bit**. This holds information by switching between one of two possible states. Each storage unit can therefore only represent two possible values and all data to be entered into memory must be organized into a format which recognizes this restriction. This means that numbers must be entered in binary format, where each digit in the number can have only one of two values, 0 or 1. The binary representation is particularly convenient for computers because bits can be represented very simply electronically as either zero or non-zero voltages. However, the conversion is tedious for humans. Starting from the right-hand side of a binary number, where the first digit represents 2^0 (i.e. 1), each successive binary digit represents progressively higher powers of two. For example, in the binary number 1111, the first digit (starting from the right-hand side) represents 1, the next 2, the next 4 and the final, leftmost digit represents 8; thus the decimal equivalent is $1 + 2 + 4 + 8 = 15$.

Example 10.1
Convert the following 8-bit binary number to its decimal equivalent: 10110011.

Solution
Starting at the right-hand side, we have:

1×2^0	$+1 \times 2^1$	$+0 \times 2^2$	$+0 \times 2^3$	$+1 \times 2^4$	$+1 \times 2^5$	$+0 \times 2^6$	$+1 \times 2^7$
$=1$	$+2$	$+0$	$+0$	$+16$	$+32$	$+0$	$+128$

$=179$

For data storage purposes, memory elements are combined into larger units known as **bytes**, which are usually considered to consist of 8 bits each. Each bit holds one binary digit, and therefore a memory unit consisting of 8 bits can store eight-digit binary numbers in the range of 00000000 to 11111111 (equivalent to decimal numbers in the range of 0 to 255). A binary number in this system of 10010011 for instance would correspond to the decimal number 147.

This range is clearly inadequate for most purposes, including measurement systems, because even if all data could be conveniently scaled the maximum resolution obtainable is only 1 part in 128. Numbers are therefore normally stored in units of either 2 or 4 bytes, which allow the storage of integer (whole) numbers in the range of 0–65 535 or 0–4 294 967 296.

No means have been suggested so far for expressing the sign of numbers, which is clearly necessary in the real world where negative as well as positive numbers occur. A simple way to do this is to reserve the most significant (left-hand) bit in a storage unit to define the sign of a number, with '0' representing a positive number and '1' a negative number. This alters the range of numbers representable in a 1 byte storage unit to -127 to $+127$, as only 7 bits are left to express the magnitude of the number, and also means that there are two representations of the value 0. In this system the binary number 10010011 translates to the decimal number -19 and 00010011 translates to $+19$. For reasons dictated by the mode of operation of the CPU, however, most computers use an alternative representation known as the two's complement form.

The **two's complement** of a number is most easily formed by going via an intermediate stage of the **one's complement**. The one's complement of a number is formed by reversing all digits in the binary representation of the magnitude of a number, changing ones to zeros and zeros to ones, and then changing the left-hand bit to a 1 if the original number was negative. The two's complement is then formed by adding 1 at the least significant (right-hand) end of the one's complement. As before for a 1 byte storage unit, only 7 bits are available for representing the magnitude of a number, but, because there is now only one representation of zero, the decimal range representable is -128 to $+127$.

Example 10.2

Find the one's and two's complement 8 bit binary representation of the following decimal numbers:

$$56 \quad -56 \quad 73 \quad 119 \quad 27 \quad -47$$

Method of solution

Take first the decimal value of 56.
Form 7 bit binary representation: 0111000
Reverse digits in this: 1000111
Add sign bit to left-hand end to form one's complement: 01000111
Form two's complement by adding one to one's complement:

$$01000111 + 1 = 01001000$$

Take next the decimal value of -56.
Form 7 bit binary representation: 0111000
Reverse digits in this: 1000111
Add sign bit to left-hand end to form one's complement: 11000111
Form two's complement by adding one to one's complement:

$$11000111 + 1 = 11001000$$

The summary of solution for all values is shown in Table 10.1.

Table 10.1

Decimal no.	Binary representation of magnitude (7 bit)	Digits reversed in 7 bit representation	One's complement (8 bit)	Two's complement (8 bit)
56	0111000	1000111	01000111	01001000
-56	0111000	1000111	11000111	11001000
73	1001001	0110110	00110110	00110111
119	1110111	0001000	00001000	00001001
27	0011011	1100100	01100100	01100101
-47	0101111	1010000	11010000	11010001

We have therefore established the binary code in which the computer stores positive and negative integers (whole numbers). However, it is frequently necessary also to handle real numbers (those with fractional parts). These are most commonly stored using the floating point representation.

The **floating point representation** divides each memory storage unit (notionally, not physically) into three fields, known as the sign field, the exponent field and the mantissa field. The sign field is always 1 bit wide but there is no formal definition for the relative sizes of the other fields. However, a common subdivision of a 32 bit (4 byte) storage unit is to have a 7 bit exponent field and a 24 bit mantissa field, as shown below:

The value contained in the storage unit is evaluated by multiplying the number in the mantissa field by two raised to the power of the number in the exponent field. Negative as well as positive exponents are obtained by biasing the exponent field by 64 (for a 7 bit field), such that a value of 64 is interpreted as an exponent of 0, a value of 65 as an exponent of 1, a value of 63 as an exponent of -1, etc. Suppose therefore that the sign bit field has a 0, the exponent field has a value of 0111110 (decimal 62) and the mantissa field has a value of 000000000000000001110111 (decimal 119), i.e. the contents of the storage unit are 00111110000000000000000001110111.

The number stored is $+119 \times 2^{-2}$. Changing the first (sign) bit to a 1 would change the number stored to -119×2^{-2}.

If a human being were asked to enter numbers in these binary forms, however, the procedure would be both highly tedious and also very prone to error, and, in consequence, simpler ways of entering binary numbers have been developed. Two such ways are to use octal and hexadecimal numbers, which are translated to binary numbers at the input–output interface to the computer.

Octal numbers use a base of eight and consist of decimal digits in the range 0–7 which each represent three binary digits. Thus a 24 bit binary number is represented by eight octal digits.

Hexadecimal numbers have a base of 16 and are used much more commonly than octal numbers. They use decimal digits in the range 0–9 and letters in the range A–F which each represent four binary digits. The decimal digits 0–9 translate directly to the decimal values 0–9 and the letters A–F translate respectively to the decimal values 10–15. A 24 bit binary number requires six hexadecimal digits to represent it. Table 10.2 shows the octal, hexadecimal and binary equivalents of decimal numbers in the range 0–15.

10.2.2 Octal/hexadecimal-to-binary conversion

Octal and hexadecimal conversion is very simple. Each octal/hexadecimal digit is taken in turn and converted to its binary representation according to Table 10.2.

Table 10.2

Decimal	Octal	Hexadecimal	Binary
0	0	0	0
1	1	1	1
2	2	2	10
3	3	3	11
4	4	4	100
5	5	5	101
6	6	6	110
7	7	7	111
8	10	8	1000 ,
9	11	9	1001
10	12	A	1010
11	13	B	1011
12	14	C	1100
13	15	D	1101
14	16	E	1110
15	17	F	1111

Example 10.3
Convert the octal number 7654 to binary.

Solution
Using Table 10.2, write down the binary equivalent of each octal digit:

$$| \ 7 \ | \ 6 \ | \ 5 \ | \ 4 \ |$$
$$= | \ 111 \ | \ 110 \ | \ 101 \ | \ 100 \ |$$

Thus, the binary code is 111110101100.

Example 10.4
Convert the hexadecimal number ABCD to binary.

Solution
Using Table 10.2, write down the binary equivalent of each hexadecimal digit:

$$| \ A \ | \ B \ | \ C \ | \ D \ |$$
$$= |1010 \ |1011 \ |1100 \ |1101 \ |$$

Thus, the binary code is 1010101111001101.

10.2.3 Binary-to-octal/hexadecimal conversion

Conversion from binary to octal or hexadecimal is also simple. The binary digits are taken in groups of three at a time (for octal) or four at a time (for hexadecimal), starting at the least significant end of the number (right-hand side) and writing down the appropriate octal or hexadecimal digit for each group.

Example 10.5

Convert the binary number 010111011001 into octal and hexadecimal.

Solution

$$| 010 | 111 | 011 | 001 |$$

$$= | 2 | 7 | 3 | 1 | = 2731 \text{ octal}$$

$$| 0101 | 1101 | 1001 |$$

$$= | 5 | D | 9 | = 5D9 \text{ hexadecimal}$$

Example 10.6

The 24 bit binary number 011111001001001101011010 is to be entered into a computer. How would it be entered using (a) octal code and (b) hexadecimal code?

Solution

(a) Divide the 24 bit number into groups of three, starting at the right-hand side:

$$| 011 | 111 | 001 | 001 | 001 | 101 | 011 | 010 | = 37111532 \text{ octal}$$

Thus, the number would be entered as 37111532 using octal code.

(b) Divide the 24 bit number into groups of four, starting at the right-hand side:

$$| 0111 | 1100 | 1001 | 0011 | 0101 | 1010 | = 7C935A \text{ hexadecimal}$$

In carrying out such conversions, it is essential that the groupings of binary digits start from the right-hand side. Groupings starting at the left-hand side give completely wrong values unless the number of binary digits happens to be an integer multiple of the grouping size. Consider a ten-digit binary number: 1011100011. Grouping digits starting at the right-hand side gives the values 1343 octal and 2E3 hexadecimal. Grouping digits starting at the left gives the (incorrect) values of 5611 octal and B83 hexadecimal.

When converting a binary number to octal or hexadecimal representation, a check must also be made that all of the binary digits represent data. In some systems, the first (left-hand) digit is used as a sign bit and the last (right-hand) digit is used as a parity bit.

Example 10.7

In a system which uses the first bit as a sign bit and the last bit as a parity bit, what is the octal and hexadecimal representation of the binary code 110111000111?

Solution

The ten data bits are 1011100011. This converts to 1343 octal and 2E3 hexadecimal.

10.2.4 **Programming and program execution**

In most modes of usage, including use as part of intelligent instruments, computers are involved in manipulating data. This requires data values to be input, processed and output according to a sequence of operations defined by the computer program.

Programming the microprocessor within an intelligent instrument is not normally the province of the instrument user; indeed, there is rarely any provision for the user to create or modify operating programs even if he/she wished to do so. There are several reasons for this. First, the signal processing needed within an intelligent instrument is usually well defined, and therefore it is more efficient for a manufacturer to produce this rather than to have each individual user produce near identical programs separately. Secondly, better program integrity and instrument operation are achieved if a standard program produced by the instrument manufacturer is used. Finally, use of a standard program allows it to be burnt into ROM, thereby protecting it from any failure of the instrument power supply. This also facilitates software maintenance and updates, by the mechanism of the manufacturer providing a new ROM which simply plugs into the slot previously occupied by the old ROM.

Whilst it is not normally therefore a task undertaken by the user of an intelligent instrument, some appreciation of microprocessor programming is useful background material. To illustrate the techniques involved in programming, consider a very simple program which reads in a value from a transducer, adds a pre-stored value to it to compensate for a bias in the transducer measurement, and outputs a corrected reading to a display device.

Let us assume that the addresses of the transducer and output display device are 00C0 and 00C1 respectively, and that the required scaling value has already been stored in memory address 0100. The instructions below are formed from the instruction set for a Z80 microprocessor and make use of CPU registers A and B:

```
IN A,C0
IN B,100
ADD A,B
OUT C1,A
```

This list of four instructions constitutes the computer program which is necessary to execute the required task. The CPU normally executes the instructions one at a time, starting at the top of the list and working downwards (though jump and branch instructions change this order). The first instruction (IN A,C0) reads in a value from the transducer at address C0 and places the value in CPU register A (often called the accumulator). The mechanics of the execution of this instruction consist of the CPU putting the required address C0 on the address bus and then putting a command on the control bus which causes the contents of the target address (C0) to be copied on to the data bus and subsequently transferred into the A register. The next instruction (IN B,100) reads in a value from address 100 (the pre-stored biasing value) and stores it in register B. The following instruction (ADD A,B) adds together the contents of registers A and B and stores the result in register A. Register A now contains the measurement read from the transducer but corrected for bias. The final instruction (OUT C1,A) transfers the contents of register A to the output device on address C1.

10.3 **Interfacing**

The input–output interface connects the computer to the outside world, and is therefore an essential part of the computer system. When the CPU puts the address of a peripheral on to the address bus, the input–output interface decodes the address and identifies the unique computer peripheral with which a data transfer operation is to be executed. The interface also has to interpret the command on the control bus so that the timing of the data transfer is correct. One further very important function of the input–output interface is to provide a physical electronic highway for the flow of data between the computer data bus and the external peripheral. In many computer applications, including their use within intelligent instruments, the external peripheral requires signals to be in analog form. Therefore the input–output interface must provide for conversion between these analogue signals and the digital signals required by a digital computer. This is satisfied by analog-to-digital and digital-to-analog conversion elements within the input–output interface.

The rest of this section presents some elementary concepts of interfacing in simple terms. A more detailed discussion follows later in Chapter 11, where the combining of intelligent instruments into larger networks is discussed.

10.3.1 **Address decoding**

A typical address bus in a microcomputer is 16 bits wide, allowing 65 536 separate addresses to be accessed in the range 0000–FFFF (in hexadecimal representation). Special commands on some computers (including the Z80) are reserved for accessing the bottom-end 256 of these addresses in the range 0000–00FF, and, if these commands are used, only 8 bits are needed to specify the required address. For the purpose of explaining address-decoding techniques, the scheme below shows how the lower 8 bits of the 16 bit address line are decoded to identify the unique address referenced by one of these special commands. Decoding of all 16 address lines follows a similar procedure but requires a substantially greater number of integrated circuit chips.

Address decoding is performed by a suitable combination of logic gates. Figure 10.3 shows a very simple hardware scheme for decoding eight address lines. This consists of 256 eight-input NAND gates, which each uniquely decode one of 256 addresses. A NAND gate is a logic element which only gives a logic level 1 output when all inputs are zero, and a logic level 0 output for any other combination of inputs. The inputs to the NAND gates are connected on to the lower eight lines of the address bus and the computer peripherals are connected to the output of the particular gates which decode their unique addresses. There are two pins for each input to the NAND gates which respectively invert and do not invert the input signal. By connecting the eight address

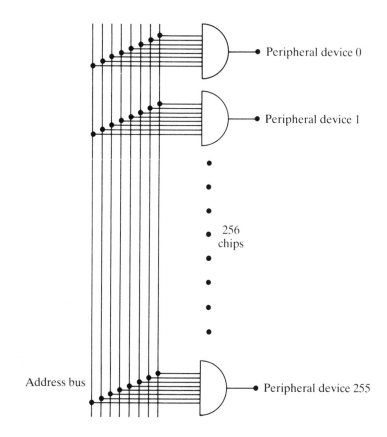

Figure 10.3 Simple hardware scheme for decoding eight address lines

lines appropriately to these two alternative pins at each input, the gate is made to decode a unique address. Consider for instance the pin connections shown in Figure 10.4. This NAND gate decodes address C5 (hexadecimal) which is 11000101 in binary. Because of the way that the input pins to the chip are connected, the NAND gate will see all zeros at its input when 11000101 is on the lower 8 bits of the address bus and therefore have an output of 1. Any other binary number on the address bus will cause this NAND gate to have a zero output.

10.3.2 **Data transfer control**

The transfer of data between the computer and peripherals is managed by control and status signals carried on the control bus which determine the exact sequencing and timing of I/O operations. Such management is necessary because of the different operating speeds of the computer and its peripherals and because of the multi-tasking

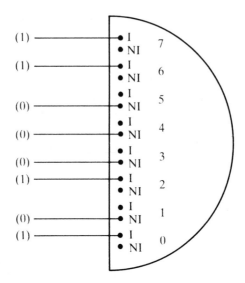

Figure 10.4 Pin connections to NAND gate to decode address C5

operation of many computers. This means that, at any particular instant when a data transfer operation is requested, either the computer or the peripheral may not be ready to take part in the transfer.

Typical control and status lines, and their meaning when set at a logic level of 1, are shown below:

    BUSY   Peripheral device busy
    READY   Peripheral device ready for data transfer
    ENABLE   CPU ready for data transfer
    ERROR   Malfunction on peripheral device

Similar control signals are set up by both the computer and the peripherals, but often different conventions are used to define the status of each device. Differing conventions occur particularly when the computer and peripherals come from different manufacturers, and might mean for instance that the computer interprets a logic level of 1 as defining a device to be busy but the peripheral device uses logic level 0 to define 'device busy' on the appropriate control line. Therefore, translation of the control lines between the computer and peripherals is required, which is achieved by a further series of logic gates within the I/O interface.

10.3.3 Analog-to-digital (A/D) conversion

Many computer inputs, particularly those from transducers within intelligent instruments, consist of analog signals which must be converted to a digital form before they can be accepted by the computer. This conversion is performed by an analog-to-digital conversion circuit within the computer interface.

Important factors in the design of an analog-to-digital converter are the speed of conversion and the number of digital bits used to represent the analog signal level. The minimum number of bits used in analog-to-digital converters is eight, which means that the analog signal can be represented to a resolution of 1 part in 256 if the input signal is carefully scaled to make full use of the converter range. It is more common, however, to use either 10 bit or 12 bit analog-to-digital converters which give resolutions respectively of 1 part in 1024 and 1 part in 4096. Several types of analog-to-digital converter exist; they differ in the technique used to effect signal conversion, in operational speed and in cost.

The simplest type of analog-to-digital converter is the **counter analog-to-digital converter**, as shown in Figure 10.5. This, like most types of analog-to-digital converter, does not convert continuously, but in a stop–start mode triggered by special signals on the computer's control bus. At the start of each conversion cycle, the counter is set to zero. The digital counter value is converted to an analog signal by a digital-to-analog converter (a discussion of digital-to-analog converters follows in the next section), and a comparator then compares this analog counter value with the unknown analog signal. The output of the comparator forms one of the inputs to an AND logic gate. The other input to the AND gate is a sequence of clock pulses. The comparator acts as a switch which can turn on and off the passage of pulses from the clock through the AND gate. The output of the AND gate is connected to the input of

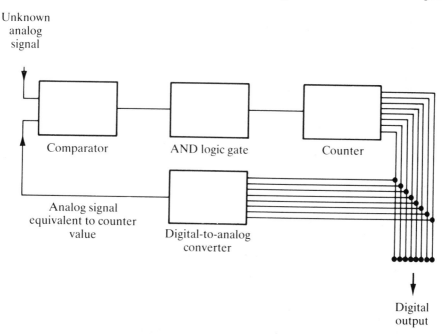

Figure 10.5 Counter analog-to-digital converter

the digital counter. Following reset of the counter at the start of the conversion cycle, clock pulses are applied continuously to the counter through the AND gate, and the analog signal at the output of the digital-to-analog converter gradually increases in magnitude. At some point in time, this analogue signal becomes equal in magnitude to the unknown signal at the input to the comparator. The output of the comparator changes state in consequence, closing the AND gate and stopping further increments of the counter. At this point in time, the value held in the counter is a digital representation of the level of the unknown analog signal.

10.3.4 Digital-to-analog (D/A) conversion

Digital-to-analog conversion is much simpler to achieve than analog-to-digital conversion and the cost of building the necessary hardware circuit is considerably less. It is required wherever the output of an intelligent instrument needs to be presented on a display device which operates in an analog manner. A common form of digital-to-analog converter is illustrated in Figure 10.6. This is shown with 8 bits for simplicity of explanation, although in practice 10 and 12 bit D/A converters are used more frequently.

This form of D/A converter consists of a resistor ladder network on the input to an operational amplifier. The analog output voltage from the amplifier is given by:

$$V_A = V_7 + \frac{V_6}{2} + \frac{V_5}{4} + \frac{V_4}{8} + \frac{V_3}{16} + \frac{V_2}{32} + \frac{V_1}{64} + \frac{V_0}{128}$$

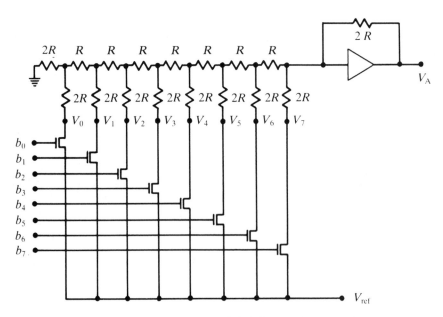

Figure 10.6 Common form of digital-to-analog converter

$V_0 \ldots V_7$ are set at either the reference voltage level V_{ref} or at 0 volts according to whether an associated switch is open or closed. Each switch is controlled by the logic level of one of the bits 0–7 of the 8 bit binary signal being converted. A particular switch is open if the relevant binary bit has a value of 0 and closed if the value is 1.

Consider for example a digital signal with a binary value of 11010100. The values of $V_7 \ldots V_0$ are therefore:

$$V_7 = V_6 = V_4 = V_2 = V_{ref} \quad V_5 = V_3 = V_1 = V_0 = 0$$

The analog output from the converter is then given by:

$$V_A = V_{ref} + \frac{V_{ref}}{2} + \frac{V_{ref}}{8} + \frac{V_{ref}}{32}$$

10.3.5 **Other interface considerations**

The foregoing discussion has presented some of the necessary elements in an input–output interface in a relatively simplistic manner which is just sufficient to give the reader the flavour of what is involved in an interface. Much fine detail has been omitted, and the amount of work involved in the practical design of a real interface should not be underestimated.

One significant omission so far is a discussion of the scaling which is generally required within the analog–digital interface of a computer. The raw analog input and output signals are generally either too large or too small for compatibility with the operating voltage levels of a digital computer and they have to be scaled upwards or downwards. This is normally achieved by operational amplifiers and/or potentiometers.

The main features of an operational amplifier are its high gain (typically $\times 1\,000\,000$) and large bandwidth (typically 1 MHz or better). When it is used at high frequencies, however, the bandwidth becomes significant. The quality of an amplifier is often measured by a criterion called the gain–bandwidth product which is the product of its gain and bandwidth. Other important attributes of the operational amplifier, particularly when used within intelligent instruments, are its distortion level, overload recovery capacity and offset level. Special instrumentation amplifiers, which are particularly good in these attributes, as described in section 5.1, have been developed for instrumentation applications.

10.4 **Intelligent instruments in use**

The intelligent instrument behaves as a black box as far as the user is concerned, and no knowledge of its internal mode of operation is required in normal measurement situations. The foregoing discussion on the operating principles of intelligent instruments is therefore presented for interest only and is not mandatory reading for potential users of such instruments.

Intelligent instruments offer many advantages over their non-intelligent counterparts, principally because of the improvement in accuracy achieved by processing the output of transducers to correct for errors inherent in the measurement process. Proper procedures must always be followed in their use, however, as discussed in Chapter 3 (sections 3.2.4 and 3.3.3), to avoid the possibility of introducing extra sources of measurement error. The solutions which intelligent instruments offer to many problems occurring in measurement systems will be discussed at various points in the chapters following in the first part of this book, and particular applications will be presented later in Part Two.

One such example of the benefit that intelligence can bring to instruments is in volume flow rate measurement, where the flow rate is inferred by measuring the differential pressure across an orifice plate placed in a fluid-carrying pipe (see Chapter 14 for more details). The flow rate is proportional to the square root of the difference in pressure across the orifice plate. For a given flow rate, this relationship is affected both by the temperature and by the mean pressure in the pipe, and changes in the ambient value of either of these cause measurement errors.

A typical intelligent flow rate measuring instrument contains three transducers: a primary one measuring the pressure difference across an orifice plate and secondary ones measuring absolute pressure and temperature. The instrument is programmed to correct the output of the primary differential pressure transducer according to the values measured by the secondary transducers, using appropriate physical laws which quantify the effect of ambient temperature and pressure changes on the fundamental relationship between flow and differential pressure. The instrument is also normally programmed to convert the square root relationship between flow and signal output into a direct one, making the output much easier to interpret. Typical accuracy levels of such intelligent flow measuring instruments are ±0.1%, compared with ±0.5% for their non-intelligent equivalents, showing an improvement by a factor of five.

Intelligent instruments usually provide many other facilities in addition to those mentioned above, such as the following:

1. Signal damping with selectable time constants.
2. Switchable ranges (using several primary transducers within the instrument which each measure over a different range).
3. Switchable output units (e.g. display in Imperial or SI units).
4. Diagnostic facilities.
5. Remote adjustment and control of instrument options from up to 1500 metres away via four-way, 20 mA signal lines.

10.5 **Exercises**

10.1 Describe briefly the three essential elements of a microcomputer system.

10.2 Write down the one's complement and two's complement 8 bit representations of the following decimal numbers: (a) 47, (b) −119, (c) −101, (d) 86, (e) 108.

10.3 Write down the binary, octal and hexadecimal representations of the following decimal numbers: (a) 57, (b) 101, (c) 175, (d) 259, (e) 999, (f) 1234.

10.4 The binary code representation of a number is 111101001101. If all twelve digits are data bits, what are (a) the octal and (b) the hexadecimal equivalents of this number?

10.5 The binary code representation of a number is 1011000110100101. If all sixteen digits are data bits, what are (a) the octal and (b) the hexadecimal equivalents of this number?

10.6 The binary code representation of a number is 110100011000. The first and last binary digits are sign and parity bits respectively, so only the middle ten digits are data bits. What are (a) the octal, (b) the hexadecimal and (c) the decimal data values?

References and further reading

Barney, G. C. (1985) *Intelligent Instruments*, Prentice Hall: Hemel Hempstead.

11 Instrumentation/computer networks

11.1 Introduction

The inclusion of computer processing power in intelligent instruments and intelligent actuators creates the possibility of building an instrumentation system where several intelligent devices collaborate together, transmit information to one another and execute process control functions. Such an arrangement is often known as a **distributed control system**. Additional computer processors can also be added to the system to provide the necessary computational power when the computation of complex control algorithms is required. Such an instrumentation system is far more fault tolerant and reliable than older control schemes where data from several discrete instruments is carried to a centralized computer controller via long instrumentation cables. This improved reliability arises from the fact that the presence of computer processors in every unit injects a degree of redundancy into the system and measurement and control action can still continue, albeit in a degraded form, if one unit fails.

In order to effect the necessary communication when two or more intelligent devices are to be connected together, some form of electronic highway must be provided between them which permits the exchange of information. Apart from data transfer, a certain amount of control information also has to be transferred. The main purpose of this control information is to make sure that the target device is ready to receive information before data transmission starts. This control information also prevents more than one device trying to send information at the same time.

The electronic highway can be a serial communication line, a parallel data bus, or a local-area network. Serial data lines are very slow and are only used where a low data transmission speed is acceptable. Parallel data buses are limited to connecting a modest number of devices spread over a small geographical area, typically a single room, but provide reasonably fast data transmission. Local-area networks are used to connect larger numbers of devices spread over larger geographical distances, typically a single building or site. They transmit data in digital format at high speed. Instrumentation networks which are geographically larger than a single building or site can also be built but these generally require transmission systems which include telephone lines as well as local networks at particular sites within the large system.

11.2 **Interfacing**

The input–output interface of an intelligent device provides the necessary connection between the device and the electronic highway. The interface can be either serial or parallel.

A **serial interface** is used to connect a device on to a serial communication line. The connection is effected physically by a multi-pin plug which fits into a multi-pin socket on the casing of the device. The pins in this plug/socket match the signals lines used in the serial communication line exactly in number and function.

Effectively, there is only one standard format for serial data transmission which enjoys international recognition. Whilst this is advantageous in avoiding compatibility problems when connecting together devices coming from different manufacturers, serial transmission is relatively slow.

A **parallel interface** is used to connect devices on to parallel instrument buses and also into all other types of network systems. Like the serial interface, the parallel interface exists physically as a multi-pin plug which fits into a multi-pin socket on the casing of the device. The pins in the plug/socket are matched exactly in number and function with the data and control lines used by a particular parallel instrument bus. Unfortunately, there are a number of different parallel instrument buses in use and thus a corresponding number of different parallel interface protocols, with little compatibility between them. Hence, whilst parallel data transmission is much faster than serial transmission, there are serious compatibility problems to be overcome when connecting together devices coming from different manufacturers because of these different parallel interface protocols used.

11.3 **Serial communication lines**

This is a very slow type of electronic highway which is only used where low data transfer rates are acceptable. Data is transferred down a single line in the highway 1 bit at a time, at a maximum speed of 19 200 bits/s. The start and finish of each item of data is denoted by special sequences of control characters which precede and follow the data bits. These control characters use the same signal line as the data and so impose a further limitation on the maximum data rate that the interface can handle.

Two alternative forms of serial communication exist, known as half duplex and full duplex. In **half-duplex** mode (HDX), the same data wire is used by a device both to send and to receive data, and thus simultaneous sending and receiving of data is not possible. In **full-duplex** operation (FDX), on the other hand, two separate data lines are used, one for send and one for receive, and simultaneous sending and receiving of data is therefore possible.

The structure of the data and control characters used in serial transmission is shown in Figure 11.1. The binary digits of 1 and 0 are represented by voltage logic levels of $+V$ and zero. The start of transmission of each character is denoted by a binary 0 digit. The following seven digits represent a coded character. The next digit

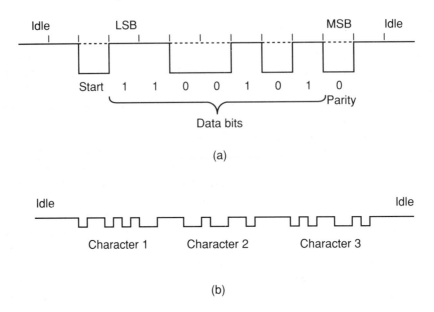

Figure 11.1 Serial data transmission: (a) data and control bits for one character (even parity); (b) a string of serially transmitted characters

is known as a parity bit. Finally, the end of transmission of the character is denoted by either one or two stop bits, which are binary 1 digits.

The parity bit is provided as an error-checking mechanism, and is set to make the total number of binary 1 digits in the character representation either odd or even, according to whether the odd or even parity system is being used (some manufacturers use odd parity and some use even parity). The seven-character digits are usually coded using the ASCII (American Standard Code for Information Interchange) system. This provides for the transmission of the full set of alphabetic, numerical, punctuation and control characters.

Whilst its data transfer rate is low, serial transmission does have the advantage that only two different standard formats exist, and these are very similar. The two formats, which have achieved international recognition, are the RS232 standard (USA) and the CCITT V24 standard (Europe). The only significant difference between these is the logic voltage level used (3 V for RS232 and 6 V for V24), and this incompatibility can be handled very easily. Within either of these standards, there are options about the type of parity (odd or even), the number of stop bits and whether data transmission is in full- or half-duplex mode. The various equipment manufacturers use the options differently but it is a relatively simple matter to accommodate these differences when connecting together devices coming from different manufacturers.

11.4 **Parallel data bus**

Parallel data buses allow much higher data rates than serial communication lines because data is transmitted in parallel, i.e. several bits are transmitted simultaneously. Control signals are transmitted on separate control lines within the bus. This means that the data lines are used solely for data transmission, thereby optimizing the capability of the data transmission rate.

There are a number of different parallel data buses in existence, but there is little compatibility between them, with differences existing in the number of data lines used, the number of control lines used, the interrupt structure used, the data timing system and the logic levels used for operation. Equipment manufacturers tend to keep to the same parallel interface protocol for all their range of devices, but different manufacturers use different protocols. Thus, while it will normally be easy to connect together a number of intelligent devices which all come from the same manufacturer, interfacing difficulties are likely to be experienced if devices from different manufacturers are connected together.

In practice, limiting the choice of devices to those available from the same manufacturer is unlikely to be acceptable. Even if all the units required can be obtained from the same manufacturer, this limitation is likely to mean having to use devices with a lower performance specification than desirable and at a cost penalty. Therefore, it is necessary to find a way of converting the different interface protocols into a common communication format at the interface between the device and the transmission medium. This is normally done using some type of network as the transmission medium rather than a single instrument bus. Many different systems exist for dealing with the different communication protocols used by different equipment manufacturers, and allow all the devices to be connected on to one particular type of instrument bus. These systems include CAMAC and MEDIA.

CAMAC was conceived some 30 years ago and was one of the first interfacing systems designed to connect together devices from different manufacturers. It provides a 24 bit parallel interface and can handle data transmission at speeds up to 5 Mbits/s. Physically, it consists of a crate containing a motherboard into which up to 24 devices can be plugged. The 24 sockets in the crate are connected together by an 86 wire bus known as the dataway. The bus accommodates 24 data lines, with all other lines being reserved for various control functions. The supervisory computer is connected to the bus by a special plug-in card. Up to eight crates can be connected together, communicating via a 66 wire bus extension cable, thus permitting the full system to have up to 192 modules.

MEDIA is a proprietary interface system developed by Imperial Chemical Industries. The system can handle up to 4096 units which are plugged into a highway consisting of 16 data lines, 16 address lines and several control lines.

Some examples of parallel interface buses are the IEEE 488 bus, Multibus and the S100 Bus. The IEEE 488 bus is now used by a large number of manufacturers and is covered in detail below.

11.4.1 **IEEE 488 bus**

The **IEEE 488** standard instrument bus was developed about 20 years ago and provides a parallel interface which facilitates the connection of intelligent instruments, actuators and controllers within a single room. This bus is also known by the alternative names of **IEC bus**, **HP-IB bus**, **GBIB bus**, **Plus bus** and **ANSI std MC1-1 bus**. The maximum length of bus allowable is about 20 m and no more than 15 instruments should be distributed along its length. The maximum distance between two particular units on the bus should not exceed about 2 m. The maximum data transfer rate permitted by the bus is 1 Mbit/s in theory, though the maximum data rate achieved in practice over a full 20 m length of bus is more likely to be in the range of 250–500 kbits/s.

Physically, the bus consists of a shielded, 24 conductor cable; 16 of the conductors are used as signal lines, 8 carrying data and 8 carrying control signals. The remaining 8 conductors are used as ground wires for the control signals, each control wire being twisted together along its full length with one of the ground wires. This minimizes cross-talk between the control wires. Normal practice is to route the eight twisted pairs carrying control signals in the centre of the cable and place the eight data wires around the periphery. This bundle of wires is surrounded by shielding and an outer insulated coating. Each end of the cable is connected to a standard 24 pin metal connector, with generally a female connector at one end and a male connector at the other. This facilitates several cables being chained from one device to another.

Figure 11.2 shows three devices connected on to an IEEE 488 bus. The bus can only carry one lot of information at a time, and which unit is sending data and which is receiving it is controlled by a supervisory computer connected to the bus. This supervisor ensures that only one unit can put data on the bus at a time, and thus prevents the corruption of data which would occur if several instruments had simultaneous access.

Having eight data lines means that the bus can transmit 8 bits of data in parallel at the same time. This was originally designed so that 8 bit computer words could be transmitted as whole words. This does not prevent the bus being used with computers of a different wordlength, for example 16 or 32 bits. However, if the wordlength is longer than 8 bits, whole words cannot be transmitted in one go: they have to be transmitted 8 bits at a time.

The eight status lines provide the necessary control to ensure that when data transmission takes place between two units, three conditions are satisfied simultaneously. These three conditions are (a) that the sender unit is ready to transmit data, (b) that the receiver unit is ready to receive data and (c) that the bus does not currently have any data on it. The functioning of the eight status lines is as follows:

DAV (Data valid) This goes to a logic 0 when the data on the eight data lines is valid.

NRFD (Not ready for data) This goes to logic 0 when the receiver unit is ready to accept data.

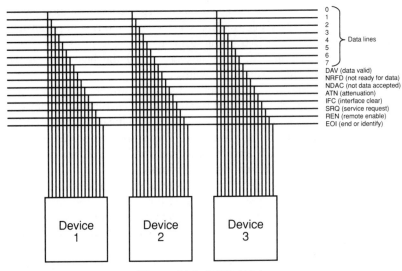

Figure 11.2 IEEE 488 bus

NDAC (Not data accepted) This goes to logic 0 when the receiver unit has finished receiving data.

ATN (Attention) This is a general control signal which is used for various purposes to control the use of data lines and specify the send and receive devices to be used.

IFC (Interface clear) The controller uses this status line to put the interface into a wait state.

SRQ (Service request) This is an interrupt status line which allows high-priority devices such as alarms to interrupt current bus traffic and get immediate access to the bus.

REN (Remote enable) This status line is used to specify which of two alternative sets of device programming data are to be used.

EOI (End of output or identify) This status line is used by the sending unit to indicate that it has finished transmitting data.

The IEEE 488 bus protocol uses a logic level of less than 0.8 V to represent a logic 0 signal and a voltage level greater than 2.0 V to represent a logic 1 signal.

11.5 **Local-area networks (LANs)**

Local-area networks transmit data serially in digital format along a simple cable. Digital transmission has particular advantages in that the possibility of signal corruption during transmission is greatly reduced compared with analog transmission. LANs have particular value in the monitoring and control of plants which are large and/or widely dispersed over a large area. Indeed, for such large instrumentation systems, a local-area network is the only viable transmission medium in terms of performance and cost. Parallel data buses, which transmit data in analog form, suffer from signal attenuation and noise pick-up over large distances, and the high cost of the long, multi-cored cables which they need is prohibitive.

The development of instrumentation networks is not without problems, however. Careful design of the network is required to prevent the corruption of data when two or more devices on the network try to access it simultaneously and perhaps put information on to the data bus at the same time. This problem is solved by designing a suitable network protocol which ensures that network devices do not access the network simultaneously, thus preventing data corruption.

In a local-area network, the electronic highway is either a simple pair of twisted wires or a coaxial cable. Fiber optic cables can also be used as the data transmission medium as discussed later. There are many different protocols for local-area networks but these are all based on one of three network structures known as star networks, bus networks and ring networks, as shown in Figure 11.3. A local-area network operates within a single building or site and can transmit data over distances up to about 500 m without signal attenuation being a problem. For transmission over greater distances, telephone lines are used in the network. Intelligent devices are interfaced to the telephone line used for data transmission via a modem. The **modem** converts the signal into a frequency-modulated analog form. In this form, it can be transmitted over either the public switched telephone network or over private lines rented from telephone companies. The latter, being dedicated lines, allow higher data transmission rates.

11.5.1 **Star networks**

In a **star network**, each instrument and actuator is connected directly to the supervisory computer by its own signal cable. One apparent advantage of a star network is that data can be transferred if necessary using a serial communication protocol such as RS232. This is an industry-standard protocol and so compatibility problems do not arise, but of course data transfer is very slow. Because of this speed problem, parallel communication is usually preferred even for star networks.

Whilst star networks are simple in structure, the central supervisory computer node is a critical point in the system and failure of this means total failure of the whole system. When any device in the network needs to communicate with another device, a request has to be made to the central supervisory computer and all data transferred is routed through this central node. If the central node is inoperational for any reason then data communication in the network is stopped.

11.5.2 **Ring and bus networks**

In contrast, both ring and bus networks have a high degree of resilience in the face of one node breaking down and so are generally preferred to star networks. If the

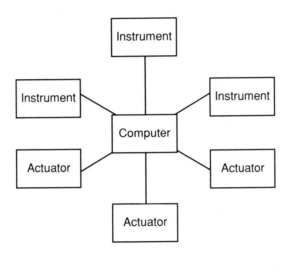

(a)

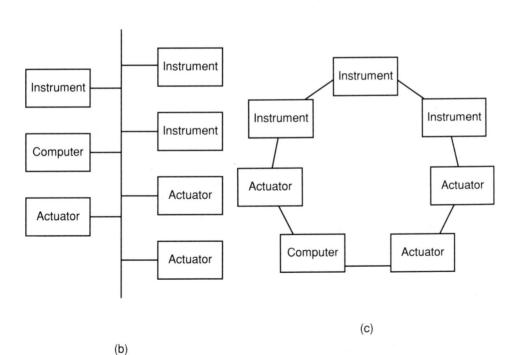

(b)

(c)

Figure 11.3 Network protocols: (a) star network; (b) bus network; (c) ring network

processor in any node breaks down, the data transmission paths in the network are still maintained and operation of the network can continue, albeit at a degraded performance level, using the remaining computational power in the other processors. Most computer and intelligent instrument/actuator manufacturers provide standard conversion modules which allow their equipment to interface to one of these standard networks.

In a **ring network**, all the intelligent devices are connected to a bus which is formed into a continuous ring. The ring protocol sends a special packet (or token) continuously round the ring to control access to the network. A station can only send data when it receives the token. During data transmission, the token is attached to the back of the message sent so that, once the information has been safely received, the token can continue on its journey round the network. A typical data transmission speed is 10 Mbits/s. Commonly used token ring protocols are the Cambridge ring, Arcnet and the IEEE 802.5 bus.

A **bus network** is similar to a ring network but the bus on to which the devices are connected is not continuous. Bus networks are also resilient to the breakdown of one node in the network. A **contention protocol** is normally used which allows any station to have immediate access to the network unless another station is using it simultaneously, in which case the protocol manages the situation and prevents data loss/corruption. Bus networks have a similar data transmission speed to ring networks of 10 Mbits/s. Ethernet, Proway and the IEEE 802.3 standard bus are all examples of bus networks.

11.5.3 Fiber optic distributed networks

The virtues of fiber optic cables as a data transmission medium have been expounded in Chapter 9. Apart from the high immunity of the signals to noise, a fiber optic transmission system can transfer data at speeds up to 100 Mbits/s. This compares with 10 Mbits/s for signal transmission along copper conductors. The reduction in signal attenuation during transmission also means that much longer transmission distances are possible without repeaters being necessary. For instance, the allowable distances between repeaters for a fiber optic Arcnet network are quoted as 1 km for half-duplex operation and between 1 and 3.5 km for full-duplex operation.

11.5.4 Fieldbus

Fieldbus is the most recent international instrumentation network standard to be proposed. It is a low-level network particularly aimed at small instrumentation/control systems.

At the time of writing, the final standard for fieldbus has not yet been agreed. However, the essential features of fieldbus have been defined, and a prominent characteristic will be that all outputs, as well as transmission, will be digital only. This means that a new range of output displays, recorders and actuators will have to be developed alongside fieldbus. Existing analog devices will be incompatible. Two standard data transfer rates are proposed, 31.25 kbits/s and 1 Mbit/s. The low speed of 31.25 kbits/s is compatible with most existing equipment and is intended for cable lengths up to 1900 m. The high speed of 1 Mbit/s is not generally compatible with existing equipment and will allow a maximum cable length of only 750 m.

11.6 Communication protocols for very large systems

Once a system gets too large to be covered by a local-area network, it is generally necessary to use telephone lines to provide communication over large distances. Public telephone lines are readily available for this, but there is a fundamental problem about their use. Whilst instrumentation networks need high bandwidths, public networks operate at the low bandwidth required to satisfy speech-based telephony. High bandwidths can be obtained by leasing private telephone lines but this solution is expensive and often uneconomic. The solution that is emerging is to extend LAN technology into public telephone networks. The LAN extended in this way is renamed a metropolitan-area network (MAN).

An IEEE standard for MAN (IEEE 802.6) was first published in 1990. Messages between nodes are organized in packets. MANs cover areas which are typically up to 50 km in diameter, but in some cases links can be several hundred kilometres long. Use of the public switched telephone network for transmission is most common although private lines are sometimes used.

Both ring and bus networks lose efficiency as the number of nodes increases and are unsuitable for adoption by MAN. Instead, a protocol known as the **distributed queue dual bus** (DQDB) is used. DQDB is a hybrid bus which carries isochronous data for the public switched telephone network as well as providing the data bus for a MAN. For handling data on a MAN, DQDB has a pair of buses on which data, preceded by the target address, circulates in fixed-size packets in opposite directions, i.e. there is a clockwise bus and an anticlockwise bus. All stations have access to both buses and the protocol establishes a distributed queue which ensures that all stations have access to the bus on a fair basis; this means that stations have their access demands satisfied in the order in which they arise (i.e. a first-come, first-served basis) but commensurate with ensuring that the bus is used efficiently. Fiber optic cables are commonly used for the buses, allowing data transmission at speeds up to 140 Mbits/s.

11.6.1 Communication network rationalization

As the size of networks increases, the diversity of protocols used in different systems becomes more and more troublesome. In response, attempts are being made to define standard protocols for these larger systems. These standard protocols now try to cover all aspects of computer communications, i.e. management and stock control information, etc., as well as instrumentation/process control networks.

This development started with the publication in 1978 by the International

Organization for Standardization of the **Open Systems Interconnection seven-layer model (ISO-7)**, covering all aspects of computer communications. The ISO seven-layer model defines a set of standard message formats, and rules for their interchange, which can be applied both within local-area networks and in much larger global networks which involve data transmission via telephone lines. Whilst the standards and protocols involved in ISO-7 are highly complex, the network builder does not need to have a detailed understanding of them as long as all devices used in the network are certified by their manufacturer as conforming to the standard.

The **Manufacturing Automation Protocol** (MAP), conceived by General Motors in 1980, conforms with ISO-7, and is a similar attempt at providing a standard computer communications protocol. MAP is now being adopted by many large companies and organizations, including the European ESPRIT research programme, and promises to realize what it set out to achieve, i.e. the provision of an industry-wide common communications standard.

Part Two
Instruments for measurement of particular physical variables

This second part of the book discusses the range of instruments which are available for measuring various physical quantities. In presenting each range of instruments, the aim has been to be as comprehensive as possible, to make the book useful as a future source of reference when choosing an instrument for any particular measurement situation. This method of treatment means that quite rare and expensive instruments as well as very common and cheap instruments are included, and therefore it is necessary to choose from the instrument ranges presented with care.

The first step in choosing an instrument for a particular measurement situation is to specify the static and dynamic characteristics required. These must always be defined carefully, as a high degree of accuracy, sensitivity, etc., in the instrument specification inevitably involves a high cost. The characteristics specified should therefore be the minimum necessary to achieve the required level of performance in the system that the instrument is providing a measurement for. Once the required static and dynamic characteristics have been defined, the range of instruments presented in each of the following chapters can be reduced to a subset containing those instruments which satisfy these defined specifications.

The second step in choosing an instrument is to consider the working conditions in which the instrument will operate. Conditions demanding special consideration are those where the instrument will be exposed to mechanical shocks, vibration, fumes, dust or fluids. Such considerations subdivide further the subset of possible instruments already identified. If at this stage a choice still exists, then the final criterion is one of cost.

The relevant criteria in choosing an instrument for measuring particular physical quantities are considered further in the following chapters.

12 Temperature measurement

Temperature measurement is very important in all spheres of life and especially so in the process industries. However, temperature measurement poses particular problems, since it cannot be related to a fundamental standard of temperature in the same way that the measurement of other quantities can be related to the primary standards of mass, length and time. If two bodies of lengths l_1 and l_2 are connected together end to end, the result is a body of length $l_1 + l_2$. A similar relationship exists between separate masses and separate times. However, if two bodies at the same temperature are connected together, the joined body has the same temperature as each of the original bodies.

This is a root cause of the fundamental difficulties which exist in establishing an absolute standard for temperature in the form of a relationship between it and other measurable quantities for which a primary standard unit exists. In the absence of such a relationship, it is necessary to establish fixed, reproducible reference points for temperature in the form of freezing and boiling points of substances where the transition between solid, liquid and gaseous states is sharply defined. The **International Practical Temperature Scale** (IPTS) uses this philosophy and defines six **primary fixed points** for reference temperatures in terms of:

the triple point of equilibrium hydrogen:	$-259.34\,°C$
the boiling point of oxygen:	$-182.962\,°C$
the boiling point of water:	$100.0\,°C$
the freezing point of zinc:	$419.58\,°C$
the freezing point of silver:	$961.93\,°C$
the freezing point of gold:	$1064.43\,°C$
(all at standard atmospheric pressure)	

The freezing points of certain other metals are also used as **secondary fixed points** to provide additional reference points during calibration procedures. These are of particular use for calibrating instruments measuring high temperatures. Some examples are:

freezing point of tin:	$231.968\,°C$
freezing point of lead:	$327.502\,°C$
freezing point of zinc:	$419.58\,°C$

freezing point of antimony:	630.74 °C
freezing point of aluminium:	660.37 °C
freezing point of copper:	1084.5 °C
freezing point of nickel:	1455 °C
freezing point of palladium:	1554 °C
freezing point of platinum:	1772 °C
freezing point of rhodium:	1963 °C
freezing point of iridium:	2447 °C
freezing point of tungsten:	3387 °C

Instruments for measuring temperature can be divided into seven separate classes according to the physical principle on which they operate. These principles are as follows:

1. Thermal expansion
2. The thermoelectric effect
3. Resistance change
4. Resonant frequency change
5. Radiative heat emission
6. Thermography
7. Acoustic thermometry
8. Fiber optic devices

12.1 Thermal expansion methods

Thermal expansion methods make use of the fact that the dimensions of all substances, whether solids, liquids or gases, change with temperature. Instruments operating on this physical principle include the liquid-in-glass thermometer, the bimetallic thermometer and the pressure thermometer.

12.1.1 Liquid-in-glass thermometers

The liquid-in-glass thermometer is a well-known temperature measuring instrument which is used in a wide range of applications. The fluid used is usually either mercury or coloured alcohol, and this is contained within a bulb and capillary tube, as shown in Figure 12.1. As the temperature rises, the fluid expands along the capillary tube and the meniscus level is read against a calibrated scale etched on the tube. The process of estimating the position of the curved meniscus of the fluid against the scale introduces some error into the measurement process and a measurement accuracy better than ±1% of full-scale reading is hard to achieve.

However, an accuracy of ±0.15% can be obtained in the best industrial instruments. Industrial versions of the liquid-in-glass thermometer are normally used to measure temperature in the range between −200 °C and +1000 °C, although

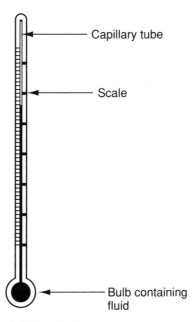

Figure 12.1 The liquid-in-glass thermometer

instruments are available to special order which can measure temperatures up to 1500 °C.

12.1.2 **Bimetallic thermometer**

The bimetallic principle is probably more commonly known in connection with its use in thermostats. It is based on the fact that if two strips of different metals are bonded together, any temperature change will cause the strips to bend, as this is the only way in which the differing rates of change of length of each metal in the bonded strip can be accommodated. In the bimetallic thermostat, this is used as a switch in control applications.

If the magnitude of bending is measured, the bimetallic device becomes a thermometer. For such purposes, the strip is often arranged in a spiral configuration, as shown in Figure 12.2, as this gives a relatively large displacement of the free end for any given temperature change. Strips in a helical shape are an alternative for this purpose. The measurement sensitivity is increased further by choosing the pair of materials carefully such that the degree of bending is maximized, with Invar (a nickel–steel alloy) and brass being commonly used.

The system used to measure the displacement of the strip must be carefully designed. Very little resistance must be offered to the end of the strip, otherwise the spiral or helix will distort and cause a false reading in the measurement of the displacement. For visual indication purposes, the end of the strip can be made to turn

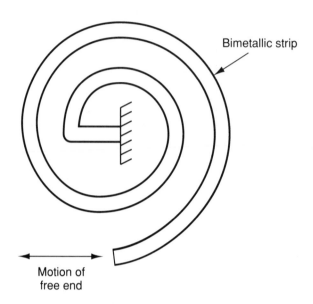

Bimetallic strip

Motion of
free end

Figure 12.2 The bimetallic thermometer

a pointer mounted in low-friction bearings which moves against a calibrated scale. If an electrical output is required, the linear variable differential transformer (LVDT) is a suitable form of translational displacement transducer. Alternatively, a fiber optic device known as the shutter sensor can be used (see Figure 9.7).

Bimetallic thermometers are used to measure temperatures between −75 °C and +1500 °C. The accuracy of the best instruments can be as good as ±0.5% but such devices are quite expensive. Many instrument applications do not require this degree of accuracy in temperature measurements, and in such cases much cheaper bimetallic thermometers with substantially inferior accuracy are used.

12.1.3 **Pressure thermometers**

The sensing element in a pressure thermometer consists of a bulb containing gas. If the gas were not constrained, temperature rises would cause its volume to increase. However, because it is constrained in a bulb and cannot expand, its pressure rises instead. As such, the pressure thermometer does not strictly belong to the thermal expansion class of instruments but is included because of the relationship between volume and pressure according to Boyle's gas law: $PV = KT$.

The change in pressure of the gas is measured by a suitable pressure transducer such as the Bourdon tube (see Chapter 13). This transducer is located remotely from the bulb and connected to it by a capillary tube as shown in Figure 12.3. The need to protect the pressure measuring instrument from the environment where the temperature is being measured can require the use of capillary tubes up to 5 m long,

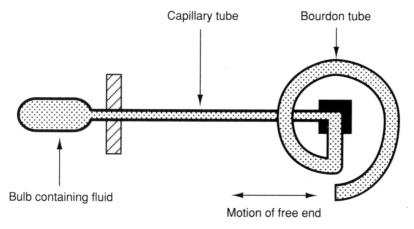

Capillary tube   Bourdon tube

Bulb containing fluid

Motion of free end

Figure 12.3 The pressure thermometer

and the temperature gradient and hence pressure gradient along the tube acts as a modifying input which can introduce a significant measurement error. Correction for this using the principle of introducing an opposing modifying input, as discussed in Chapter 3, can be carried out according to the scheme shown in Figure 12.4. This includes a second, dummy capillary tube whose temperature gradient is measured by a second Bourdon tube. The outputs of the two Bourdon tubes are connected together in such a manner that the output from the second tube is subtracted from the output of the first, thus eliminating the error due to the temperature gradient along the tube.

Pressure thermometers are used to measure temperatures in the range between $-250\,°C$ and $+2000\,°C$. Their typical accuracy is $±0.5\%$ of full-scale reading.

12.2 **Thermoelectric-effect instruments (thermocouples)**

Thermoelectric-effect instruments rely on the physical principle that, when any two different metals are connected together, an e.m.f., which is a function of the temperature, is generated at the junction between the metals. The general form of this relationship is:

$$e = a_1 T + a_2 T^2 + a_3 T^3 + \ldots + a_n T^n$$

This is clearly non-linear, which is inconvenient for measurement applications. Fortunately, for certain pairs of materials, the terms involving squared and higher powers of T ($a_2 T^2$, $a_3 T^3$, etc.) are approximately zero and the e.m.f.–temperature relationship is approximately linear according to:

$$e \approx a_1 T$$

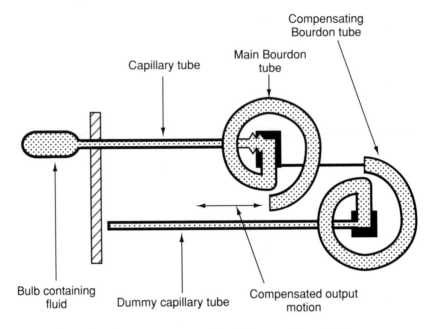

Figure 12.4 Correction for temperature gradient along capillary tube in pressure thermometer

Wires of such pairs of materials are connected together at one end, and in this form are known as **thermocouples**. Thermocouples are a very important class of device as they provide the most commonly used method of measuring temperatures in industry.

Thermocouples are manufactured from various combinations of the base metals copper and iron, the base-metal alloys of alumel (Ni/Mn/Al/Si), chromel (Ni/Cr), constantan (Cu/Ni), nicrosil (Ni/Cr/Si) and nisil (Ni/Si/Mn), the noble metals platinum and tungsten, and the noble-metal alloys of platinum/rhodium and tungsten/rhenium. Only certain combinations of these are used as thermocouples and each standard combination is known by an internationally recognized type letter; for instance, type K is chromel–alumel. The temperature–e.m.f. characteristics for some of these standard thermocouples are shown in Figure 12.5. Certain of these show reasonable linearity over short temperature ranges and their characteristic can therefore be approximated by a series of straight-line relationships for use in intelligent instruments containing thermocouples. In general, however, the temperature indicated by a given e.m.f. output measurement has to be calculated from tables. The set of these tables corresponding to all the standard types of thermocouple available are known as thermocouple tables.

A typical thermocouple, made from one chromel wire and one constantan wire, is shown in Figure 12.6(a). For analysis purposes, it is useful to represent the

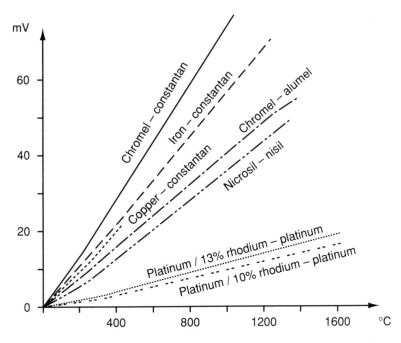

Figure 12.5 Temperature–e.m.f. characteristics for some standard thermocouple materials

thermocouple by its equivalent electrical circuit, shown in Figure 12.6(b). The e.m.f. generated at the junction is represented by a voltage source, E_1, and it is customary also to show the temperature of the junction on the diagram. This is known as the **hot junction** and is represented by T_h. The junction between the open ends of the thermocouple and the voltage measuring instrument is known as the **reference junction**, and thermocouple tables are calibrated assuming that this is maintained at a temperature of $0\,°C$. This is achieved by immersing the reference junction in an ice bath.

Tables for a range of standard thermocouples are given in Appendix 4. In these tables, a range of temperatures is given in the left-hand column and the e.m.f. output for each standard type of thermocouple is given in the columns to the right.

Example 12.1

If the e.m.f. output measured from a chromel–constantan thermocouple is 13.419 mV with the reference junction at 0 °C, the appropriate column in the tables shows that this corresponds to a hot-junction temperature of 200 °C.

In practice, any general e.m.f. output measurement taken at random would not be found exactly in the tables, and interpolation would be necessary between quoted values.

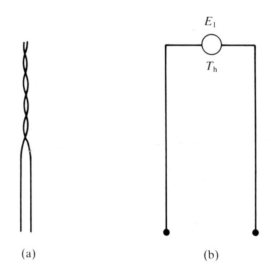

Figure 12.6 (a) Typical chromel–alumel thermocouple; (b) equivalent circuit

Example 12.2

If the measured output e.m.f. for a chromel–constantan thermocouple (reference junction at 0 °C) is 10.65 mV, it is necessary to carry out linear interpolation between the temperature of 160 °C corresponding to an e.m.f. of 10.501 mV shown in the tables and the temperature of 170 °C corresponding to an e.m.f. of 11.222 mV. This interpolation procedure gives an indicated hot-junction temperature of 162 °C.

12.2.1 Extension leads

In order to make a thermocouple conform to some precisely defined e.m.f.– temperature characteristic described by standard tables, it is necessary that all metals used are refined to a high degree of pureness and all alloys are manufactured to an exact specification. This makes the materials used expensive, and consequently thermocouples are typically only a few centimetres long. It is clearly impractical to connect a voltage measuring instrument to measure the thermocouple output in close proximity to the environment whose temperature is being measured, and therefore extension leads up to several metres long are normally connected between the thermocouple and the measuring instrument. This modifies the equivalent circuit to that shown in Figure 12.7. There are now three junctions in the system and consequently three voltage sources, E_1, E_2 and E_3, which add to give a net e.m.f. output E_n given by:

$$E_n = E_1 + E_2 + E_3$$

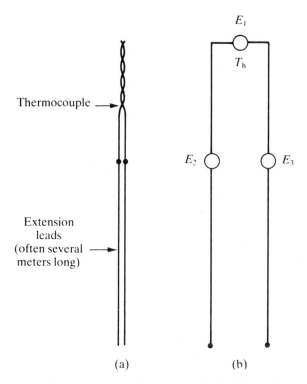

Figure 12.7 (a) Thermocouple with extension leads; (b) equivalent circuit

The e.m.f. of interest for calculating the unknown temperature is E_1. However, only E_n is measured and therefore E_2 and E_3 must be calculated to obtain E_1, which requires that the temperature at the junction between the thermocouples and extension wires be measured. This difficulty can be avoided if the extension wires are manufactured from materials which make the magnitudes of E_2 and E_3 approximately zero. This is most easily achieved by choosing the extension leads of the same basic materials as the thermocouple but manufactured to a lower specification which significantly reduces their cost. Such a solution is still prohibitively expensive in the case of noble-metal thermocouples, and it is necessary in this case to search for base-metal extension leads which have a similar thermoelectric behaviour to the noble-metal thermocouple. In this form, the extension leads are known as **compensating leads**. A typical example of this is the use of nickel/copper–copper extension leads connected to a platinum/rhodium–platinum thermocouple.

12.2.2 **Connection of voltage measuring instrument**

No account so far has been taken of the actual connection between the thermocouple extension leads and the voltage measuring instrument. This introduces two further junctions between dissimilar metals and hence two extra voltage sources E_4 and E_5, and the total number of voltage sources in the equivalent circuit is increased to five, as shown in Figure 12.8. Provided that these extra junctions are maintained at the same temperature, this is unimportant as E_4 and E_5 cancel out ($E_4 = -E_5$) and the net effect on the temperature indicated by the thermocouple tables is zero. However, in various circumstances, the junctions can be at different temperatures and correction then has to be made for E_4 and E_5. The junction between a copper connection lead and, say, a chromel extension lead does not involve a standard pair of thermocouple materials and is therefore not covered by thermocouple tables. Use of a thermoelectric law known as the **law of intermediate metals** has to be used in these circumstances. This states that the e.m.f. generated at the junction between two metals or alloys A and C is equal to the sum of the e.m.f. generated at the junction

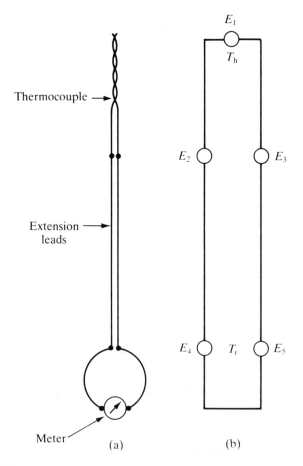

Figure 12.8 (a) Thermocouple with extension leads and connections to a meter; (b) equivalent circuit

between metals or alloys A and B and the e.m.f. generated at the junction between metals or alloys B and C, where all junctions are at the same temperature. This can be expressed more simply as:

$$e_{AC} = e_{AB} + e_{BC} \tag{12.1}$$

Example 12.3

Suppose that the junction between a copper connection lead and a chromel extension lead is maintained at a temperature of 80 °C. Equation (12.1) can be applied by introducing the intermediate alloy, constantan. Then:

$$e_{copper-chromel} = e_{copper-constantan} + e_{constantan-chromel}$$
$$= e_{copper-constantan} - e_{chromel-constantan}$$

Therefore, from thermocouple tables:

$$e_{copper-chromel} = 3.357 - 4.983 = -1.626 \text{ mV}$$

12.2.3 Reference junction temperature

Whilst thermocouple tables are calibrated assuming a zero reference junction temperature, it is not always possible in real applications to maintain such a low reference temperature. In many practical applications of thermocouples, it is necessary to locate the reference junction in a controlled environment maintained by an electrical heating element at a temperature greater than 0 °C. Correction for this different reference junction temperature then has to be made using a second thermoelectric law known as the **law of intermediate temperatures**. This states that:

$$E_{(T_h, T_0)} = E_{(T_h, T_r)} + E_{(T_r, T_0)} \tag{12.2}$$

where $E_{(T_h, T_0)}$ is the e.m.f. with the junctions at temperatures T_h and T_0, $E_{(T_h, T_r)}$ is the e.m.f. with the junctions at temperatures T_h and T_r, $E_{(T_r, T_0)}$ is the e.m.f. with the junctions at temperatures T_r and T_0, T_h is the hot-junction measured temperature, T_0 is 0 °C and T_r is the non-zero reference junction temperature which is somewhere between T_0 and T_h.

Example 12.4

Suppose that the reference junction of a chromel–constantan thermocouple is maintained at a temperature of 80 °C and the output e.m.f. measured is 40.102 mV when the hot junction is immersed in a fluid.
The quantities given are $T_r = 80$ °C and $E_{(T_h, T_r)} = 40.102$ mV
From the tables, $E_{(T_r, T_0)} = 4.983$ mV
Now applying Equation (12.2), $E_{(T_h, T_0)} = 40.102 + 4.983 = 45.085$ mV
Again referring to the tables, this indicates a fluid temperature of 600 °C.

In using thermocouples, it is essential that they are connected correctly. Large errors can result if they are connected incorrectly, for example by interchanging the extension leads or by using incorrect extension leads. Such mistakes are particularly serious because they do not prevent some sort of output being obtained, which may look sensible even though it is incorrect, and so the mistake may go unnoticed for a long period of time. The following examples illustrate the sort of errors which may arise.

Example 12.5
In a particular industrial situation, a chromel–alumel thermocouple with chromel–alumel extension wires is used to measure the temperature of a fluid. In connecting up this measurement system, the instrumentation engineer responsible has inadvertently interchanged the extension wires from the thermocouple. The ends of the extension wires are held at a reference temperature of 0 °C and the output e.m.f. measured is 12.1 mV. If the junction between the thermocouple and extension wires is at a temperature of 40 °C, what temperature of fluid is indicated and what is the true fluid temperature?

Solution
The initial step necessary in solving a problem of this type is to draw a diagrammatical representation of the system and to mark on this the e.m.f. sources, temperatures, etc., as shown in Figure 12.9. The first part of the problem is solved very simply by looking up in the thermocouple tables what temperature the e.m.f. output of 12.1 mV indicates for a chromel–alumel thermocouple. This is 297.4 °C.
  Then, summing e.m.f.s around the loop:

$$V = 12.1 = E_1 + E_2 + E_3 \quad \text{or} \quad E_1 = 12.1 - E_2 - E_3$$

$$E_2 = E_3 = \text{e.m.f. (alumel–chromel)}_{40} = -\text{e.m.f. (chromel–alumel)}_{40}{}^*$$
$$= -1.611 \text{ mV}$$

Hence $E_1 = 12.1 + 1.611 + 1.611 = 15.322$ mV. Interpolating from the thermocouple tables, this indicates that the true fluid temperature is 374.5 °C.

Example 12.6
An iron–constantan thermocouple measuring the temperature of a fluid is connected by mistake with copper–constantan extension leads (such that the two constantan wires are connected together and the copper extension wire is connected to the iron thermocouple wire). If the fluid temperature was actually 200 °C, and the junction between the thermocouple and extension wires was at 50 °C, what e.m.f. would be measured at the open ends of the extension wires if the reference junction is

---

*The thermocouple tables quote e.m.f.s using the convention that going from chromel to alumel is positive. Hence the e.m.f. going from alumel to chromel is minus the e.m.f. going from chromel to alumel.

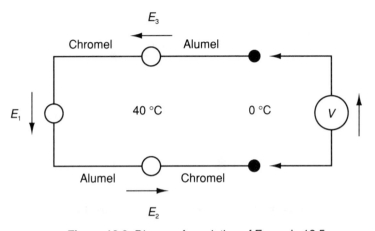

Figure 12.9 Diagram for solution of Example 12.5

maintained at 0 °C? What fluid temperature would be deduced from this (assuming that the connection mistake was not known about)?

Solution
Again, the initial step necessary is to draw a diagram showing the junctions, temperatures and e.m.f.s, as shown in Figure 12.10. The various quantities can then be calculated:

$$E_2 = \text{e.m.f. (iron–copper)}_{50}$$

By the law of intermediate metals:

$$\text{e.m.f. (iron–copper)}_{50} = \text{e.m.f. (iron–constantan)}_{50} - \text{e.m.f. (copper–constantan)}_{50}$$
$$= 2.585 - 2.035 \quad \text{(from thermocouple tables)}$$
$$= 0.55 \, \text{mV}$$
$$E_1 = \text{e.m.f. (iron–constantan)}_{200} = 10.777 \quad \text{(from thermocouple tables)}$$
$$V = E_1 - E_2 = 10.777 - 0.55 = 10.227$$

Using tables and interpolating to find the temperature, 10.227 mV indicates a temperature of:

$$\left(\frac{10.227 - 10.222}{10.777 - 10.222} \right) 10 + 190 = 190.1 \, ^\circ\text{C}$$

12.2.4 Thermocouple materials

The five standard base-metal thermocouples are chromel–constantan (type E), iron–constantan (type J), chromel–alumel (type K), nicrosil–nisil (type N) and

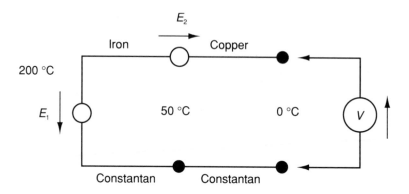

Figure 12.10 Diagram for solution of Example 12.6

copper–constantan (type T). These are all relatively cheap to manufacture but they become inaccurate with age and have a short life.

Chromel–constantan devices give the highest measurement sensitivity of 80 μV/°C, with an accuracy of ±0.75% and a useful measuring range of −200 °C up to 900 °C. Unfortunately, whilst they can operate satisfactorily in oxidizing environments, their performance and life are seriously affected by reducing atmospheres. Iron–constantan thermocouples have a sensitivity of 60 μV/°C and are the preferred type for general-purpose measurements in the temperature range of −150 °C to +1000 °C, where the typical measurement accuracy obtained is ±1%. Their performance is little affected by either oxidizing or reducing atmospheres. Copper–constantan devices have a similar measurement sensitivity of 60 μV/°C and find their main application in measuring subzero temperatures down to −200 °C, with an accuracy of ±0.5%. They can also be used in both oxidizing and reducing atmospheres to measure temperatures up to 350 °C. Chromel–alumel thermocouples have a measurement sensitivity of only 45 μV/°C, although their characteristic is particularly linear over the temperature range between 700 °C and 1200 °C and this is therefore their main application. Like chromel–constantan devices, they are suitable for oxidizing atmospheres but not for reducing ones. The measurement accuracy given by them is ±0.75%. Nicrosil–nisil thermocouples are a recent development which resulted from attempts to improve the performance and stability of chromel–alumel thermocouples. Their thermoelectric characteristic has a very similar shape to type K devices with equally good linearity over the high-temperature measurement range and a measurement sensitivity of 40 μV/°C. The operating environment limitations are the same as for chromel–alumel devices but their long-term stability and life are at least three times better.

Noble-metal thermocouples are always expensive but enjoy high stability and long life in conditions of high temperature and oxidizing environments. They are chemically inert except in reducing atmospheres. Thermocouples made from platinum and a platinum/rhodium alloy have a high accuracy of ±0.2% and can measure temperatures up to 1500 °C, but their measurement sensitivity is only 10 μV/°C. Alternative devices made from tungsten and a tungsten/rhenium alloy have a better sensitivity of 20 μV/°C and can measure temperatures up to 2300 °C.

12.2.5 Thermocouple protection

Thermocouples are delicate instruments which must be treated carefully if their specified operating characteristics are to be reproduced. One major source of error is induced strain in the hot junction which reduces the e.m.f. output, and precautions are normally taken to minimize this by mounting the thermocouple horizontally rather than vertically. In some operating environments, no further protection is necessary to obtain satisfactory performance from the instrument. However, thermocouples are prone to contamination by various metals and protection is often necessary to minimize this. Such contamination alters the thermoelectric behaviour of the device, such that its characteristic varies from that published in standard tables. It also becomes brittle and its life is therefore shortened.

Protection takes the form of enclosing the thermocouple in a sheath. Some common sheath materials and their maximum operating temperatures are shown in Table 12.1. Whilst the thermocouple is an instrument with a naturally first-order type of step response characteristic, the time constant is usually so small as to be negligible when the thermocouple is used unprotected. When enclosed in a sheath, however, the time constant of the combination of thermocouple and sheath is significant. The size of the thermocouple and hence the diameter required for the sheath has a large effect on the importance of this. The time constant of a thermocouple in a 1 mm diameter sheath is only 0.15 s and this has little practical effect in most measurement situations, whereas a larger sheath of 6 mm diameter gives a time constant of 3.9 s which cannot be ignored so easily.

Table 12.1 Common sheath materials for thermocouples

Material	Maximum operating temperature (°C)*
Mild steel	900
Nickel–chromium	900
Fused silica	1000
Special steel	1100
Mullite	1700
Recrystallized alumina	1850
Beryllia	2300
Magnesia	2400
Zirconia	2400
Thoria	2600

*The maximum operating temperatures quoted assume oxidizing or neutral atmospheres. For operation in reducing atmospheres, the maximum allowable temperature is usually reduced.

12.2.6 **Thermocouple manufacture**

Thermocouples are manufactured by connecting together two wires of different materials, where each material is produced so as to conform precisely with some defined composition specification. This ensures that its thermoelectric behaviour accurately follows that for which standard thermocouple tables apply. The connection between the two wires is effected by welding, soldering or in some cases just by twisting the wire ends together. Welding is the most common technique generally used, with silver soldering being reserved for copper–constantan instruments.

The diameter of wire used to construct thermocouples is usually in the range between 0.4 mm and 2 mm. The larger diameters are used where ruggedness and long life are required, although these advantages are gained at the expense of increasing the measurement time constant. In the case of noble-metal thermocouples, the use of large-diameter wire incurs a substantial cost penalty. Some special applications have a requirement for a very fast response time in the measurement of temperature, and in such cases wire diameters as small as 0.1 μm (0.1 microns) can be used.

12.2.7 **The thermopile**

The thermopile is the name given to a temperature measuring device which consists of several thermocouples connected together in series, such that all the reference junctions are at the same cold temperature and all the hot junctions are exposed to the temperature being measured, as shown in Figure 12.11. The effect of connecting n thermocouples together in series is to increase the measurement sensitivity by a factor of n. A typical thermopile manufactured by connecting together 25 chromel–constantan thermocouples gives a measurement resolution of 0.001 °C.

12.2.8 **The continuous thermocouple**

The continuous thermocouple is one of a class of devices which detect and respond to heat. Other devices in this class include the **line-type heat detector** and **heat-sensitive cable**. The basic construction of all these devices consists of two or more strands of wire separated by insulation within a long, thin cable. Whilst they sense temperature, they do not in fact provide an output measurement of temperature. Their function is to respond to abnormal temperature rises and thus prevent fires, equipment damage, etc.

The advantages of continuous thermocouples become more apparent if the problems with other types of heat detector are considered. The insulation in the line-type heat detector and heat-sensitive cable consists of a plastic or ceramic material with a negative temperature coefficient (i.e. the resistance falls as the temperature rises). An alarm signal can be generated when the measured resistance falls below a certain level. Alternatively, in some instruments, the insulation is allowed to break down completely, in which case the device acts as a switch. The major limitation of these devices is that the temperature change has to be relatively large,

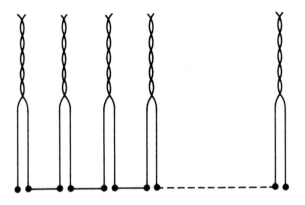

Figure 12.11 The thermopile

typically 50–200 °C above ambient temperature, before they respond. Also, it is not generally possible for such devices to give an output which indicates that an alarm condition is developing before it actually happens, and thus to allow preventative action. Furthermore, after a device has generated an alarm it usually has to be replaced. This is particularly irksome because there is a large variation in the characteristics of detectors coming from different batches and so replacement of the device requires extensive on-site recalibration of the system.

In contrast, the continuous thermocouple suffers from very few of these problems. It differs from other types of heat detector in that the two strands of wire inside it are a pair of thermocouple materials* separated by a special, patented mineral insulation and contained within a stainless steel protective sheath. If any part of the cable is subjected to heat, the resistance of the insulation at that point is reduced and a hot junction is created between the two wires of dissimilar metals. An e.m.f. is generated at this hot junction according to normal thermoelectric principles.

The continuous thermocouple can detect temperature rises as small as 1 °C above normal. Unlike other types of heat detector, it can also monitor abnormal rates of temperature rise and provide a warning of alarm conditions developing before they actually happen. Replacement is only necessary if a great degree of insulation breakdown has been caused by a substantial hot spot at some point along the detector's length. Even then, the use of thermocouple materials of standard characteristics in the detector means that recalibration is not needed if it is replaced. Calibration is not affected either by cable length, and so a replacement cable may be of a different length to the one it is replacing. One further advantage of continuous thermocouples over earlier forms of heat detector is that no power supply is needed, thus significantly reducing installation costs.

---

*Normally type E, chromel–constantan, or type K, chromel–alumel.

12.3 **Varying-resistance devices**

Varying-resistance devices rely on the physical principle of the variation of resistance with temperature. The instruments are known as either resistance thermometers or thermistors according to whether the material used for their construction is a metal or a semiconductor. The normal method of measuring resistance is to use a d.c. bridge. The excitation voltage of the bridge has to be chosen very carefully because, although a high value is desirable for achieving high measurement sensitivity, the self-heating effect of high currents flowing in the temperature transducer creates an error by increasing the temperature of the device and so changing the resistance value.

12.3.1 **Resistance thermometers (resistance–temperature devices)**

Resistance thermometers, which are alternatively known as resistance–temperature devices (or RTDs), rely on the principle that the resistance of a metal varies with temperature according to the relationship

$$R = R_0(1 + a_1 T + a_2 T^2 + a_3 T^3 + \ldots + a_n T^n) \qquad (12.3)$$

This equation is non-linear and so is inconvenient for measurement purposes. The equation becomes linear if all the terms in $a_2 T^2$ and higher powers of T are negligible. This is approximately true for some metals over a limited temperature range and, in such cases, the resistance and temperature are related according to:

$$R \approx R_0(1 + a_1 T)$$

Platinum is one such metal where the resistance–temperature relationship is linear within $\pm 0.4\%$ over the temperature range between $-200\,°C$ and $+40\,°C$. Even at $+1000\,°C$, the quoted accuracy is $\pm 1.2\%$. The characteristics of platinum and other metals are summarized in Figure 12.12.

Its good linearity and chemical inertness make platinum the first choice for resistance thermometers in many applications. Platinum is very expensive, however, and consequently the cheaper but less accurate alternatives of nickel and copper are sometimes used. These two metals are very susceptible to oxidation and corrosion and so the range of applications where they can be used is strictly limited even if their reduced accuracy is acceptable. Another metal, tungsten, is also used in some circumstances, particularly for high-temperature measurements. The working ranges of each of these four types of resistance thermometer are as shown below:

Platinum: $-270\,°C$ to $+1000\,°C$ (though use above $650\,°C$ is uncommon)
Copper: $-200\,°C$ to $+260\,°C$
Nickel: $-200\,°C$ to $+430\,°C$
Tungsten: $-270\,°C$ to $+1100\,°C$

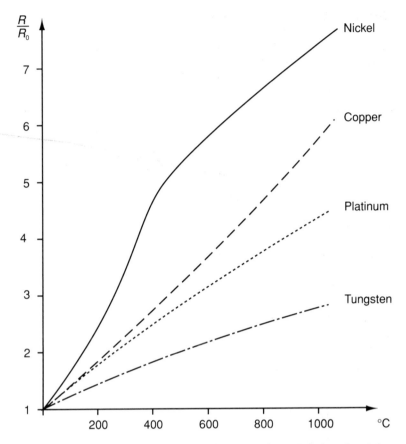

Figure 12.12 Typical resistance–temperature characteristics of metals

In the case of non-corrosive and non-conducting environments, resistance thermometers are used without protection. In all other applications, they are protected inside a sheath. As in the case of thermocouples, such protection affects the speed of response of the system to rapid changes in temperature. A typical time constant for a sheathed platinum resistance thermometer is 0.4 seconds.

The resistance thermometer exists physically as a coil of resistance wire of the chosen metal. Different instruments available have resistance elements ranging from 10 Ω right up to 25 kΩ. The devices with high resistance have several operational advantages. The first of these is that any connection resistances within the circuit become negligible in their effect. The second advantage is that the relatively high-voltage output produced by these instruments makes any induced e.m.f.s produced by thermoelectric behaviour at the junction with connection leads negligible in magnitude.

12.3.2 **Thermistors**

Thermistors are manufactured from beads of semiconductor material prepared from oxides of the iron group of metals such as chromium, cobalt, iron, manganese and nickel. The resistance of such materials varies with temperature according to the following expression:

$$R = R_0 \exp[\beta(1/T - 1/T_0)] \qquad (12.4)$$

This relationship, as illustrated in Figure 12.13, exhibits a large negative temperature coefficient (i.e. the resistance decreases as the temperature increases), and so is

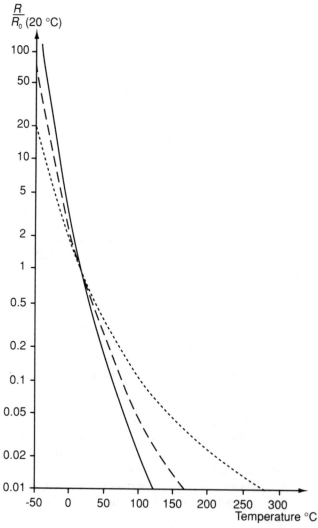

Figure 12.13 Typical resistance–temperature characteristics of thermistor materials

fundamentally different from the relationship for the resistance thermometer, which shows a positive temperature coefficient. The form of Equation (12.4) is such that it is not possible to make a linear approximation to the curve over even a small temperature range, and hence the thermistor is very definitely a non-linear instrument.

The major advantages of thermistors are their relatively low cost and their small size. This size advantage means that the time constant of thermistors operated in sheaths is small. However, the size reduction also decreases its heat dissipation capability, and so makes the self-heating effect greater. In consequence, thermistors have to be operated at generally lower current levels than resistance thermometers and so the measurement sensitivity afforded is less. Like resistance thermometers, the resistance of a thermistor is usually measured with a d.c. bridge.

12.4 **Quartz thermometers**

The quartz thermometer makes use of the principle that the resonant frequency of a material such as quartz is a function of temperature, thus enabling temperature changes to be translated into frequency changes. The temperature-sensing element consists of a quartz crystal enclosed within a probe (sheath). The crystal is connected electrically so as to form the resonant element within an electronic oscillator. Measurement of the oscillator frequency therefore allows the measured temperature to be calculated.

The instrument has a very linear output characteristic over the temperature range between $-40\,°C$ and $+230\,°C$. The measurement resolution given is $0.1\,°C$. The characteristics of the instrument are generally very stable over long periods of time and therefore only infrequent calibration is necessary.

12.5 **Radiation thermometers**

All bodies emit electromagnetic radiation as a function of their temperature above absolute zero. The total rate of radiation emitted per second is given by:

$$E = KT^4 \tag{12.5}$$

The power spectral density of this emission varies with temperature in the manner shown in Figure 12.14. The major part of the frequency spectrum lies within the band of wavelengths between $0.3\,\mu m$ and $40\,\mu m$, which corresponds to the visible (0.3–$0.72\,\mu m$) and infrared (0.72–$1000\,\mu m$) ranges. As the magnitude of the radiation varies with temperature, measurement of the emission from a body allows the temperature of the body to be calculated. Choice of the best method of measuring the emitted radiation depends on the temperature of the body. At low temperatures, the peak of the power spectral density function (Figure 12.14) lies in the infrared region,

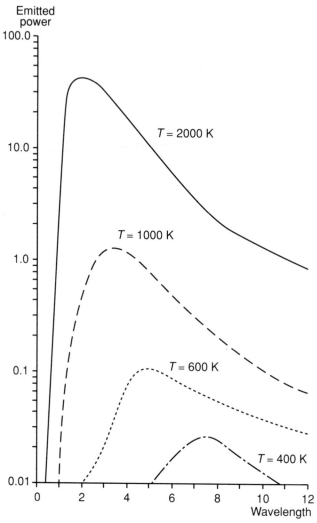

Figure 12.14 Power spectral density of radiated energy emission at various temperatures

whereas at higher temperatures it moves towards the visible part of the spectrum. This phenomenon is observed as the red glow which a body begins to emit as its temperature is increased beyond 600 °C.

Radiation thermometers have one major advantage in that they do not require to be in contact with the hot body in order to measure its temperature. Thus, there is no disturbance at all of the measured system. Furthermore, there is no possibility of contamination, which is particularly important in the food and many other process industries. They are especially suitable for measuring high temperatures that are beyond the capabilities of contact instruments such as thermocouples, resistance thermometers and thermistors. They are also capable of measuring moving bodies, for instance the temperature of steel bars in a rolling mill. Their use is not as straightforward as the discussion so far might have suggested, however, because the radiation from a body varies with the composition and surface condition of the body as well as with temperature. This dependence on surface condition is quantified by the **emissivity** of the body. The use of radiation thermometers is further complicated by absorption and scattering of the energy between the emitting body and the radiation detector. This energy is scattered by atmospheric dust and water droplets and absorbed by carbon dioxide, ozone and water vapour molecules. Therefore, all radiation thermometers have to be carefully calibrated for each particular body whose temperature they are required to monitor.

Various types of radiation thermometer exist, as described below. The optical pyrometer can only be used to measure high temperatures, but various types of radiation pyrometers are available which between them cover the whole temperature spectrum.

12.5.1 **Optical pyrometer**

The optical pyrometer, illustrated in Figure 12.15, is designed to measure temperatures where the peak radiation emission is in the red part of the visible spectrum, i.e. where the measured body glows a certain shade of red according to the temperature. This limits the instrument to measuring temperatures above 600 °C. The instrument contains a heated tungsten filament within its optical system. The current in the filament is increased until its colour is the same as the hot body: under these conditions the filament apparently disappears when viewed against the background of

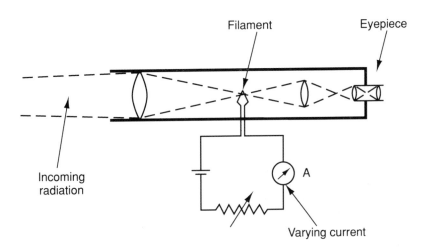

Figure 12.15 The optical pyrometer

the hot body. Temperature measurement is therefore obtained in terms of the current flowing in the filament. As the brightness of different materials at any particular temperature varies according to the emissivity of the material, the calibration of the optical pyrometer must be adjusted according to the emissivity of the target. Manufacturers provide tables of standard material emissivities to assist with this.

The inherent measurement inaccuracy of an optical pyrometer is $\pm 5\,°C$. However, in addition to this error, there can be a further operator-induced error of $\pm 10\,°C$ arising from the difficulty in judging the moment when the filament 'just' disappears. Measurement accuracy can be improved somewhat by employing an optical filter within the instrument which passes a narrow band of frequencies of wavelength around $0.65\,\mu m$ corresponding to the red part of the visible spectrum.

The instrument cannot be used in automatic temperature control schemes because the eye of the human operator is an essential part of the measurement system. The reading is also affected by fumes in the path of sight. Because of these difficulties and its low accuracy, the optical pyrometer is being rapidly overtaken by hand-held radiation pyrometers in popularity, although the instrument is still widely used in industry for measuring temperatures in furnaces and similar applications.

12.5.2 Radiation pyrometers

All the alternative forms of radiation pyrometer described below have an optical system which is similar to that in the optical pyrometer and focuses the energy emitted from the measured body. They differ, however, by omitting the filament and eyepiece and using instead an energy detector in the same focal plane as the eyepiece was. The radiation detector is either a thermal detector, which measures the temperature rise in a black body at the focal point of the optical system, or a photon detector. Thermal detectors respond equally to all wavelengths in the frequency spectrum whereas photon detectors respond selectively to a particular band within the full spectrum.

Thermopiles, resistance thermometers and thermistors are all used as thermal detectors in different versions of these instruments. These typically have time constants of several milliseconds, because of the time taken for the black body to heat up and the temperature measuring instrument to respond to the temperature change.

Photon detectors are usually of the photoconductive or photovoltaic type. Both of these types respond very much faster to temperature changes than thermal detectors because they involve atomic processes, and typical measurement time constants are a few microseconds.

The size of objects measured by a radiation pyrometer is limited by the optical resolution, which is defined as the ratio of target size to distance. A ratio of 1:300 is regarded as good, and this would allow temperature measurement of a 1 mm sized object at a range of 300 mm. With large distance/target size ratios, accurate aiming and focusing of the pyrometer at the target are essential. It is now common to find 'through the lens' viewing provided in pyrometers, using a principle similar to SLR camera technology, as focusing the instrument for visible light automatically focuses it for infrared light.

Various forms of electrical output are available from the radiation detector: these are functions of the incident energy on the detector and are therefore functions of the temperature of the measured body. Whilst this therefore makes such instruments of use in automatic control systems, their accuracy is often inferior to optical pyrometers. This reduced accuracy arises first because a radiation pyrometer is sensitive to a wider band of frequencies than the optical instrument and the relationship between emitted energy and temperature is less well defined. Secondly, the magnitude of energy emitted at low temperatures gets very small, according to Equation (12.5), increasing the difficulty of accurate measurement.

The forms of radiation pyrometer described below differ mainly in the technique used to measure the emitted radiation. They also differ in the range of energy wavelengths which each is designed to measure and hence in the temperature range measured. One further difference is the material used to construct the energy-focusing lens. Outside the visible part of the spectrum, glass becomes almost opaque to infrared wavelengths, and other lens materials such as arsenic trisulphide are used.

Broad-band (unchopped) radiation pyrometers

The form of the radiation detector used in this instrument is shown in Figure 12.16. The instrument finds wide application in industry and has a measurement accuracy which varies from ±0.05% of full scale in the best instruments to ±0.5% in the cheapest. However, the level of accuracy deteriorates significantly over a period of time, and an error of 10 °C is common after 1–2 years of operation at high temperatures. As its name implies, the instrument measures radiation across the whole frequency spectrum and so uses a thermal detector. This consists of a blackened platinum disk to which a thermopile* is bonded. The temperature of the detector increases until the heat gain from the incident radiation is balanced by the heat loss due to convection and radiation. For high-temperature measurement, a two-couple thermopile gives acceptable measurement sensitivity and has a fast time constant of about 0.1 s. At lower measured temperatures, where the level of incident

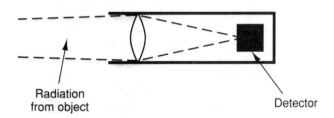

Radiation
from object

Detector

Figure 12.16 Structure of the radiation pyrometer

*Typically manganin–constantan.

radiation is much less, thermopiles constructed from a greater number of thermocouples must be used to get sufficient measurement sensitivity. This increases the measurement time constant to as much as 2 s. Standard instruments of this type are available to measure temperatures between $-20\,°C$ and $+1800\,°C$, although in theory much higher temperatures could be measured by this method.

Chopped broad-band radiation pyrometers

The construction of this form of pyrometer is broadly similar to that shown in Figure 12.16 except that a rotary mechanical device is included which periodically interrupts the radiation reaching the detector. The voltage output from the thermal detector thus becomes an alternating quantity which switches between two levels. This form of a.c. output can be amplified much more readily than the d.c. output coming from an unchopped instrument. This is particularly important when amplification is necessary to achieve an acceptable measurement resolution in situations where the level of incident radiation from the measured body is low. For this reason, this form of instrument is the more common when measuring body temperatures associated with peak emission in the infrared part of the frequency spectrum. For such chopped systems, the time constant of thermopiles is too long. Instead, thermistors are generally used, giving a time constant of 0.01 s. Standard instruments of this type are available to measure temperatures between $+20\,°C$ and $+1300\,°C$. This form of pyrometer suffers a similar drift in accuracy to unchopped forms. Its life is also limited to about 2 years because of motor failures.

Narrow-band radiation pyrometers

Narrow-band radiation pyrometers are highly stable instruments which suffer a drift in accuracy which is typically only $1\,°C$ in 10 years. They are also less sensitive to emissivity changes than other forms of radiation pyrometer. They use photodetectors of either the photoconductive or photovoltaic form whose performance is unaffected by carbon dioxide or water vapour in the path between the target object and the instrument. A photoconductive detector exhibits a change in resistance as the incident radiation level changes whereas a photovoltaic cell exhibits an induced voltage across its terminals which is also a function of the incident radiation level. All photodetectors are preferentially sensitive to a particular narrow band of wavelengths in the range $0.5–1.2\,\mu m$ and all have a form of output which varies in a highly non-linear fashion with temperature; thus a microcomputer inside the instrument is highly desirable. Four commonly used materials for photodetectors are cadmium sulphide, lead sulphide, indium antimonide and lead tin telluride. Each of these is sensitive to a different band of wavelengths and therefore all find application in measuring the particular temperature ranges corresponding to each of these bands.

The output from a narrow-band radiation pyrometer is normally chopped into an a.c. signal in the same manner as used in the chopped broad-band pyrometer. This simplifies the amplification of the output signal which is necessary to achieve an acceptable measurement resolution. The typical time constant of a photon detector is only $5\,\mu s$ which allows high chopping frequencies up to $20\,kHz$. This gives such instruments an additional advantage in being able to measure fast transients in temperature as short as $10\,\mu s$.

Two-colour pyrometer (ratio pyrometer)

As stated earlier, the emitted radiation–temperature relationship for a body depends on the emissivity. This is very difficult to calculate, and therefore in practice all pyrometers have to be calibrated to the particular body they are measuring. The two-colour pyrometer (alternatively known as a ratio pyrometer) is a system which largely overcomes this problem by using the arrangement shown in Figure 12.17. Radiation from the body is split equally into two parts which are applied to separate narrow-band filters. The outputs from the filters consist of radiation within two narrow bands of wavelength λ_1 and λ_2. Detectors sensitive to these frequencies produce output voltages V_1 and V_2 respectively. The ratio of these outputs, V_1/V_2, can be shown (see Bentley 1983) to be a function of temperature and to be independent of the emissivity provided that the two wavelengths λ_1 and λ_2 are close together.

The theoretical basis of the two-colour pyrometer is that the output is independent of emissivity because the emissivities at the two wavelengths λ_1 and λ_2 are equal. This is based on the assumption that λ_1 and λ_2 are very close together. In practice, this assumption does not hold and therefore the accuracy of the two-colour pyrometer tends to be relatively poor. However, the instrument is still of great use in conditions where the target is obscured by fumes or dust, which is a common problem in the cement and mineral processing industries. Two-colour pyrometers typically cost 50–100% more than other types of pyrometer.

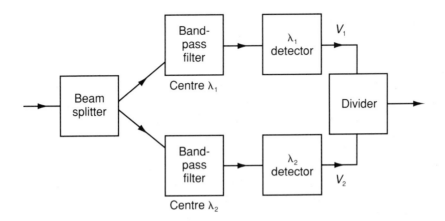

Figure 12.17 Two-colour pyrometer system

Selected-waveband pyrometer

The selected-waveband pyrometer is sensitive to one waveband only, e.g. 5 μm, and is dedicated to particular, special situations where other forms of pyrometer are inaccurate. One example of such a situation is measuring the temperature of steel billets which are being heated in a furnace. If an ordinary radiation pyrometer is aimed through the furnace door at a hot billet, it receives radiation from the furnace walls (by reflection off the billet) as well as radiation from the billet itself. If the temperature of the furnace walls is measured by a thermocouple, a correction can be made for the reflected radiation, but variations in transmission losses inside the furnace through fumes, etc., make this correction inaccurate. However, if a carefully chosen selected-waveband pyrometer is used, this transmission loss can be minimized and the measurement accuracy is thereby greatly improved.

12.6 **Thermography (thermal imaging)**

Thermography, or thermal imaging, involves scanning an infrared radiation detector across an object. The information gathered is then processed and an output in the form of the temperature distribution across the object is produced. Temperature measurement over the range from $-20\,°C$ up to $+1500\,°C$ is possible. Elements of the system are shown in Figure 12.18.

The radiation detector uses the same principles of operation as a radiation pyrometer in inferring the temperature of the point that the instrument is focused on from a measurement of the incoming infrared radiation. However, instead of providing a measurement of the temperature of a single point at the focal point of the instrument, the detector is scanned across a body or scene, and thus provides information about temperature distributions.

Because of the scanning mode of operation of the instrument, radiation detectors with a very fast response are required, and only photoconductive or photovoltaic sensors are suitable. These are sensitive to the portion of the infrared spectrum between the wavelengths of 2 μm and 14 μm.

Simpler versions of thermal imaging instruments consist of hand-held viewers which are pointed at the object of interest. The output from an array of infrared detectors is directed on to a matrix of red light-emitting diodes assembled behind a glass screen, and the output display thus consists of different intensities of red on a black background, with the different intensities corresponding to different temperatures. Measurement resolution is high, with temperature differences as small as 0.1 °C being detectable. Such instruments are used in a wide variety of applications such as monitoring product flows through pipework, detecting insulation faults, and detecting hot spots in furnace linings, electrical transformers, machines, bearings, etc. The number of applications is extended still further if the instrument is carried in a helicopter, where uses include scanning electrical transmission lines for faults, searching for lost or injured people and detecting the source and spread pattern of forest fires.

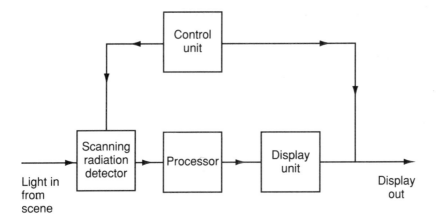

Figure 12.18 Thermography (thermal imaging) system

More complex thermal imaging systems comprise a tripod-mounted detector connected to a desktop computer and display system. Multi-colour displays are commonly used in such systems, where up to 16 different colours represent different bands of temperature across the measured range. The heat distribution across the measured body or scene is thus displayed graphically as a contoured set of coloured bands representing the different temperature levels. Such colour–thermography systems find many applications such as inspecting electronic circuit boards and monitoring production processes. There are also medical applications in body scanning.

12.7 **Acoustic thermometers**

The principle of acoustic thermometry was discovered as long ago as 1873, but until very recently it had only been used for measuring cryogenic (very low) temperatures. Acoustic thermometers use the fact that the velocity of sound through a gas varies with temperature according to the equation:

$$v = (\alpha RT/M)^{1/2} \tag{12.6}$$

where v is the velocity of sound, T is the gas temperature, M is the molecular weight of the gas and both R and α are constants. The various versions of acoustic thermometer which are available differ according to the technique used for generating sound and measuring its velocity in the gas. Further information can be found in Moore (1984).

12.8 **Fiber optic temperature sensors**

Fiber optic cables can be used as temperature sensors in various ways, as discussed in section 9.3, although special attention has to be paid to providing a suitable protective coating when high temperatures are measured. Two out of many examples of fiber optic temperature sensors are the fiber optic cross-talk sensor and special types of fiber where the temperature-sensitive refractive indices of the core and cladding are almost equal. Further examples are the phase-modulating fiber optic temperature sensor, and forms of monomode fiber which modulate the polarization of the light transmitted according to the temperature.

One extremely accurate form of extrinsic fiber optic sensor is a device known as the Accufibre temperature sensor. This is a form of radiation pyrometer which has a black-box cavity at the focal point of the lens system. A fiber optic cable is used to transmit radiation from the black-box cavity to a spectrometric device which computes the temperature. This has a measurement range of 500 °C to 2000 °C, a resolution of 10^{-5} °C and an inaccuracy of only $\pm 0.0025\%$ of full scale.

Fiber optic cables are also frequently used to transmit light into standard radiation pyrometers from a remote targetting lens. The cables can be used in all types of radiation pyrometer, including the two-colour version. However, it is not possible to use fiber optics for measuring very low temperatures because of the attenuation along the fiber of the very small radiation levels which exist at low temperatures. The minimum temperature which can be measured is about 50 °C, and the light guide for this must not exceed 600 mm in length. At temperatures exceeding 1000 °C, lengths of fiber up to 20 m can be successfully used as a light guide.

12.9 **Intelligent temperature measuring instruments**

Intelligent temperature transmitters have now been introduced into the catalogues of most instrument manufacturers, and they bring about the usual benefits associated with intelligent instruments. Such transmitters are separate boxes designed for use with transducers which have either a d.c. voltage output in the millivolt range or an output in the form of a resistance change. They are therefore suitable for use in conjunction with thermocouples, thermopiles, resistance thermometers, thermistors and broad-band radiation pyrometers. All of the transmitters presently available have non-volatile memories where all constants used in correcting output values for modifying inputs, etc., are stored, thus enabling the instrument to survive power failures without losing such information.

The transmitters now available include adjustable damping, noise rejection, self-adjustment for zero and sensitivity drifts and expanded measurement range. These features allow an accuracy level of $\pm 0.05\%$ of full scale to be specified.

The cost of intelligent temperature transducers is significantly more than their non-intelligent counterparts, and justification purely on the grounds of their superior accuracy is hard to make. However, their expanded measurement range means immediate savings are made in terms of the reduction in the number of spare instruments needed to cover a number of measurement ranges. Their capability for self-diagnosis and self-adjustment means that they require attention much less frequently, giving additional savings in maintenance costs.

12.10 Summary of temperature transducers

The suitability of different instruments in any particular measurement situation depends substantially on whether the medium to be measured is a solid or a fluid. For measuring the temperature of solids, it is essential that good contact is made between the body and the transducer unless a radiation thermometer is used. This restricts the range of suitable transducers to thermocouples, thermopiles, resistance thermometers and thermistors. On the other hand, fluid temperatures can be measured by any of the instruments described in this chapter, with the exception of radiation thermometers.

The range of instruments working on the thermal expansion principle are mainly used as temperature-indicating devices rather than as components within automatic control schemes. Mercury-in-glass thermometers are used up to $+1000\,°C$, bimetallic thermometers to $+1500\,°C$ and pressure thermometers to $+2000\,°C$. The displacement form of output in these last two devices is frequently connected mechanically to the pointer of a chart recorder in process monitoring applications. The usual measurement accuracy is in the range of $±0.5\%$ to $±1.0\%$.

The most common instrument used in industry for temperature measurement is the base-metal thermocouple. This is relatively cheap, has a typical accuracy of $±0.5\%$ of full scale and can measure temperatures in the range of $-250\,°C$ to $+1200\,°C$. Noble-metal thermocouples are much more expensive, but are chemically inert and can measure temperatures up to $2300\,°C$ with an accuracy of $±0.2\%$ of full scale.

Resistance thermometers and thermistors are another relatively common class of devices used for measuring temperature. Resistance thermometers are used in the temperature range of $-270\,°C$ to $+1100\,°C$ to give a measurement accuracy of $±0.5\%$. Thermistors are much smaller and cheaper, and give a fast output response to temperature changes, but they have a lower measurement sensitivity than resistance thermometers.

Dual diverse sensors are a new development which include a thermocouple and a resistance thermometer inside the same sheath. Both of these devices are affected by various factors in the operating environment, but each tends to be sensitive to different things in different ways. Thus, comparison of the two outputs means that any change in characteristics is readily detected, and appropriate measures to replace or recalibrate the sensors can be taken.

Pulsed sensors are a further recent development. They consist of a water-cooled thermocouple or resistance thermometer, and enable temperature measurement to be made well above the normal upper temperature limit for these devices. At the measuring instant, the water cooling is temporarily stopped, causing the temperature in the sensor to rise towards the process temperature. Cooling is restarted before the sensor temperature rises to a level where the sensor would be damaged, and the process temperature is then calculated by extrapolating from the measured temperature according to the exposure time.

The quartz thermometer is a relatively new development. It is fairly expensive because of the complex electronics required to analyze the frequency-change form of output. It only operates over the limited temperature range of $-40\,°C$ to $+230\,°C$, but gives a high measurement accuracy of $±0.1\%$ within this range.

Radiation thermometers are of considerable benefit in many measurement situations because of their non-invasive mode of measurement. In some applications, the disturbance imposed by putting a transducer into a medium to measure its temperature can be significant. Radiation thermometers are non-contact devices and so avoid such disturbance. Optical pyrometers are commonly used to monitor temperatures above $600\,°C$ in industrial furnaces, etc., but only give a measurement accuracy of $±5\%$. Various forms of radiation pyrometer are used over the temperature range between $-20\,°C$ and $+1800\,°C$ and can give measurement accuracies as good as $±0.05\%$. One particular merit of narrow-band radiation pyrometers is their ability to measure fast temperature transients of duration as small as $10\,\mu s$. No other instrument can measure transients anywhere near as fast as this.

The serious difficulty usually encountered in applying radiation thermometers is the need to calibrate the instrument to the body being measured, so correcting for the variable emissivity. This problem is avoided in the two-colour pyrometer.

12.11 **Exercises**

12.1 The output e.m.f. from a chromel–alumel thermocouple (type K), with its reference junction maintained at $0\,°C$, is $12.207\,mV$. What is the measured temperature?

12.2 The output e.m.f. from a nicrosil–nisil thermocouple (type N), with its reference junction maintained at $0\,°C$, is $4.21\,mV$. What is the measured temperature?

12.3 A platinum/10% rhodium–platinum (type S) thermocouple is used to measure the temperature of a furnace. The output e.m.f., with the reference junction maintained at $50\,°C$, is $5.975\,mV$. What is the temperature of the furnace?

12.4 In a particular industrial situation, a nicrosil–nisil thermocouple with nicrosil–nisil extension wires is used to measure the temperature of a fluid. In connecting up this measurement system, the instrumentation engineer responsible has inadvertently interchanged the extension wires from the thermocouple. The ends of the extension wires are held at a reference temperature of $0\,°C$ and the output e.m.f. measured is $21.0\,mV$. If the junction between the thermocouple and extension wires is at a temperature of $50\,°C$, what temperature of fluid is indicated and what is the true fluid temperature?

12.5 A chromel–constantan thermocouple measuring the temperature of a fluid is connected by mistake with copper–constantan extension leads (such that the two constantan wires are connected together and the copper extension wire is connected to the chromel thermocouple wire). If the fluid temperature was actually 250 °C, and the junction between the thermocouple and extension wires was at 80 °C, what e.m.f. would be measured at the open ends of the extension wires if the reference junction is maintained at 0 °C? What fluid temperature would be deduced from this (assuming that the connection mistake was not known about)?

References and further reading

Bentley, J. P. (1983) *Principles of Measurement Systems*, Longman: London.
Moore, G. (1984) 'Acoustic thermometry', *Electronics and Power*, **30**, 675–7.

13 Pressure measurement

Pressure measurement is a very common requirement for most industrial process control systems and many different types of pressure-sensing and pressure measurement systems are available. Before considering these in detail, however, it is important to explain some terms used in pressure measurement and to define clearly the difference between absolute pressure, gauge pressure and differential pressure. The **absolute pressure** of a fluid defines the difference between the pressure of the fluid and the absolute zero of pressure, whereas **gauge pressure** describes the difference between the pressure of a fluid and atmospheric pressure. Absolute and gauge pressure are therefore related by the expression:

$$\text{absolute pressure} = \text{gauge pressure} + \text{atmospheric pressure}$$

The term **differential pressure** is used to describe the difference between two absolute pressure values, such as the pressures at two different points within the same fluid (often between the two sides of a flow restrictor in a system measuring volume flow rate).

The range of pressures for which measurement is commonly required is the span from 1.013 bar to 7000 bar (1 atmosphere to 6910 atmospheres). In the following discussion, therefore, devices used for measuring pressures in this common mid-range are described first, and this is followed by separate consideration of the techniques used for measuring pressure outside this range.

Apart from certain types of diaphragm-type instrument which use either a piezoelectric crystal or a capacitive transducer to measure the pressure-induced displacement, all the instruments discussed in this chapter are restricted to measuring only static pressures. Instruments and techniques for measuring dynamic pressures have not been included because their measurement is a very specialized area which is not of general interest.

13.1 Measurement of mid-range pressures (1.013–7000 bar)

The only instrument commonly used for measuring absolute pressures in the mid-range of 1.013 bar to 7000 bar is the U-tube manometer, in the form where one end of the U-tube is sealed and evacuated. The U-tube contains a fluid, and the unknown pressure is applied to the open end of the tube (see Figure 13.1). The absolute pressure is measured in terms of the difference between the mercury levels in the two halves of the tube. Quite apart from the difficulty of judging exactly where the meniscus levels in the mercury are, such an instrument cannot give a perfect measurement because of the impossibility of achieving a total vacuum at the sealed end of the tube. Although it is possible by modern techniques to design an instrument which does give a reasonably accurate measurement of absolute pressure, the problem is usually avoided in practice by measuring gauge pressure instead of absolute pressure.

Gauge pressure is generally measured in one of two ways, either by comparison with a known weight acting on a known area, or by deflection of elastic elements. Instruments belonging to the former class are a second form of the U-tube manometer and the dead-weight gauge, whereas the latter class consists of either a diaphragm or some form of Bourdon tube. Apart from these two standard classes of measurement device, modern developments in electronics now allow the use of other principles in pressure measurement, such as in the resonant-wire device.

Choosing between the various types of instrument available for measuring mid-range pressures is usually strongly influenced by the intended application. U-tube manometers are very commonly used where a visual indication of pressure levels is required, and dead-weight gauges, because of their superior accuracy, are used in calibrating other pressure measuring devices.

Where compatibility with automatic control schemes is required, the choice of transducer is usually either a diaphragm type or a Bourdon tube, with the former now being predominant. Bellows-type instruments are also sometimes used for this purpose (but much less frequently), mainly in applications where their greater measurement sensitivity is required.

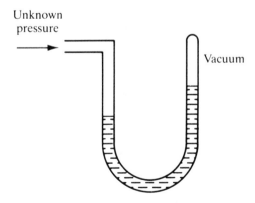

Figure 13.1 Sealed U-tube measuring absolute pressure

13.1.1 **U-tube manometer**

The U-tube manometer consists of a glass vessel in the shape of a letter U. When the manometer is used for measuring gauge pressure, both ends of the tube are open, with an unknown pressure being applied at one end and the other end being open to the atmosphere, as shown in Figure 13.2(a). The unknown gauge pressure of the fluid (p_1) is then related to the difference (h) in the levels of fluid in the two halves of the tube by the expression:

$$p_1 = h\rho g$$

where ρ is the specific gravity of the fluid.

In a third way of using the U-tube manometer, as shown in Figure 13.2(b), each open end of the tube is connected to different unknown pressures (p_1 and p_2), and the instrument thereby measures the differential pressure ($p_1 - p_2$) according to the expression:

$$(p_1 - p_2) = h\rho g$$

U-tube manometers are typically used to measure gauge and differential pressures up to about 2 bar. The type of liquid used in the instrument depends on the pressure and characteristics of the fluid being measured. Water is a convenient and certainly cheap choice, but it evaporates easily and is difficult to see. It is nevertheless used

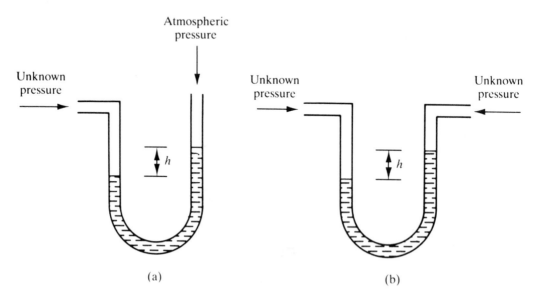

Figure 13.2 (a) U-tube measuring gauge pressure; (b) U-tube measuring differential pressure

extensively, with the major obstacles to its use being overcome by using coloured water and regularly topping up the tube to counteract evaporation.

Situations where water is definitely not used as the U-tube manometer fluid include the measurement of fluids which react with or dissolve in water, and also where higher-magnitude pressure measurements are required. In such circumstances, liquids such as aniline, carbon tetrachloride, bromoform, mercury or transformer oil are used.

The U-tube manometer, in one of its various forms, is an instrument commonly used in industry to give a visual measurement of pressure which can be acted upon by a human operator. It is not normally possible to transform the output of a U-tube manometer into an electrical signal, however, and so this instrument is not suitable for use as part of automatic control systems.

13.1.2 **Dead-weight gauge**

The dead-weight gauge, as shown in Figure 13.3, is a null-reading type of measuring instrument in which weights are added to the piston platform until the piston is adjacent to a fixed reference mark, at which time the downward force of the weights on top of the piston is balanced by the pressure exerted by the fluid beneath the piston. The fluid pressure is therefore calculated in terms of the weight added to the platform and the known area of the piston. The instrument offers the ability to measure pressures to a high degree of accuracy but is inconvenient to use. Its major application is as a reference instrument against which other pressure measuring devices are calibrated. Various versions are available which allow measurement of pressures up to 7000 bar.

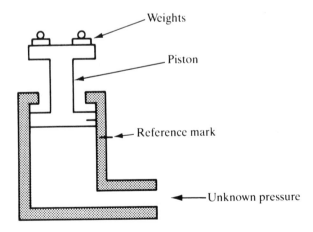

Figure 13.3 Dead-weight gauge

13.1.3 **Well-type manometer (cistern manometer)**

The well-type manometer, shown in Figure 13.4, is effectively a U-tube manometer in which one-half of the tube is made very large and takes the form of a well. The change in the level of the well as the measured pressure varies is negligible. Therefore, the liquid level in only one tube has to be measured, which makes the instrument much easier to use than the U-tube manometer. The gauge pressure is given by:

$$p_1 = h\rho$$

It might appear that the instrument would give a better measurement accuracy than the U-tube manometer because the need to subtract two liquid-level measurements in order to arrive at the pressure value is avoided. However, this benefit is swamped by errors which arise due to the typical variations in cross-sectional area in the glass used to make the tube. Such variations do not affect the accuracy of the U-tube manometer.

13.1.4 **Inclined manometer (draft gauge)**

The inclined manometer, shown in Figure 13.5, is a variation on the well-type manometer in which one leg of the tube is inclined to increase measurement sensitivity. Similar comments to those above apply about accuracy.

13.1.5 **Diaphragm**

The diaphragm is one of three types of elastic-element pressure transducer, and is shown schematically in Figure 13.6. Diaphragm-type instruments predominate in the measurement of pressures up to 10 bar. Applied pressure causes displacement of the diaphragm and this movement is measured by a displacement transducer. Both gauge

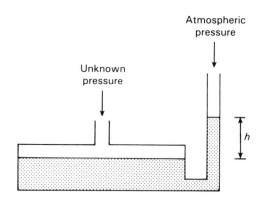

Figure 13.4 Well-type manometer

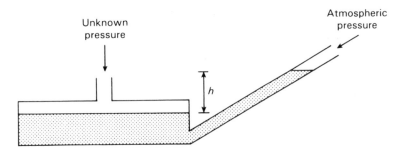

Figure 13.5 Inclined manometer

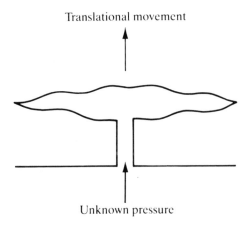

Figure 13.6 Diaphragm

pressure and differential pressure can be measured by different versions of diaphragm-type instruments. In the case of differential pressure, the two pressures are applied to either side of the diaphragm and the displacement of the diaphragm corresponds to the pressure difference. The magnitude of displacement in either version is typically 0.1 mm, which is well suited to a strain gauge type of measuring transducer.

Four strain gauges are normally used in a bridge configuration, in which an excitation voltage is applied across two opposite points of the bridge. The output voltage measured across the other two points of the bridge is then a function of the resistance change due to the strain in the diaphragm. This arrangement automatically provides compensation for environmental temperature changes. Older pressure transducers of this type used metallic strain gauges bonded to a diaphragm typically made of stainless steel. Apart from manufacturing difficulties arising from the problem of bonding the gauges, metallic strain gauges have a low gauge factor, which means that the low output from the strain gauge bridge has to be amplified by an expensive d.c. amplifier. The development of semiconductor (piezoresistive) strain gauges provided a solution to the low-output problem, as they have gauge factors up to 100 times greater than metallic gauges. However, the difficulty of bonding gauges to the diaphragm remained and a new problem emerged regarding the highly non-linear characteristic of the strain–output relationship.

The problem of strain gauge bonding was solved with the emergence of monolithic piezoresistive pressure transducers some 10 years ago, and these are now the most commonly used type of diaphragm pressure transducer. The monolithic cell consists of a diaphragm made of a silicon sheet into which resistors are diffused during the manufacturing process. Such pressure transducers can be made to be very small and are often known as **microsensors**. Also, besides avoiding the difficulty with bonding, such monolithic silicon measuring cells have the advantage of being very cheap to manufacture in large quantities. Although the inconvenience of a non-linear characteristic remains, this is normally overcome by processing the output signal with an active linearization circuit or incorporating the cell into a microprocessor-based intelligent measuring transducer. The latter usually provides analog-to-digital conversion and interrupt facilities within a single chip and gives a digital output which is readily integrated into computer control schemes. Such instruments can also offer automatic temperature compensation, built-in diagnostics and simple calibration procedures. These features allow measurement accuracies of up to ±0.1% of full-scale reading.

As an alternative to strain-gauge-type displacement measurement, capacitive transducers are sometimes used. These can also be diffused into a silicon chip, and thus enable the fabrication of very small microsensors. As a further option, developments in optical fibers have been exploited in the Fotonic sensor (Figure 13.7), which is a diaphragm-type device in which the displacement is measured by optoelectronic means. In this device, light travels from a light source down an optical fiber, is reflected back from the diaphragm and travels back along a second fiber to a photodetector. There is a characteristic relationship between the light reflected and the distance from the fiber ends to the diaphragm, thus making the amount of reflected light dependent upon the measured pressure.

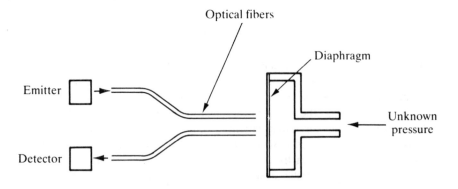

Figure 13.7 Fotonic sensor

Whilst the discussion so far has been about measuring static pressures, diaphragm-type instruments can also be used to measure dynamic pressures. The principal displacement measuring device used in such circumstances is the piezoelectric crystal. However, the strain gauge in its various forms can be used as an alternative, and there are particular advantages in its use in respect of its greater measurement sensitivity and also its ability to measure static pressures. A further alternative for measuring dynamic pressures is the diaphragm-based capacitive displacement transducer.

13.1.6 **Bellows**

The bellows, illustrated in Figure 13.8, is a device which operates on a very similar principle to the diaphragm. The sensitivity of a bellows is greater than for a diaphragm, which is its principal attribute. Pressure changes within the bellows produce translational motion of the end of the bellows which can be measured by capacitive, inductive (LVDT) or potentiometric transducers according to the range of movement produced. A typical measuring range for a bellows-type instrument is 0–1 bar (gauge pressure). The bellows is now rarely used because of its high manufacturing cost and proneness to failure. Its traditional advantage of being able to measure low pressures with a high sensitivity can now be satisfied more cheaply by diaphragm-type devices, using recent developments in electronics.

13.1.7 **Bourdon tube**

The Bourdon tube is the third type of elastic-element pressure transducer and is a very common industrial measuring instrument used for measuring the pressure of both gaseous and liquid fluids. It consists of a specially shaped piece of oval-section flexible tube which is fixed at one end and free to move at the other. When pressure is applied at the open, fixed end of the tube, the oval cross-section becomes more circular. As the cross-section of the tube tends towards a circular shape, a deflection of the closed, free end of the tube is caused. This displacement is measured by some form of displacement transducer, which is commonly a potentiometer or LVDT, or less often a capacitive sensor. In yet another version, the displacement is measured optically.

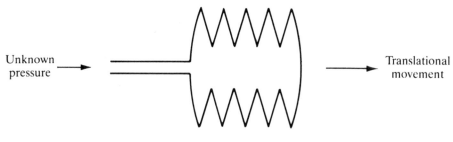

Unknown pressure →   → Translational movement

Figure 13.8 Bellows

The three common shapes of Bourdon tube are shown in Figure 13.9. The maximum possible deflection of the free end of the tube is proportional to the angle subtended by the arc through which the tube is bent. For a C-type tube, the maximum value for this arc is somewhat less than 360 degrees. Where greater measurement sensitivity and resolution are required, spiral and helical tubes are used, where the possible magnitude of the arc subtended is limited only by a practical limit on how many turns it is convenient to have in the helix or spiral. This increased measurement performance is only gained, however, at the expense of a substantial increase in manufacturing difficulty and cost compared with C-type tubes, and is also associated with a large decrease in the maximum pressure which can be measured.

C-type tubes are available for measuring pressures up to 6000 bar. A typical C-type tube of 25 mm radius has a maximum displacement travel of 4 mm, giving a moderate level of measurement resolution. Measurement accuracy is typically quoted at ±1% of full-scale deflection. Similar accuracy is available from helical and spiral types, but whilst the measurement resolution is higher, the maximum pressure measurable is only 700 bar.

The existence of one potentially major source of error in Bourdon-tube pressure measurement has not been widely documented and few manufacturers of Bourdon tubes make any attempt to warn users of their products. The problem concerns the relationship between the fluid being measured and the fluid used for calibration. The

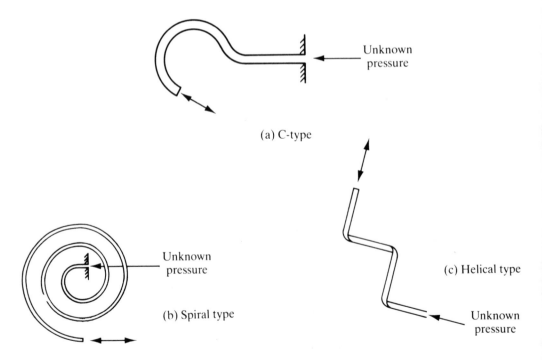

(a) C-type

(b) Spiral type

(c) Helical type

Figure 13.9 Bourdon tubes

pointer of Bourdon tubes is normally set at zero during manufacture, using air as the calibration medium. If, however, a different fluid, especially a liquid, is subsequently used with a Bourdon tube, the fluid in the tube will cause a non-zero deflection according to its weight compared with air, resulting in a reading error of up to 6%. This can be avoided by calibrating the Bourdon tube with the fluid to be measured instead of with air, assuming of course that the user is aware of the problem. Alternatively, correction can be made according to the calculated weight of the fluid in the tube. Unfortunately, difficulties arise with both of these solutions if air is trapped in the tube, since this will prevent the tube being filled completely by the fluid. Then, the amount of fluid actually in the tube, and its weight, will be unknown.

In conclusion, therefore, Bourdon tubes only have guaranteed accuracy limits when measuring gaseous pressures. Their use for accurate measurement of liquid pressures poses great difficulty unless the gauge can be totally filled with liquid during both calibration and measurement, a condition which is very difficult to fulfil practically.

13.1.8 **Resonant-wire devices**

The resonant-wire device is a relatively new instrument which has emanated from recent advances in the electronics field. A typical device is shown schematically in Figure 13.10. Wire is stretched across a chamber containing fluid at unknown

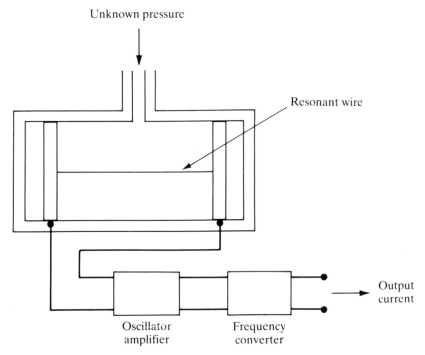

Figure 13.10 Resonant-wire device

pressure subjected to a magnetic field. The wire resonates at its natural frequency according to its tension, which varies with pressure. Thus pressure is calculated by measuring the frequency of vibration of the wire. Such frequency measurement is normally carried out by electronics integrated into the cell. Such devices are highly accurate, with $\pm 0.2\%$ of full-scale reading being typical, and they are particularly insensitive to changes in ambient conditions.

13.1.9 Fiber optic pressure sensors

In fiber optic pressure sensors, known as microbend sensors, the refractive index of the fiber and hence the intensity of light transmitted varies according to the mechanical deformation of the fibers caused by pressure. A simple form of this is sketched in Figure 9.6(a). The sensitivity of pressure sensors can be optimized by applying the pressure via a roller chain such that the bending is applied periodically (see Figure 9.6(b)). The optimal pitch for the chain varies according to the radius, refractive index and type of cable involved. Microbend sensors are typically used to measure the small pressure changes generated in vortex-shedding flowmeters. Vortices can be detected more simply in the alternative version shown in Figure 13.11, where a fiber optic cable is merely stretched across the pipe.

The shutter sensor described earlier in Chapter 9 (Figure 9.7) is a further type of fiber optic device used for measuring pressure. It is typically used to measure the displacement of transducers such as Bourdon tubes and diaphragms.

Phase-modulating fiber optic pressure sensors also exist. Their mode of operation was discussed in Chapter 9.

13.2 Low-pressure measurement (less than 1.013 bar)

Adaptations of most of the common types of pressure transducer already described can be used for absolute pressure measurement in the vacuum range (less than atmospheric pressure). Special forms of Bourdon tubes measure pressures down to 10 mbar, manometers and bellows-type instruments measure pressures down to 0.1 mbar, and diaphragms can be designed to measure pressures down to 0.001 mbar. Other more specialized instruments are also used to measure vacuum pressures. These instruments include the thermocouple gauge, the Pirani gauge, the thermistor gauge, the McLeod gauge and the ionization gauge, and are covered in more detail in the following subsections.

As for the case of mid-range pressures, the choice between instruments suitable for measuring low pressures depends very much on the application requirements. At the upper end of the low-pressure measurement range (0.001 mbar to 1 bar), Bourdon tubes, manometers, bellows-type and diaphragm instruments are all used depending on whether the application is in process monitoring, automatic control schemes or calibration duties. At greater cost, but generally offering greater sensitivity and accuracy, thermocouple/thermistor, Pirani and McLeod gauges all find application in measuring this range of pressures.

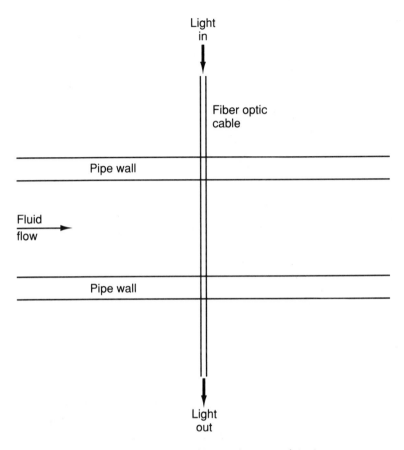

Figure 13.11 Simple fiber optic vortex detector

Below pressures of 0.001 mbar, only the more specialized instruments are applicable. Thermocouple/thermistor types measure down to 10^{-4} mbar, Pirani gauges to 10^{-5} mbar and ionization gauges down to 10^{-13} mbar.

13.2.1 **Thermocouple gauge**

The thermocouple gauge is one of the group of gauges working on the thermal conductivity principle. The Pirani and thermistor gauges also belong to this group. At low pressure, the kinematic theory of gases predicts a linear relationship between pressure and thermal conductivity. Thus measurement of thermal conductivity gives an indication of pressure.

Figure 13.12 shows a sketch of a thermocouple gauge. Operation of the gauge depends on the thermal conduction of heat between a thin, hot metal strip in the centre and the cold outer surface of a glass tube (which is normally at room temperature). The metal strip is heated by passing a current through it and its

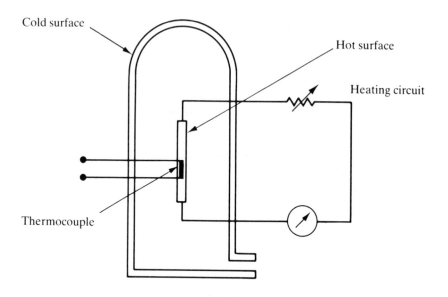

Figure 13.12 Thermocouple gauge

temperature is measured by a thermocouple. The temperature measured depends on the thermal conductivity of the gas in the tube and hence on its pressure. A source of error in this instrument is the fact that heat is also transferred by radiation as well as conduction. This is of a constant magnitude, independent of pressure, and so could be measured and corrected for. However, it is usually more convenient to design for low radiation loss by choosing a heated element with low emissivity. Thermocouple gauges are used to measure pressures typically in the range 10^{-4} mbar up to 1 mbar.

13.2.2 Pirani gauge

A typical form of Pirani gauge is shown in Figure 13.13. This is similar to a thermocouple gauge but has a heated element which consists of four coiled tungsten wires connected in parallel. Two identical tubes are normally used, connected in a bridge circuit as shown in Figure 13.14, with one containing the gas at unknown pressure and the other evacuated to a very low pressure. Current is passed through the tungsten element, which attains a certain temperature according to the thermal conductivity of the gas. The resistance of the element changes with temperature and causes an imbalance of the measurement bridge. Thus the Pirani gauge avoids the use of a thermocouple to measure temperature (as in the thermocouple gauge) by effectively using a resistance thermometer as the heated element. Such gauges cover the pressure range 10^{-5} mbar to 1 mbar.

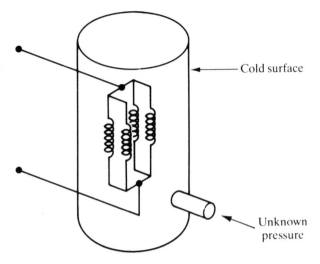

Figure 13.13 Pirani gauge

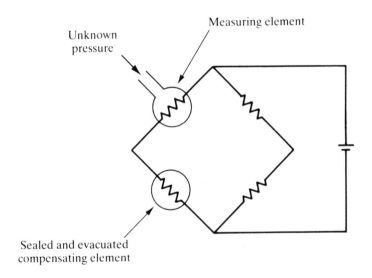

Figure 13.14 Wheatstone bridge circuit to measure output of Pirani gauge

13.2.3 **Thermistor gauge**

Thermistor gauges operate on identical principles to the Pirani gauge but use semiconductor materials for the heated elements instead of metals. The normal pressure range covered is 10^{-4} mbar to 1 mbar.

13.2.4 **McLeod gauge**

Figure 13.15 shows the general form of a McLeod gauge, in which low-pressure fluid is compressed to a higher pressure which is then read by manometer techniques. In essence, the gauge can be visualized as a U-tube manometer, sealed at one end, and where the bottom of the U can be blocked at will. To operate the gauge, the piston is first withdrawn, causing the level of mercury in the lower part of the gauge to fall below the level of the junction J between the two tubes in the gauge marked Y and Z. Fluid at unknown pressure p_u is then introduced via the tube marked Z, from where it also flows into the tube marked Y, of cross-sectional area A. Next, the piston is pushed in, moving the mercury level up to block the junction J. At the stage where J is just blocked, the fluid in tube Y is at pressure p_u and contained in a known volume V_u. Further movement of the piston compresses the fluid in tube Y and this process continues until the mercury level in tube Z reaches a zero mark. Measurement of the height (h) above the mercury column in tube Y then allows calculation of the compressed volume of the fluid V_c as:

$$V_c = hA$$

Then, by Boyle's law:

$$p_u V_u = p_c V_c$$

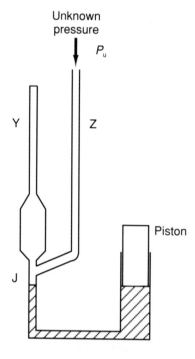

Figure 13.15 McLeod gauge

Also, applying the normal manometer equation:

$$p_c = p_u + h\rho g$$

where ρ is the mass density of mercury. Hence:

$$p_u = \frac{Ah^2\rho g}{V_u - Ah} \tag{13.1}$$

The compressed volume V_c is often very much smaller than the original volume, in which case Equation (13.1) approximates to:

$$p_u = \frac{Ah^2\rho g}{V_u} \quad \text{for } Ah \ll V_u \tag{13.2}$$

Although the best accuracy achievable with McLeod gauges is $\pm 1\%$, this is still better than most other gauges for measuring pressures in this range, and they are therefore often used as a standard against which other gauges are calibrated. The minimum pressure normally measurable is 10^{-4} bar, although lower pressures can be measured if pressure-dividing techniques are applied.

13.2.5 Ionization gauge

The ionization gauge is a special type of instrument used for measuring very low pressures in the range 10^{-13} to 10^{-3} bar. Gas of unknown pressure is introduced into a glass vessel containing free electrons discharged from a heated filament, as shown in Figure 13.16. The gas pressure is determined by measuring the current flowing

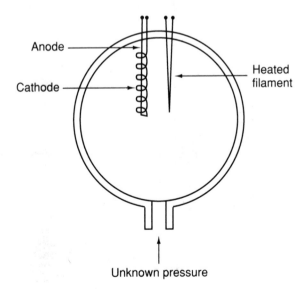

Figure 13.16 The ionization gauge

between an anode and cathode within the vessel. This current is proportional to the number of ions per unit volume, which in turn is proportional to the gas pressure.

13.3 High-pressure measurement (greater than 7000 bar)

Measurement of pressures above 7000 bar is normally carried out electrically by monitoring the change of resistance of wires of special materials. Materials having resistance–pressure characteristics which are suitably linear and sensitive include gold–chromium alloys and manganin. A coil of such wire is enclosed in a sealed, kerosene-filled, flexible bellows, as shown in Figure 13.17. The unknown pressure is applied to one end of the bellows, which transmits the pressure to the coil. The magnitude of the applied pressure is then determined by measuring the coil resistance.

13.4 Intelligent pressure transducers

Adding microprocessor power to pressure transducers brings about substantial improvements in their characteristics. Improved measurement sensitivity, extended measurement range, compensation for hysteresis and other non-linearities, and correction for ambient temperature and pressure changes are just some of the facilities offered by intelligent pressure transducers. Accuracies of ±0.1% can be achieved with piezoresistive-bridge silicon devices, for instance. In view of their much superior characteristics, it is perhaps surprising that intelligent pressure instruments represent only about 1% of the total number of pressure measuring devices sold at the

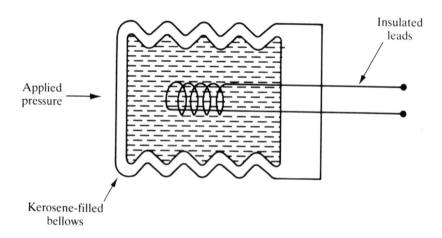

Figure 13.17 High-pressure measurment – wire coil in bellows

present time. The significantly higher cost compared with non-intelligent devices has been cited as the reason for this, but this hardly seems adequate to explain such a very low level of market penetration.

Some recent microprocessor-based pressure transducers make use of novel techniques of displacement measurement. For example, both diaphragm and helical Bourdon-tube devices are now available which use an optical method of displacement measurement of the form shown in Figure 13.18. In this, the motion is transmitted to a vane which progressively shades one of two monolithic photodiodes that are exposed to infrared radiation. The second photodiode acts as a reference, enabling the microprocessor to compute a ratio signal which is linearized and is available as either an analog or digital measurement of pressure. The measurement accuracy is typically ±0.1%.

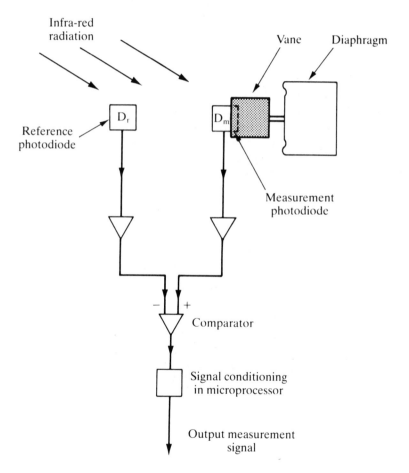

Figure 13.18 One type of intelligent pressure measuring instrument

14 Flow measurement

Flow measurement is extremely important in all the process industries. The manner in which the flow rate is quantified depends on whether the quantity flowing is a solid, liquid or gas. In the case of solids, it is appropriate to measure the rate of mass flow, whereas in the case of liquids and gases, flow is usually measured in terms of the volume flow rate. In a few cases, such as measuring the rate of usage of liquid fuel in a rocket, it is necessary to measure the mass flow rate of a liquid, and special techniques are available for this.

14.1 Mass flow rate

Measurement of the mass flow rate of solids in the process industries is normally concerned with solids which are in the form of small particles produced by a crushing or grinding process. Such materials are usually transported by some form of conveyor, and this allows the mass flow rate to be calculated in terms of the mass of the material on a given length of conveyor multiplied by the speed of the conveyor.

Figure 14.1 shows a typical system for mass flow measurement. A load cell measures the mass M of material distributed over a length L of the conveyor. If the conveyor velocity is v, the mass flow rate, Q, is given by:

$$Q = Mv/L$$

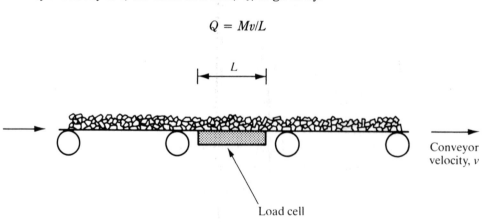

Figure 14.1 Measurement of mass flow rate

The mass flow rate of a fluid is usually determined by simultaneous measurement of the volume flow rate and the fluid density, although a recently available alternative is the Coriolis meter. As the measurement of the mass flow rate of fluids is not a common requirement, the instruments suitable for this are not covered by this text but full details can be found elsewhere (e.g. Medlock 1983, Furness and Heritage 1986, Medlock and Furness 1990).

14.2 **Volume flow rate**

Volume flow rate is the appropriate way of quantifying the flow of all materials which are in a gaseous, liquid or semi-liquid slurry form (where solid particles are suspended in a liquid host). Materials in these forms are carried in pipes, and the common classes of instrument used for measuring the volume flow rate can be summarized as follows:

1. Differential pressure meters
2. Variable area meters
3. Positive-displacement meters
4. Turbine flowmeters
5. Electromagnetic flowmeters
6. Vortex-shedding flowmeters
7. Gate-type meters
8. Ultrasonic flowmeters
9. Cross-correlation flowmeters
10. Laser Doppler flowmeters.

14.3 **Differential pressure meters**

Differential pressure meters involve the insertion of some device into a fluid-carrying pipe which causes an obstruction and creates a pressure difference on either side of the device. Such devices include the **orifice plate**, the **venturi tube**, the **flow nozzle** and the **Dall flow tube**. When such a restriction is placed in a pipe, the velocity of the fluid through the restriction increases and the pressure decreases. The volume flow rate is then proportional to the square root of the pressure difference across the obstruction. The manner in which this pressure difference is measured is important. Measuring the two pressures with different instruments and calculating the difference between the two measurements is not satisfactory because of the large measurement error which can arise when the pressure difference is small, as explained in Chapter 7. The normal procedure is therefore to use a diaphragm-based differential pressure transducer.

The **pitot tube** is a further device which measures flow by creating a pressure difference within a fluid-carrying pipe. However, in this case, there is negligible obstruction of flow in the pipe. The pitot tube is a very thin tube which obstructs only a small part of the flowing fluid and thus measures flow at a single point across the cross-section of the pipe. This measurement only equates to average flow velocity in the pipe for the case of uniform flow. The **annubar** is a type of multi-port pitot tube which does measure the average flow across the cross-section of the pipe by forming the mean value of several local flow measurements across the cross-section of the pipe.

All applications of this method of flow measurement assume that flow conditions upstream of the obstruction device are in steady state, and a certain minimum length of straight run of pipe ahead of the measurement point is specified to ensure this. The minimum lengths required for various pipe diameters are specified in British Standards tables (and also in alternative but equivalent national standards used in other countries), but a useful rule of thumb widely used in the process industries is to specify a length of ten times the pipe diameter. If physical restrictions make this impossible to achieve, special flow-smoothing vanes can be inserted immediately ahead of the measurement point.

Flow-restriction-type instruments are popular because they have no moving parts and are therefore robust, reliable and easy to maintain. One disadvantage of this method is that the obstruction causes a permanent loss of pressure in the flowing fluid. The magnitude and hence importance of this loss depends on the type of obstruction element used, but where the pressure loss is large, it is sometimes necessary to recover the lost pressure by an auxiliary pump further down the flow line. This class of device is not normally suitable for measuring the flow of slurries as the tappings into the pipe to measure the differential pressure are prone to blockage, although the venturi tube can be used to measure the flow of dilute slurries.

Figure 14.2 illustrates approximately the way in which the flow pattern is interrupted when an orifice plate is inserted into a pipe. The other obstruction devices also have a similar effect to this. Of particular interest is the fact that the minimum cross-sectional area of flow occurs not within the obstruction but at a point downstream of there. Knowledge of the pattern of pressure variation along the pipe, as shown in Figure 14.3, is also of importance in using this technique of measurement of volume flow rate. This shows that the point of minimum pressure coincides with the point of minimum cross-section flow, a little way downstream of the obstruction. Figure 14.3 also shows that there is a small rise in pressure immediately before the obstruction. It is therefore important not only to position the instrument measuring P_2 exactly at the point of minimum pressure, but also to measure the pressure P_1 at a point upstream of the point where the pressure starts to rise before the obstruction.

In the absence of any heat transfer mechanisms, and assuming frictionless flow of an incompressible fluid through the pipe, the theoretical volume flow rate of the fluid, Q, is given by (see Bentley (1983) for the derivation):

$$Q = \frac{A_2}{[1 - (A_2/A_1)^2]^{1/2}} [2(P_1 - P_2)/\rho]^{1/2} \tag{14.1}$$

where A_1 and P_1 are the cross-sectional area and pressure of the fluid flow before the

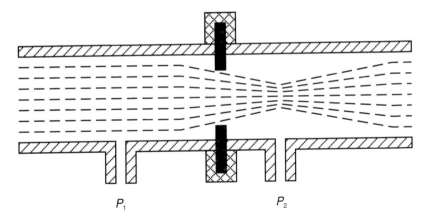

Figure 14.2 Profile of flow across orifice plate

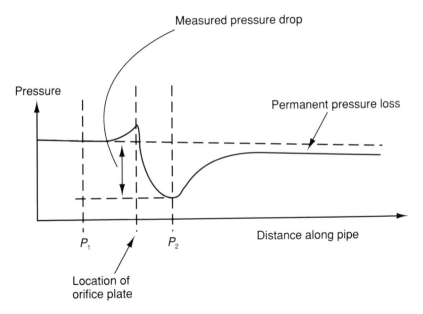

Figure 14.3 Pattern of pressure variation along pipe obstructed by an orifice plate

obstruction, A_2 and P_2 are the cross-sectional area and pressure of the fluid flow at the narrowest point of the flow beyond the obstruction, and ρ is the fluid density.

Equation (14.1) is never applicable in practice for several reasons. First, frictionless flow is never achieved. However, in the case of turbulent flow through smooth pipes, friction is low and it can be adequately accounted for by a variable called the Reynolds number, which is a measurable function of the flow velocity and the viscous friction.

The other reasons for the non-applicability of Equation (14.1) are that the initial cross-sectional area of the fluid flow is less than the diameter of the pipe carrying it and that the minimum cross-sectional area of the fluid is less than the diameter of the obstruction. Therefore, neither A_1 nor A_2 can be measured. These problems are taken account of by modifying Equation (14.1) to the following:

$$Q = \frac{C_D A_2'}{[1 - (A_2'/A_1')^2]^{1/2}} [2(P_1 - P_2)/\rho]^{1/2} \tag{14.2}$$

where A_1' and A_2' are the pipe diameters before and at the obstruction and C_D is a constant, known as the discharge coefficient, which accounts for the Reynolds number and the difference between the pipe and flow diameters.

Before Equation (14.2) can be evaluated, the discharge coefficient must be calculated. As this varies between each measurement situation, it would appear at first sight that the discharge coefficient must be determined by practical experimentation in each case. However, provided that certain conditions are met, standard tables can be used to obtain the value of the discharge coefficient appropriate to the pipe diameter and fluid involved.

It is particularly important in applications of flow restriction methods to choose an instrument whose range is appropriate to the magnitudes of flow rate being measured. This requirement arises because of the square-root type of relationship between the pressure difference and the flow rate, which means that as the pressure difference decreases, the error in flow rate measurement can become very large. In consequence, restriction-type flowmeters are only suitable for measuring flow rates between 30% and 100% of the instrument range.

14.3.1 Orifice plate

The orifice plate is a metal disk with a hole in it, as shown in Figure 14.4, inserted into the pipe carrying a flowing fluid. This hole is normally concentric with the disk. Over 50% of the instruments used in industry for measuring volume flow rate are of this type. The use of the orifice plate is so widespread because of its simplicity, cheapness and availability in a wide range of sizes. However, the best accuracy obtainable with this type of obstruction device is only $\pm 2\%$ and the permanent pressure loss caused in the flow is very high, being between 50% and 90% of the pressure difference $(P_1 - P_2)$ in magnitude. Other problems with the orifice plate are a gradual change in the discharge coefficient over a period of time as the sharp edges of the hole wear away, and a tendency for any particles in the flowing fluid to stick behind the hole and gradually build up and reduce its diameter. The latter problem can be minimized by using an orifice plate with an eccentric hole. If this hole is close to the bottom of the pipe, solids in the flowing fluid tend to be swept through, and the build-up of particles behind the plate is minimal.

A very similar problem arises if there are any bubbles of vapour or gas in the flowing fluid when liquid flow is involved. These also tend to build up behind an orifice

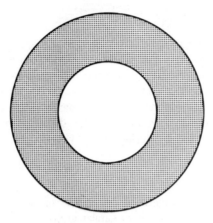

Figure 14.4 The orifice plate

plate and distort the pattern of flow. This difficulty can be avoided by mounting the orifice plate in a vertical run of pipe.

14.3.2 Flow nozzle

The form of a flow nozzle is shown in Figure 14.5. This is not prone to solid particles or bubbles of gas in a flowing fluid sticking in the flow restriction, and so in this respect it is superior to the orifice plate. Its useful working life is also greater because it does not get worn away in the same way as an orifice plate. These factors give the instrument a greater measurement accuracy. However, as the engineering effort involved in fabricating a flow nozzle is greater than that required to make an orifice plate, the instrument is somewhat more expensive. In terms of the permanent pressure loss imposed on the measured system, the flow nozzle is very similar to the orifice plate. A typical application of the flow nozzle is in measurement of steam flow.

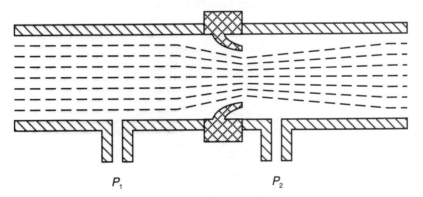

P_1 P_2

Figure 14.5 The flow nozzle

14.3.3 **Venturi**

The venturi is a precision-engineered tube of a special shape, as shown in Figure 14.6. It is a very expensive instrument but offers very good accuracy (approximately ±1%) and imposes a permanent pressure loss on the measured system of only 10–15% of the pressure difference $(P_1 - P_2)$ across it. The smooth internal shape of this type of restriction means that it is unaffected by solid particles or gaseous bubbles in the flowing fluid, and in fact can even cope with dilute slurries. It has almost no maintenance requirements and its working life is very long.

14.3.4 **Dall flow tube**

The Dall flow tube, shown in Figure 14.7, consists of two conical reducers inserted into the fluid-carrying pipe. It has a very similar internal shape to the venturi, except that it lacks a throat. This construction is much easier to manufacture than a venturi (which requires complex machining), and this gives the Dall flow tube an advantage in cost over the venturi, although the measurement accuracy obtained is not quite as good (approximately ±1.5%). Another advantage of the Dall flow tube is its shorter length, which makes the engineering task of inserting it into the flow line easier. The Dall tube has one further operational advantage, in that the permanent pressure loss

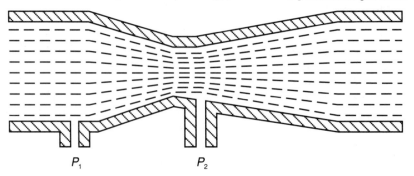

P_1 P_2

Figure 14.6 The venturi

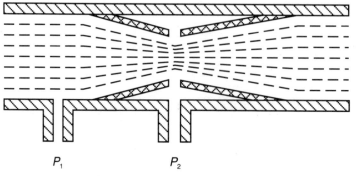

P_1 P_2

Figure 14.7 The Dall flow tube

imposed on the system is only about 5% of the measured pressure difference $(P_1 - P_2)$, and thus is only half that due to a venturi. In other respects, the two instruments are very similar, with their low maintenance requirement and long life.

14.3.5 **Pitot tube**

The pitot static tube is mainly used for making temporary measurements of flow, although it is also used in some instances for permanent flow monitoring. It measures the local velocity of flow at a particular point within a pipe rather than the average flow velocity as measured by other types of flowmeter. This may be very useful where there is a requirement to measure local flow rates across the cross-section of a pipe in the case of non-uniform flow. Multiple pitot tubes are normally used to do this.

The instrument depends on the principle that a tube placed with its open end in a stream of fluid, as shown in Figure 14.8, will bring to rest that part of the fluid which impinges on it, and the loss of kinetic energy will be converted to a measurable increase in pressure inside the tube. This pressure (P_1), as well as the static pressure of the undisturbed free stream of flow (P_2), is measured.

The flow velocity can then be calculated from the formula:

$$v = C[2g(P_1 - P_2)]^{1/2}$$

The constant C, known as the pitot-tube coefficient, corrects for the fact that not all fluid incident on the end of the tube will be brought to rest: a proportion will slip around it according to the design of the tube.

Having calculated v, the volume flow rate can then be calculated by multiplying v by the cross-sectional area of the flow pipe, A.

Pitot tubes have the advantage that they cause negligible pressure loss in the flow. They are also cheap, and the installation procedure consists of the very simple process of pushing them down a small hole drilled in the flow-carrying pipe.

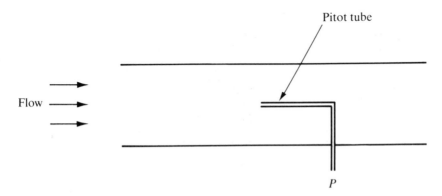

Figure 14.8 The pitot tube

Their main failing is that the measurement accuracy given is normally only about ±5%, and sensitive pressure measuring devices are needed to achieve even this limited level of accuracy, as the pressure difference created is very small. More recently, measurement capabilities with the uncertainty down to ±1% have been claimed for specially designed pitot tubes (Britton and Mesnard 1982).

The **annubar** is a development of the pitot tube which has multiple sensing ports distributed across the cross-section of the pipe. It thus provides an approximate measurement of the mean flow rate across the pipe.

14.4 **Variable area flowmeters**

In this class of flowmeter, the differential pressure across a variable aperture is used to adjust the area of the aperture. The aperture area is then a measure of the flow rate. This type of instrument normally only gives a visual indication of flow rate, and so is of no use in automatic control schemes. However, it is reliable and cheap and used extensively throughout industry. In fact, variable area meters account for 20% of all flowmeters sold. Recently, fiber optics have been incorporated into variable area flowmeters, where the position of the float is detected by a row of fibers which sense the reflection of light from the float.

In its simplest form, shown in Figure 14.9, the instrument consists of a tapered glass tube containing a float which takes up a stable position where its submerged

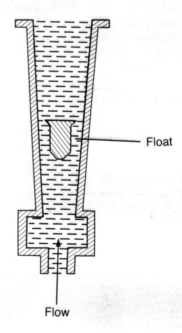

Figure 14.9 The variable area flowmeter

weight is balanced by the upthrust due to the differential pressure across it. The position of the float is a measure of the effective annular area of the flow passage and hence of the flow rate. The accuracy of the cheapest instruments is only ±3%, but more expensive versions offer measurement accuracies as high as ±0.2%. The normal measurement range is between 10% and 100% of the full-scale reading for any particular instrument.

14.5 **Positive-displacement flowmeters**

All positive-displacement meters operate by using mechanical divisions to displace discrete volumes of fluid successively. Whilst this principle of operation is common, many different mechanical arrangements exist for putting the principle into practice. All versions of positive-displacement meter are low-friction, low-maintenance and long-life devices, although they do impose a small permanent pressure loss on the flowing fluid. Low friction is especially important when measuring gas flows, and meters with special mechanical arrangements to satisfy this requirement have been developed.

The rotary piston meter is the most common type of positive-displacement meter, and this is illustrated in Figure 14.10. It uses a cylindrical piston which is displaced around a cylindrical chamber by the flowing fluid. Rotation of the piston drives an output shaft. This can either be used with a pointer-and-scale system to give a visual flow reading or be converted into an electrical output signal.

Positive-displacement flowmeters account for nearly 10% of the total number of flowmeters used in industry. Such devices are used in large numbers for metering

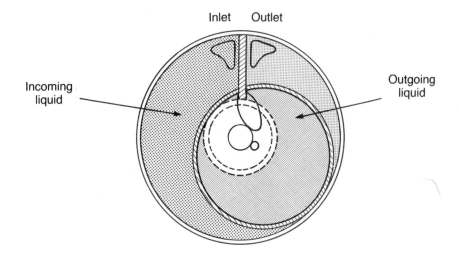

Figure 14.10 Rotary piston form of positive-displacement flowmeter

domestic gas and water consumption. The cheapest instruments have an accuracy of about ±1.5%, but the accuracy in more expensive ones can be as good as ±0.2%. These higher-quality instruments are used extensively within the oil industry, as such applications can justify the high cost of such instruments.

14.6 **Turbine meters**

A turbine flowmeter consists of a multi-bladed wheel mounted in a pipe along an axis parallel to the direction of fluid flow in the pipe, as shown in Figure 14.11. The flow of fluid past the wheel causes it to rotate at a rate which is proportional to the volume flow rate of the fluid. This rate of rotation has traditionally been measured by constructing the flowmeter such that it behaves as a variable reluctance tachogenerator. This is achieved by fabricating the turbine blades from a ferromagnetic material and placing a permanent magnet and coil inside the meter housing. A voltage pulse is induced in the coil as each blade on the turbine wheel moves past it, and if these pulses are measured by a pulse counter, the pulse frequency and hence flow rate can be deduced. In recent instruments, fiber optics are also sometimes used to count the rotations by detecting reflections off the tip of the turbine blades.

Provided that the turbine wheel is mounted in low-friction bearings, measurement accuracy can be as high as ±0.1%. However, turbine flowmeters are less rugged and reliable than flow-restriction-type instruments, and are badly affected by any particulate matter in the flowing fluid. Bearing wear is a particular problem and they also impose a permanent pressure loss on the measured system.

Turbine meters are particularly prone to large errors when there is any significant second phase in the fluid measured. For instance, using a turbine meter calibrated on

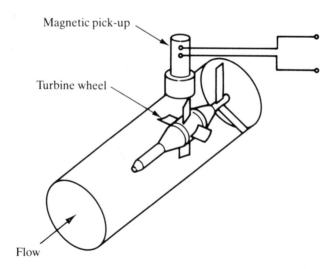

Figure 14.11 The turbine flowmeter

pure liquid to measure a liquid containing 5% air produces a 50% measurement error. As an important application of the turbine meter is in the petrochemical industries, where gas/oil mixtures are common, special procedures are being developed to avoid such large measurement errors. The most promising approach is to homogenize the two gas/oil phases prior to flow measurement (King 1988).

Turbine meters have a similar cost and market share to positive-displacement meters, and compete for many applications, particularly in the oil industry. Turbine meters are smaller and lighter than the latter and are preferred for low-viscosity, high-flow measurements. However, positive-displacement meters are superior in conditions of high viscosity and low flow rate.

14.7 **Electromagnetic flowmeters**

Electromagnetic flowmeters are limited to measuring the volume flow rate of electrically conductive fluids. A reasonable measurement accuracy, around $\pm 1.5\%$, is given, although the instrument is expensive both in terms of the initial purchase cost and also in running costs, mainly because of its electricity consumption. A further reason for its high cost is the need for careful calibration of each instrument individually during manufacture, as there is considerable variation in the properties of the magnetic materials used.

The instrument, shown in Figure 14.12, consists of a stainless steel cylindrical tube, fitted with an insulating liner, which carries the measured fluid. Typical lining materials used are neoprene, polytetrafluoroethylene (PTFE) and polyurethane. A magnetic field is created in the tube by placing mains-energized field coils either side of it, and the voltage induced in the fluid is measured by two electrodes inserted into opposite sides of the tube. The ends of these electrodes are usually flush with the inner surface of the cylinder. The electrodes are constructed from a material which is unaffected by most types of flowing fluid, such as stainless steel, platinum–iridium alloys, Hastelloy, titanium and tantalum. In the case of the rarer metals in this list, the electrodes account for a significant part of the total cost of the instrument.

By Faraday's law of electromagnetic induction, the voltage E induced across a length L of the flowing fluid moving at velocity v in a magnetic field of flux density B is given by:

$$E = BLv \tag{14.3}$$

where L is the distance between the electrodes, which is the diameter of the tube, and B is a known constant. Hence, measurement of the voltage E induced across the electrodes allows the flow velocity v to be calculated from Equation (14.3). Having thus calculated v, it is a simple matter to multiply v by the cross-sectional area of the tube to obtain a value for the volume flow rate. The typical voltage signal measured across the electrodes is 1 mV when the fluid flow rate is 1 m/s.

The internal diameter of magnetic flowmeters is normally the same as that of the

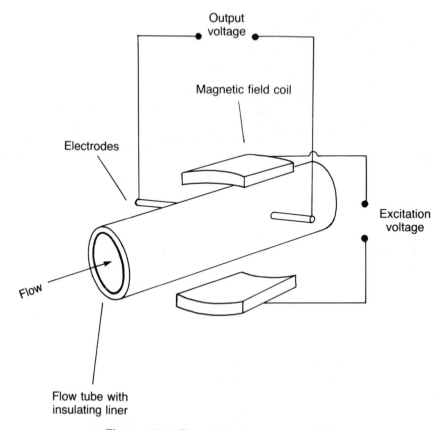

Output
voltage

Magnetic field coil

Electrodes

Excitation
voltage

Flow

Flow tube with
insulating liner

Figure 14.12 The electromagnetic flowmeter

rest of the flow-carrying pipework in the system. Therefore, there is no obstruction to the fluid flow and consequently no pressure loss associated with measurement. Like other forms of flowmeter, the magnetic type requires a minimum length of straight pipework immediately prior to the point of flow measurement in order to guarantee the accuracy of measurement, although a length equal to five pipe diameters is usually sufficient.

Whilst the flowing fluid must be electrically conductive, the method is of use in many applications and is particularly useful for measuring the flow of slurries in which the liquid phase is electrically conductive. Corrosive fluids can be handled providing a suitable lining material is used. At the present time, magnetic flowmeters account for about 15% of new flowmeters sold and this total is slowly growing. One operational problem is that the insulating lining is subject to damage when abrasive fluids are being handled, and this can give the instrument a limited life.

Current developments in electromagnetic flowmeters are producing physically smaller instruments and employing better coil designs which reduce electricity consumption and make battery-powered versions feasible (these are now commercially available). Also, whereas conventional electromagnetic flowmeters require a minimum fluid conductivity of $10\,\mu$mho/cm^3, some new versions are becoming available which can cope with fluid conductivities as low as $1\,\mu$mho/cm^3.

14.8 Vortex-shedding flowmeters

The vortex-shedding flowmeter is a relatively new type of instrument which is rapidly gaining in popularity and being used as an alternative to traditional differential pressure meters in more and more applications. The operating principle of the instrument is based on the natural phenomenon of vortex shedding, created by placing an unstreamlined obstacle (known as a bluff body) in a fluid-carrying pipe, as indicated in Figure 14.13. When fluid flows past the obstacle, boundary layers of viscous, slow-moving fluid are formed along the outer surface. Because the obstacle is not streamlined, the flow cannot follow the contours of the body on the downstream side, and the separate layers become detached and roll into eddies or vortices in the low-pressure region behind the obstacle. The shedding frequency of these alternately shed vortices is proportional to the fluid velocity past the body. Various thermal, magnetic, ultrasonic and capacitive vortex detection techniques are employed in different instruments.

Such instruments have no moving parts, operate over a wide flow range, have a low power consumption and require little maintenance. They can measure both liquid and gas flows and a common inaccuracy figure quoted is $\pm1\%$ of full-scale reading, though this can be seriously downgraded in the presence of flow disturbances upstream of the measurement point, and a straight run of pipe before the measurement point of 50 pipe diameters is recommended. Another problem with the instrument is its

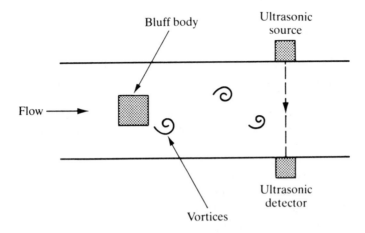

Figure 14.13 The vortex-shedding flowmeter

susceptibility to pipe vibrations, although new designs are becoming available which have a better immunity to such vibrations.

14.9 **Gate-type meters**

The gate meter was the earliest device in this class. It consists of a spring-loaded, hinged flap mounted at right angles to the direction of fluid flow in the fluid-carrying pipe. The flap is connected to a pointer outside the pipe. The fluid flow deflects the flap and pointer and the flow rate is indicated by a graduated scale behind the pointer. The major difficulty with such devices is in preventing leaks at the hinge point.

A variation on this principle is the air-vane meter which measures deflection of the flap by a potentiometer inside the pipe. This is commonly used to measure air flow within automotive fuel-injection systems.

Another device in this class is the target meter. This consists of a circular disk-shaped flap in the pipe. Fluid flow rate is inferred from the force exerted on the disk measured by strain gauges bonded to it. This meter is very useful for measuring the flow of dilute slurries but it does not find wide application elsewhere as it has a relatively high cost.

14.10 **Ultrasonic flowmeters**

The ultrasonic technique of measurement of volume flow rate is, like the magnetic flowmeter, a non-invasive method. It is not restricted to conductive fluids, however, and is particularly useful for measuring the flow of corrosive fluids and slurries. A further advantage over magnetic flowmeters is that the instrument is one which clamps on externally to existing pipework rather than being inserted as an integral part of the flow line, as in the case of the magnetic flowmeter. As the procedure of breaking into a pipeline to insert a flowmeter can be as expensive as the cost of the flowmeter itself, the ultrasonic flowmeter has enormous cost advantages. Its clamp-on mode of operation has significant safety advantages in avoiding the possibility of personnel installing flowmeters coming into contact with hazardous fluids such as poisonous, radioactive, flammable or explosive ones. Also, any contamination of the fluid being measured (e.g food substances and drugs) is avoided. The introduction of this type of flowmeter is comparatively recent and its present market share is only about 1% of flowmeters sold. In view of its distinct advantages, however, this proportion is likely to increase over the next few years.

Two different types of ultrasonic flowmeter exist which employ distinct technologies, one based on the Doppler shift and the other on transit time. In the past, the existence of these alternative technologies has not always been readily understood, and has resulted in ultrasonic technology being rejected entirely when one of these two forms has been found to be unsatisfactory in a particular application. This is unfortunate, because the two technologies have distinct characteristics and areas of application, and many situations exist where one form is very suitable and the other unsuitable. To reject both, having only tried out one, is therefore a serious mistake.

Particular care has to be taken to ensure a stable flow profile in ultrasonic flowmeter applications. It is usual to increase the normal specification of the minimum length of straight run of pipe prior to the point of measurement, expressed as a number of pipe diameters, from a figure of 10 up to 20 or in some cases even 50 diameters. Analysis of the reasons for poor performance in many instances of ultrasonic flowmeter application has shown failure to meet this stable flow profile requirement to be a significant factor.

14.10.1 Doppler shift ultrasonic flowmeter

The principle of operation of the Doppler shift flowmeter is shown in Figure 14.14. A fundamental requirement of these instruments is the presence of scattering elements within the flowing fluid which deflect the ultrasonic energy output from the transmitter such that it enters the receiver. These elements can be provided by solid particles, gas bubbles or eddies in the flowing fluid. The scattering elements cause a frequency shift between the transmitted and reflected ultrasonic energy, and measurement of this shift enables the fluid velocity to be inferred.

The instrument consists essentially of an ultrasonic transmitter–receiver pair clamped on to the outside wall of a fluid-carrying vessel. Ultrasonic energy consists of

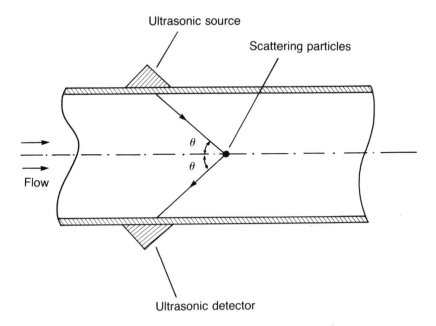

Figure 14.14 Doppler shift ultrasonic flowmeter

a train of short bursts of sinusoidal waveforms at a frequency between 0.5 MHz and 20 MHz. This frequency range is described as ultrasonic because it is outside the range of human hearing. The flow velocity, v, is given by:

$$v = \frac{c(f_t - f_r)}{2f_t \cos(\theta)} \tag{14.4}$$

where f_t and f_r are the frequencies of the transmitted and received ultrasonic waves respectively, c is the velocity of sound in the fluid being measured, and θ is the angle that the incident and reflected energy waves make with the axis of flow in the pipe. Volume flow rate is then readily calculated by multiplying the measured flow velocity by the cross-sectional area of the fluid-carrying pipe.

The electronics involved in Doppler shift flowmeters is relatively simple and therefore cheap. Ultrasonic transmitters and receivers are also relatively inexpensive, being based on piezoelectric oscillator technology. As all of its components are cheap, the Doppler shift flowmeter itself is inexpensive. The measurement accuracy obtained depends on many factors such as the flow profile, the constancy of pipe-wall thickness, the number, size and spatial distribution of scatterers, and the accuracy with which the speed of sound in the fluid is known. Consequently, accurate measurement can only be achieved by the tedious procedure of carefully calibrating the instrument in each particular flow measurement application. Otherwise measurement errors can approach $\pm 10\%$ of the reading, and for this reason Doppler shift flowmeters are often used merely as flow indicators, rather than for accurate quantification of the volume flow rate.

Versions are now available which avoid the problem of variable pipe thickness by being fitted inside the flow pipe, flush with its inner surface. An accuracy of $\pm 0.5\%$ is claimed for such devices. Other recent developments are the use of multiple-path ultrasonic flowmeters which use an array of ultrasonic elements to obtain an average velocity measurement which substantially reduces the error due to non-uniform flow profiles. There is a substantial cost penalty involved in this, however.

14.10.2 Transit time ultrasonic flowmeter

The transit time ultrasonic flowmeter is an instrument designed for measuring the volume flow rate in clean liquids or gases. It consists of a pair of ultrasonic transducers mounted along an axis aligned at an angle θ with respect to the fluid flow axis, as shown in Figure 14.15. Each transducer consists of a transmitter–receiver pair, with the transmitter emitting ultrasonic energy which travels across to the receiver on the opposite side of the pipe. These ultrasonic elements are normally piezoelectric oscillators of the same type as used in Doppler shift flowmeters. Fluid flowing in the pipe causes a time difference between the transit times of the beams travelling upstream and downstream, and measurement of this difference allows the flow velocity to be calculated. The typical magnitude of this time difference is 100 ns in a total transit time of 100 μs, and high-precision electronics are therefore needed to

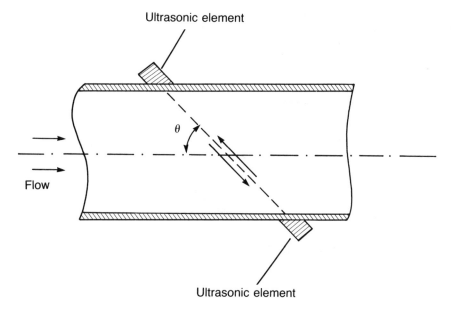

Figure 14.15 Transit time ultrasonic flowmeter

measure it. There are three distinct ways of measuring the time shift. These are direct measurement, conversion to a phase change and conversion to a frequency change. The third of these options is particularly attractive as it obviates the need to measure the speed of sound in the measured fluid which the first two methods require. A scheme applying this third option is shown in Figure 14.16. This scheme also multiplexes the transmitting and receiving functions, so that only one ultrasonic element is needed in each transducer.

The forward and backward transit times across the pipe, T_f and T_b, are given by:

$$T_f = \frac{L}{c + v\cos(\theta)} \quad T_b = \frac{L}{c - v\cos(\theta)}$$

where c is the velocity of sound in the fluid, v is the flow velocity, L is the distance between the ultrasonic transmitter and receiver, and θ is the angle of the ultrasonic beam with respect to the fluid flow axis. The time difference δT is given by:

$$\delta T = T_b - T_f = \frac{2vL\cos(\theta)}{c^2 - v^2\cos^2(\theta)}$$

This requires knowledge of c before it can be solved. However, a solution can be found much more simply if the receipt of a pulse is used to trigger the transmission of

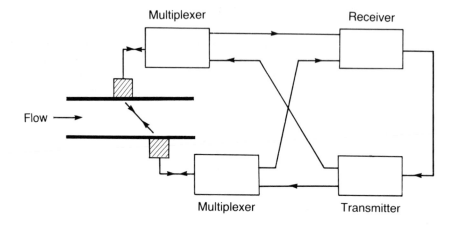

Figure 14.16 Transit time measurement system

the next ultrasonic energy pulse. Then, the frequencies of the forward and backward pulse trains are given by:

$$F_f = \frac{1}{T_f} = \frac{c - v \cos(\theta)}{L} \qquad F_b = \frac{1}{T_b} = \frac{c + v \cos(\theta)}{L}$$

If the two frequency signals are now multiplied together, the resulting beat frequency is given by:

$$\delta F = F_b - F_f = \frac{2v \cos(\theta)}{L}$$

c has now been eliminated and v can be calculated from a measurement of δF as:

$$v = \frac{L \delta F}{2 \cos(\theta)}$$

This is often known as the **sing-around flowmeter**.

Transit time flowmeters are of more general use than Doppler shift flowmeters, particularly where the pipe diameter involved is large and hence the transit time is consequently sufficiently large to be measured with reasonable accuracy. It is possible then to achieve an accuracy of ±0.5%. The instrument costs more than a Doppler shift flowmeter, however, because of the greater complexity of the electronics needed to make accurate transit time measurements.

14.11 **Cross-correlation flowmeters**

Cross-correlation flowmeters are a type of flowmeter which has not yet achieved widespread practical use in industry. Much development work is still going on, and they therefore mainly exist only as prototypes in research laboratories. However, they are included here because their use is likely to become much more widespread in the future.

Such instruments require some detectable random variable to be present in the flowing fluid. This can take forms such as velocity turbulence and temperature fluctuations. When such a stream of variables is detected by a sensor, the output signal generated consists of noise with a wide frequency spectrum.

Cross-correlation flowmeters use two such sensors placed a known distance apart in the fluid-carrying pipe and cross-correlation techniques are applied to the two output signals from these sensors. This procedure compares one signal with progressively time-shifted versions of the other signal until the best match is obtained between the two waveforms. If the distance between the sensors is divided by this time shift, a measurement of the flow velocity is obtained. A digital processor is an essential requirement to calculate the cross-correlation function, and therefore the instrument must be properly described as an intelligent one.

In practice, the existence of random disturbances in the flow is unreliable, and their detection is difficult. To answer this problem, ultrasonic cross-correlation flowmeters are under development. These use ultrasonic transducers to inject disturbances into the flow and also to detect the disturbances further downstream.

Further information about cross-correlation flowmeters can be found in Medlock (1985).

14.12 **Laser Doppler flowmeter**

This instrument gives direct measurements of flow velocity for liquids containing suspended particles flowing in a transparent pipe. Light from a laser is focused by an optical system to a point in the flow, with fiber optic cables being commonly used to transmit the light. The movement of particles causes a Doppler shift of the scattered light and produces a signal in a photodetector which is related to the fluid velocity. A very wide range of flow velocities between $10 \, \mu\text{m/s}$ and $10^5 \, \text{m/s}$ can be measured by this technique.

Sufficient particles for satisfactory operation are normally present naturally in most liquid and gaseous fluids, and the introduction of artificial particles is rarely needed. The technique is advantageous in measuring flow velocity directly rather than inferring it from a pressure difference. It also causes no interruption in the flow and, as the instrument can be made very small, it can measure velocity in confined areas. One limitation is that it measures local flow velocity in the vicinity of the focal point of the light beam, which can lead to large errors in the estimation of mean volume flow rate if the flow profile is not uniform. However, this limitation is often used constructively in applications of the instrument where the flow profile across the cross-section of a pipe is determined by measuring the velocity at a succession of points.

14.13 Intelligent flowmeters

All the usual benefits associated with intelligent instruments are potentially applicable to many types of flowmeter. However, their availability in the market place is currently very limited.

Intelligent differential pressure measuring instruments can be used to good effect in conjunction with obstruction-type flow transducers. One immediate benefit of this in the case of the commonest flow-restriction device, the orifice plate, is to extend the lowest flow measurable with acceptable accuracy down to 20% of the maximum flow value.

In positive-displacement meters, intelligence allows compensation for the thermal expansion of meter components and temperature-induced viscosity changes. Correction for variations in flow pressure is also provided for.

Intelligent electromagnetic flowmeters are also available, and these have a self-diagnosis and self-adjustment capability. The usable instrument range is typically from 3% to 100% of full-scale reading and quoted accuracy is ±0.5%. It is also normal to include a non-volatile memory to protect constants used for correcting for modifying inputs, etc., against power supply failures.

Intelligent turbine meters are able to detect their own bearing wear and also report deviations from initial calibration due to blade damage, etc. Some versions also have a self-adjustment capability.

The trend is now moving towards total flow computers which can process inputs from almost any type of transducer. Such devices allow the input of parameters like specific gravity, fluid density, viscosity, pipe diameters, thermal expansion coefficients, discharge coefficients, etc. Auxiliary inputs from temperature transducers are also catered for. After processing the raw flow transducer output with this additional data, flow computers are able to produce measurements of flow to a very high degree of accuracy.

14.14 Choice between flowmeters for particular applications

The number of relevant factors to be considered when specifying a flowmeter for a particular application is very large. These factors include the temperature and pressure of the fluid, its density, viscosity, chemical properties and abrasiveness, whether it contains particles, whether it is a liquid or gas, etc. This narrows the field to a subset of instruments which are physically capable of making the measurement. Next, the required performance factors of accuracy, rangeability, acceptable pressure drop, output signal characteristics, reliability and service life must be considered. Finally, the economic viability must be assessed and this must take account not only of purchase cost, but also of reliability, installation difficulties, maintenance requirements and service life.

Where only a visual indication of flow rate is needed, the most popular instrument is the variable area meter. This requirement is common enough for this type of instrument to account for 20% of all flowmeters sold. Where a flow measurement in the form of an electrical signal is required, the choice of available instruments is very large. The orifice plate is extremely common for such purposes and accounts for almost 50% of the instruments currently in use in industry. Beyond this very general guidance, the reader is referred elsewhere for more specific information (Furness and Heritage 1986, Instrument Society of America 1988).

References and further reading

Benedict, R. P. (1984) *Fundamentals of Temperature, Pressure and Flow Measurement*, Wiley: New York.

Bentley, J. P. (1983) *Principles of Measurement Systems*, Longman: London.

Britton, C. and Mesnard, D. (1982) 'A performance summary of averaging pitot-type primaries', *Measurement and Control*, **15**, 341–50.

Furness, R. A. and Heritage, J. E. (1986) 'Commercially available flowmeters and future trends', *Measurement and Control*, **19**, 25–35.

Instrument Society of America (1988) *Flowmeters – A comprehensive survey and guide to selection*, Pittsburgh.

King, N. W. (1988) 'Multi-phase flow measurement at NEL', *Measurement and Control*, **21**, 237–9.

Medlock, R. S. (1983) 'The techniques of flow measurement', *Measurement and Control*, **16**, 9–13.

Medlock, R. S. (1985) 'Cross-correlation flow measurement', *Measurement and Control*, **18**, 293–8.

Medlock, R. S. and Furness, R. A. (1990) 'Mass flow measurement – a state of the art review', *Measurement and Control*, **23**, 100–13.

15 Level measurement

A wide variety of instruments are available for measuring the level of liquids and in some cases solids which are in the form of powders or small particles. In most applications of such instruments, only a rough indication of level is needed, and so the accuracy demands are not high.

15.1 Dipsticks

15.1.1 Ordinary dipstick

The ordinary dipstick is perhaps the simplest and cheapest device available for measuring liquid level. It consists of a metal bar, on which a scale is etched as shown in Figure 15.1(a), which is fixed at a known position in the liquid-containing vessel. A level measurement is made by removing the instrument from the vessel and reading off how far up the scale the liquid has wetted. As a human operator is required to remove and read the dipstick, this method is limited to use in relatively small and shallow vessels.

15.1.2 Optical dipstick

The optical dipstick, illustrated in Figure 15.1(b), allows a reading to be obtained without removing the dipstick from the vessel, and so is applicable to larger, deeper tanks. Light from a source is reflected from a mirror, passes round the chamfered end of the dipstick, and enters a light detector after reflection by a second mirror. When the chamfered end comes into contact with liquid, its internal reflection properties are altered and light no longer enters the detector. By using a suitable mechanical drive system to move the instrument up and down and measure its position, the liquid level can be monitored.

15.2 Float systems

Measuring the position of a float on the surface of a liquid by means of a suitable transducer is another fairly simple and cheap method of liquid level measurement. The

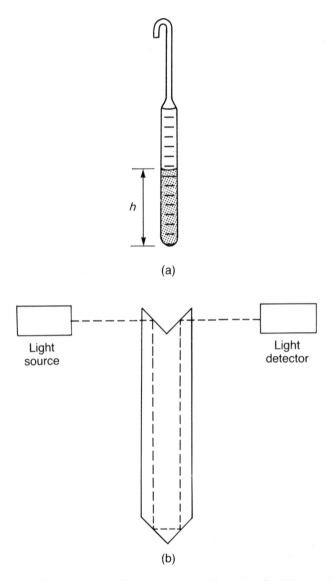

(a)

(b)

Figure 15.1 (a) Simple dipstick; (b) optical dipstick

system using a potentiometer, shown in Figure 15.2, is very common, and is well known for its application to monitoring the level in motor vehicle fuel tanks.

An alternative system, which is used in greater numbers, is called the **float and tape gauge** (or **tank gauge**). This has a tape attached to the float which passes round a pulley situated vertically above the float. The other end of the tape is attached to either a counterweight or a negative-rate counterspring. The amount of rotation of

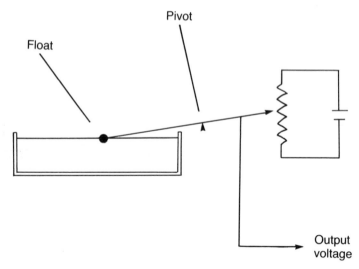

Figure 15.2 Float system

the pulley, measured by either a synchro or a potentiometer, is then proportional to the liquid level.

These two essentially mechanical systems of measurement can achieve quite high levels of accuracy, but their maintenance requirements are always high.

15.3 **Pressure measuring devices (hydrostatic systems)**

The hydrostatic pressure due to a liquid is directly proportional to its depth and hence to the level of its surface. Several instruments which use this principle for measuring liquid level are available and are widely used in many industries, particularly in harsh chemical environments.

In the case of open-topped vessels (or covered ones which are vented to the atmosphere), the level can be measured by inserting an appropriate pressure transducer at the bottom of the vessel, as shown in Figure 15.3(a). The liquid level, h, is then related to the measured pressure, P, according to:

$$h = \frac{P}{\rho g}$$

where ρ is the liquid density and g is the acceleration due to gravity.

One source of error in this method can be imprecise knowledge of the liquid density. This can be a particular problem in the case of liquid solutions and mixtures (especially hydrocarbons), and in some cases only an estimate of density is available. Even with single liquids, the density is subject to variation with temperature, and

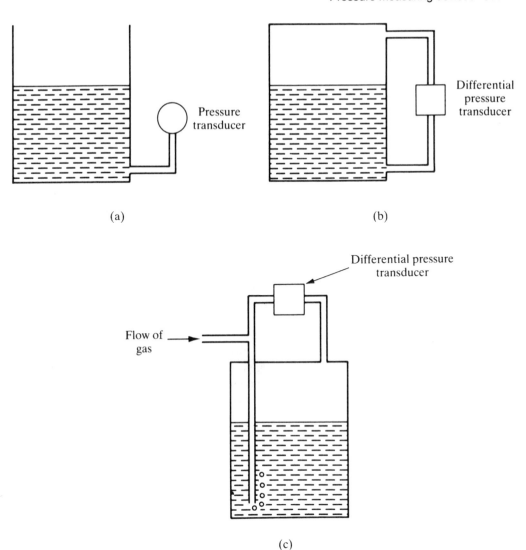

Figure 15.3 (a) Open-topped vessel; (b) sealed vessel; (c) bubbler unit

therefore temperature measurement may be required if very accurate level measurements are needed.

Where liquid-containing vessels are totally sealed, the liquid level can be calculated
by measuring the differential pressure between the top and bottom of the tank, as
shown in Figure 15.3(b). The differential pressure transducer used is normally a
standard diaphragm type, although silicon-based microsensors (see section 14.1) are

being used in increasing numbers. The liquid level is related to the differential pressure measured, δP, according to:

$$h = \frac{\delta P}{\rho g}$$

The same comments as for the case of the open vessel apply regarding uncertainty in the value of ρ. An additional problem which can occur is an accumulation of liquid on the side of the differential pressure transducer which is measuring the pressure at the top of the vessel. This can arise because of temperature fluctuations, which allow liquid alternately to vaporize from the liquid surface and then condense in the pressure tapping at the top of the vessel. The effect of this on the accuracy of the differential pressure measurement is severe, but the problem is easily avoided by placing a drain pot in the system. This should of course be drained regularly.

A final pressure-related system of level measurement is the bubbler unit shown in Figure 15.3(c). This uses a dip pipe which reaches to the bottom of the tank and is purged free of liquid by a steady flow of gas through it. The rate of flow is adjusted until gas bubbles are just seen to emerge from the end of the tube. The pressure in the tube, measured by a pressure transducer, is then equal to the liquid pressure at the bottom of the tank. It is important that the gas used is inert with respect to the liquid in the vessel. Nitrogen or sometimes just air is suitable in most cases. Gas consumption is low, and a cylinder of nitrogen may typically last for 6 months. The method is suitable for measuring the liquid pressure at the bottom of both open and sealed tanks. It is particularly advantageous in avoiding the large maintenance problem associated with leaks at the bottom of tanks at the site of the pressure tappings required by alternative methods.

15.4 **Capacitive devices**

Capacitive devices are now widely used for measuring the level of both liquids and solids in powdered or granular form. They are suitable for use in extreme conditions measuring liquid metals (high temperatures), liquid gases (low temperatures), corrosive liquids (acids, etc.) and high-pressure processes. Two versions are used according to whether the measured substance is conducting or not. For non-conducting (less than $0.1 \, \mu$mho/cm^3) substances, two bare-metal capacitor plates in the form of concentric cylinders are immersed in the substance, as shown in Figure 15.4. The substance behaves as a dielectric between the plates according to the depth of the substance. For concentric cylinder plates of radius a and b ($b > a$), and total height L, the depth of the substance h is related to the measured capacitance C by:

$$h = \frac{C \log_e(b/a) - 2\pi\varepsilon_0}{2\pi\varepsilon_0(\varepsilon - 1)} \tag{15.1}$$

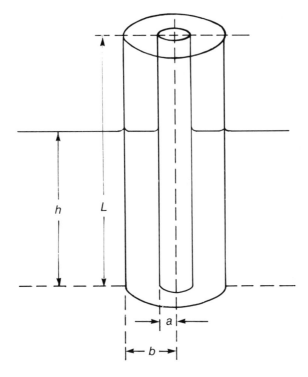

Figure 15.4 Capacitive level sensor

where ε is the relative permittivity of the measured substance and ε_0 is the permittivity of free space. The capacitance is measured by one of the methods discussed in chapter 7.

In the case of conducting substances, exactly the same measurement techniques are applied, but the capacitor plates are encapsulated in an insulating material. The relationship between C and h in Equation (15.1) then has to be modified to allow for the dielectric effect of the insulator.

Capacitive devices are useful in many applications, but become inaccurate if the measured substance is prone to contamination by agents which change its dielectric constant. Ingress of moisture into powders is one such example of this.

15.5 **Ultrasonic level gauge**

The principle of the ultrasonic level gauge is illustrated in Figure 15.5. Energy from an ultrasonic source above the liquid is reflected back from the liquid surface into an ultrasonic energy detector. Measurement of the time of flight allows the liquid level to be inferred. In alternative versions, the ultrasonic source is placed at the bottom of

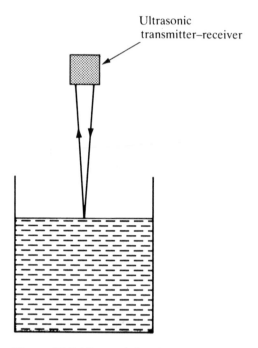

Ultrasonic
transmitter–receiver

Figure 15.5 Ultrasonic level gauge

the vessel containing the liquid, and the time of flight between emission, reflection off the liquid surface and detection back at the bottom of the vessel is measured.

Ultrasonic techniques are especially useful in measuring the position of the interface between two immiscible liquids contained in the same vessel, or measuring the sludge or precipitate level at the bottom of a liquid-filled tank. In either case, the method employed is to fix the ultrasonic transmitter–receiver transducer at a known height in the upper liquid, as shown in Figure 15.6. This establishes the level of the liquid/liquid or liquid/sludge level in absolute terms.

When using ultrasonic instruments, it is essential that proper compensation is made for the working temperature if this differs from the calibration temperature. The speed of ultrasound through air varies with temperature at the rate of 0.607 m/s per °C. The speed of ultrasound also has a small sensitivity to humidity, air pressure and carbon dioxide concentration, but these factors are usually insignificant.

Temperature compensation can be achieved in two ways. First, the operating temperature can be measured and an appropriate correction made. Secondly, and preferably, a comparison method can be used in which the system is calibrated each time it is used by measuring the transit time of ultrasonic energy between two known reference points. This second method takes account of variations in humidity, pressure and carbon dioxide concentration as well as providing temperature compensation.

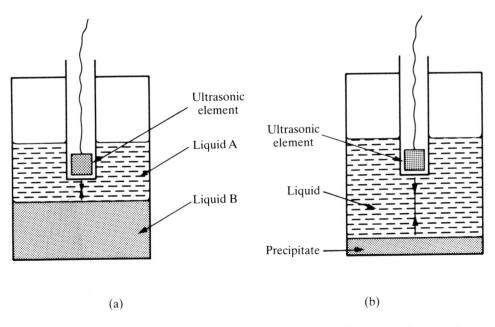

Figure 15.6 (a) Measuring liquid/liquid interface; (b) measuring liquid/precipitate interface

15.6 **Radiation methods**

The radiation method is an expensive technique which uses a radiation source and detector system located in a liquid-filled tank in the manner shown in Figure 15.7. The absorption of both beta rays and gamma rays varies with the amount of liquid between the source and detector, and hence is a function of liquid level. Caesium-137 is a gamma-ray source which is commonly used for this purpose. The radiation level measured by the detector, I, is related to the length of liquid in the path, x, according to:

$$I = I_0 \exp(-\mu\rho x) \tag{15.2}$$

where I_0 is the intensity of radiation which would be received by the detector in the absence of any liquid, μ is the mass absorption coefficient for the liquid and ρ is the mass density of the liquid.

In the arrangement shown in Figure 15.7, the radiation follows a diagonal path across the liquid, and therefore some trigonometrical manipulation has to be carried out to determine the liquid level, h, from x. In some applications, the radiation source can be located in the centre of the bottom of the tank, with the detector vertically above it. Where this is possible, the relationship between the radiation detected and the liquid level is obtained by directly substituting h in place of x in Equation (15.2).

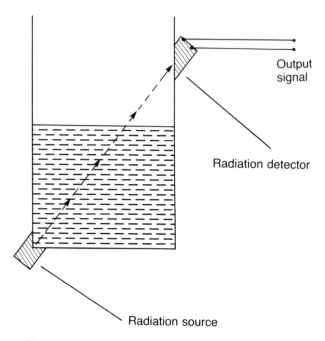

Figure 15.7 Using a radiation source to measure level

Apart from use with liquids at normal temperatures, this method is commonly used for measuring the level of hot, liquid metals.

The non-invasive nature of this technique in using a source and detector system outside the tank is particularly attractive. However, because of the obvious dangers associated with using radiation sources, very strict safety regulations have to be satisfied when applying this technique.

Very low-activity radiation sources are used in some systems to overcome safety problems but the system is then sensitive to background radiation and special precautions have to be taken regarding the provision of adequate shielding.

15.7 **Vibrating level sensor**

The principle of the vibrating level sensor is illustrated in Figure 15.8. The instrument consists of two piezoelectric oscillators fixed to the inside of a hollow tube which generate flexural vibrations in the tube at its resonant frequency. The resonant frequency of the tube varies according to the depth of its immersion in the liquid. A phase-locked loop circuit is used to track these changes in resonant frequency and adjust the excitation frequency applied to the tube by the piezoelectric oscillators. Liquid level measurement is therefore obtained in terms of the output frequency of the oscillator when the tube is resonating.

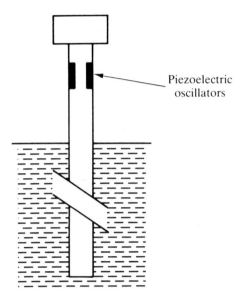

Figure 15.8 Vibrating level sensor

15.8 **Hot-wire elements**

Figure 15.9 shows a level measurement system which uses a series of hot-wire elements placed at regular intervals along a vertical line up the side of a tank. The heat transfer coefficient of such elements differs substantially depending upon whether the element is immersed in air or in the liquid in the tank. Consequently, elements in the liquid have a different temperature and therefore a different resistance to those in air. This method of level measurement is a simple one, but the measurement resolution is limited to the distance between sensors. Carbon resistors are sometimes used in place of hot-wire elements.

15.9 **Radar methods**

Level measurement using microwave radar has traditionally been a very expensive technique but recent developments in microprocessor technology have brought about sharp reductions in equipment cost. Radar techniques can provide successful level measurement in applications which are otherwise very difficult, such as measurement in closed tanks, measurement where the liquid is turbulent, and measurement in the presence of obstructions and steam condensate.

The technique involves directing a constant amplitude, frequency-modulated microwave signal at the liquid surface. A receiver measures the phase difference

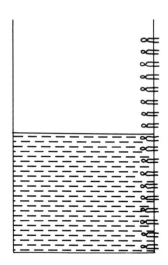

Figure 15.9 Hot-wire-element level sensor

between the reflected signal and the original signal transmitted directly through air to it, as shown in Figure 15.10. This measured phase difference is linearly proportional to the liquid level. The system is similar in principle to ultrasonic level measurement, but has the important advantage that the transmission time of radar through air is almost totally unaffected by ambient temperature and pressure fluctuations.

As the microwave frequency is within the band used for radio communications, strict conditions on amplitude levels have to be satisfied, and the appropriate licences have to be obtained.

15.10 **Laser methods**

One laser-based method is the **reflective level sensor**. This sensor uses light from a laser source which is reflected off the surface of the liquid being measured into a line array of charge-coupled devices, as shown in Figure 15.11. Only one of these will sense light, according to the level of the liquid. Recently, a further laser-based technique has been announced. This operates on the same general principles as the radar method described above but uses laser-generated pulses of infrared light directed at the liquid surface. It is immune to environmental conditions and can be used with sealed vessels provided that a glass window is provided in the top of the vessel.

15.11 **Fiber optic level sensors**

The fiber optic cross-talk sensor, as described in Chapter 9, is one example of a fiber optic sensor which can be used to measure liquid level.

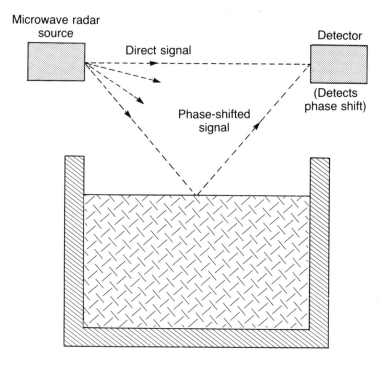

Figure 15.10 Radar level detector

Another light-loss fiber optic level sensor is the simple loop sensor shown in Figure 15.12. The amount of light loss depends on the proportion of cable which is submerged in the liquid. This effect is magnified if the alternative arrangement shown in Figure 15.13 is used. Light is reflected from an input fiber, round a prism, and then into an output fiber. Light is lost from this path into the liquid according to the depth of liquid surrounding the prism.

15.12 **Thermography**

Thermal imaging instruments, as discussed in Chapter 12, are a further means of detecting the level of liquids in tanks. Such instruments are capable of discriminating temperature differences as small as 0.1 °C. Differences of this magnitude will normally be present at the interface between the liquid, which tends to remain at a constant temperature, and the air above, which constantly fluctuates in temperature by small amounts.

The upper level of solids stored in hoppers is often detectable on the same principles.

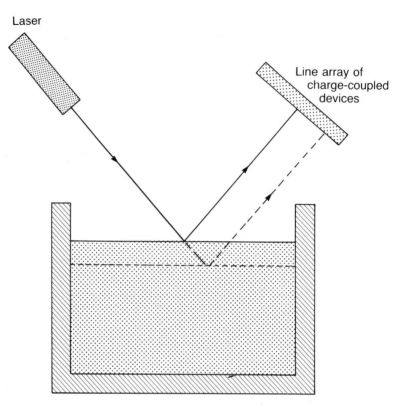

Figure 15.11 Reflective level sensor

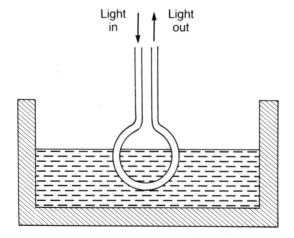

Figure 15.12 Loop level sensor

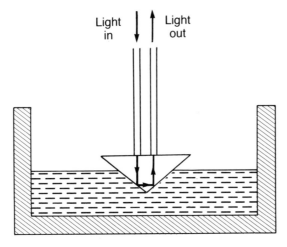

Figure 15.13 Prism level sensor

15.13 **Intelligent level measuring instruments**

Only certain types of level measuring device are suitable for use as intelligent instruments. The pressure measuring devices (section 15.3) are obvious candidates for inclusion within intelligent level measuring instruments, and versions claiming ±0.05% accuracy are now on the market. These instruments can also carry out additional functions such as providing automatic compensation for liquid density variations. Apart from this, there is little activity at present directed towards developing other forms of intelligent level measuring systems. This is mainly the result of market forces, as most applications of liquid measurement do not demand high degrees of accuracy but require instead just a rough indication of level.

16 Dimension measurement

16.1 Introduction

Dimension measurement includes the measurement of length, height, depth (of holes and slots) and angles of components. For many such measurements, it is necessary to have a reference flat plane on which the components being measured are located. This is provided by a surface plate or table as described below.

The range of instruments available for measuring length includes the steel tape (to 30 m), ultrasonic rule (to 10 m), steel rule (to 1 m), standard calipers (to 600 mm), vernier calipers (to 200 mm) and micrometer (to 50 mm). Gauge blocks and length bars, although primarily intended for calibration duties, are also used for the measurement of larger lengths when very high accuracy is required. Height and depth are measured by the height gauge, depth gauge or dial gauge. Angles are normally measured by an angle protractor, a bevel protractor or a spirit level. Further information on dimension measurement can be found in Anthony (1986) and Hume (1970).

16.2 Surface planes and tables

A flat and level reference plane is often an essential component in dimension measurement, especially where the line of measurement and the dimension being measured are not coincident. Such reference planes are available in a range of standard sizes, the smallest having nominal dimensions of 100 mm × 160 mm and the largest 1600 mm × 2500 mm. The smaller sizes exist as a surface plate resting on a supporting table, whereas the larger sizes take the form of free-standing tables, as shown in Figure 16.1. Larger sizes have a projection at the edge to facilitate the clamping of components. They are normally used in conjunction with box cubes and vee blocks (see Figure 16.2) which locate components in a fixed position. The plate or table normally has three feet,* with provision for levelling to make the surface exactly horizontal. In modern tables, granite has tended to supersede iron as the preferred material for the plate, although iron plates are still available. Granite is ideal for this purpose as it does not corrode, is dimensionally very stable and does not form burrs

*Plates with dimensions of 1 m × 1.6 m and larger often have more than three feet.

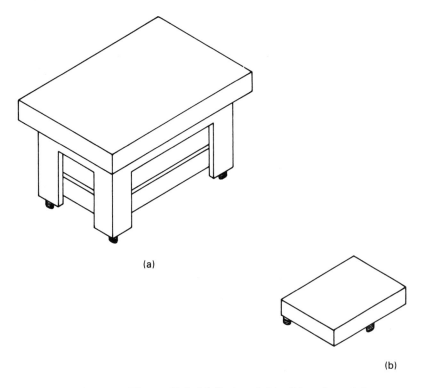

(a)

(b)

Figure 16.1 (a) Surface table; (b) surface plate

when damaged. Iron plates, on the other hand, are prone to rusting and susceptible to damage: this results in burrs on the surface which interfere with measurement procedures.

Both forms of plate are available in four standard grades, from grade 0 to grade 3. Grade 0 is the best and is used for calibration purposes. Grades 1 to 3 are recommended for inspection, marking out and lower-grade marking out respectively. Flatness is defined in terms of the distance between two parallel planes which just contain all points in the surface. Standards of flatness (defined in BS 817 (1988)) vary according to the size of the plate or table. For a $2\,\text{m} \times 1\,\text{m}$ table, the maximum permitted deviations from flatness for grades 0 to 3 are $8.5\,\mu\text{m}$, $17\,\mu\text{m}$, $34\,\mu\text{m}$, $68\,\mu\text{m}$ respectively.

16.3 **Rules and tapes**

Rules and tapes are the standard way of measuring larger dimensions. Steel rules are generally only available to measure dimensions up to 1 m. Beyond this, steel tapes and the more recent ultrasonic rule are used. Further details about these are given below.

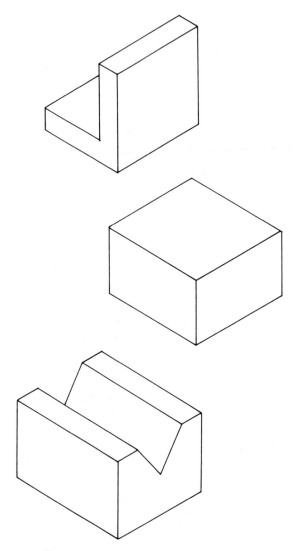

Figure 16.2 Box cubes and vee blocks

16.3.1 **Steel rule**

The steel rule is undoubtedly the simplest instrument available for measuring length. Rules are manufactured according to quality standards (see BS 4372 (1968)) which define the ruling accuracy, squareness of the end and edge straightness. The best rules have rulings at 0.05 mm intervals and a measurement resolution of 0.02 mm. When used by placing the rule against an object, the measurement accuracy is much dependent upon the skill of the human measurer and, at best, the inaccuracy is likely to be at least ±0.5%.

16.3.2 **Steel tape**

Retractable steel tapes are another well-known instrument. The end of the tape is usually provided with a flat hook which is loosely fitted so as to allow for automatic compensation of the hook thickness when the rule is used for both internal and external measurements. Again, measurement accuracy is governed by human skill, but, with care, the measurement inaccuracy can be made to be as low as $\pm 0.01\%$ of full-scale reading. Formal specifications for measuring tapes are defined in ISO 8322 (1989) and BS 4035 (1966).

16.3.3 **Ultrasonic rule**

The ultrasonic rule is a recent addition to the measurement scene. It consists of an ultrasonic energy source, an ultrasonic energy detector and battery-powered, electronic circuitry housed within a hand-held box, as shown in Figure 16.3. Both source and detector often consist of the same type of piezoelectric crystal excited at a typical frequency of 40 kHz. Energy travels from the source to a target object and is then reflected back into the detector. The time of flight of this energy is measured and this is converted into a distance reading by the enclosed electronics. Maximum measurement inaccuracy of $\pm 1\%$ of full-scale reading is claimed. This is only a modest level of accuracy which is of little use for most engineering measurements. However, it is sufficient for such purposes as measuring rooms by estate agents prior to producing sales literature, where the ease and speed of making measurements are of great value.

A fundamental problem in the use of ultrasonic energy of this type is the limited measurement resolution (7 mm) imposed by the 7 mm wavelength of sound at this frequency. Further problems are caused by the variation in the speed of sound with humidity (variations of $\pm 0.5\%$ are possible) and the temperature-induced variation of 0.2% per °C. Therefore, the conditions of use must be carefully controlled if the claimed accuracy figure is to be met.

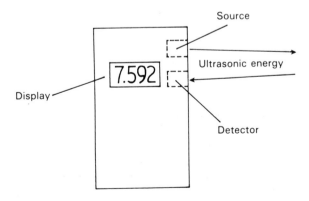

Figure 16.3 Ultrasonic rule

16.4 **Calipers**

Calipers are generally used in situations where the measurement of dimensions using only a rule or tape is inadequate. Two types exist, the standard caliper and the vernier caliper.

16.4.1 **Standard caliper**

Figure 16.4 shows two types of standard caliper. These are used to transfer the measured dimension from the workpiece to a steel rule. This avoids the necessity of aligning the end of the rule exactly with the edge of the workpiece and reduces the measurement inaccuracy by a factor of two. In the basic caliper, careless use can allow the setting of the caliper to be changed during transfer from the workpiece to the rule. Hence, the spring-loaded type, which prevents this happening, is preferable.

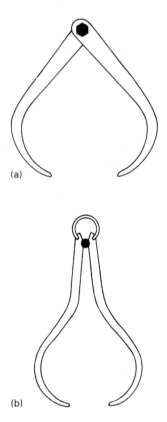

(a)

(b)

Figure 16.4 (a) Standard caliper (b) spring-loaded caliper

16.4.2 **Vernier caliper**

Vernier calipers (see BS 887 (1982), ISO 6906 (1984)), shown in Figure 16.5, are effectively the combination of standard calipers with a steel rule. The main body of the instrument includes the main scale with a fixed anvil at one end. This carries a sliding anvil which is provided with a second, vernier scale. This second scale is shorter than the main scale and is divided into units which are slightly smaller than the main scale units but related to them by a fixed factor. Determination of the point where the two scales coincide enables very accurate measurements to be made, with typical inaccuracy levels down to $\pm 0.01\%$.

Figure 16.6 shows details of a typical combination of main and vernier scales. The main scale is ruled into 1 mm units. The vernier scale is 49 mm long and divided into 50 units, thereby making each unit 0.02 mm smaller than the main scale units. Each group of five units on the vernier scale thus differs from the main scale by 0.1 mm and the numbers marked on the scale thus refer to these larger units of 0.1 mm. In the particular position shown in Figure 16.6, the zero on the vernier scale is indicating a measurement between 8 and 9 mm. Both scales coincide at a position of 6.2 (large units). This defines the interval between 8 and 9 mm to be $6.2 \times 0.1 = 0.62$ mm, i.e. the measurement is 8.62 mm.

Intelligent digital calipers are now available which give a measurement resolution of 0.01 mm and an accuracy of ± 0.03 mm. These have automatic compensation for wear, and hence calibration checks have to be very infrequent. In some of these instruments, the digital display can be directly interfaced to an external computer monitoring system.

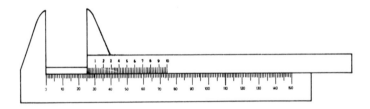

Figure 16.5 Vernier caliper

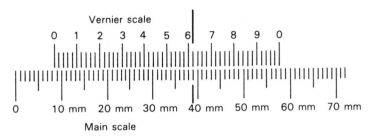

Figure 16.6 Details of vernier caliper scale

16.5 **Micrometers**

Micrometers provide a means of measuring dimensions to quite a high standard of accuracy. Different forms provide measurement of both internal and external dimensions of components and holes, slots, etc., within the components.

   In the standard micrometer, shown in Figure 16.7(a), measurement is made between two anvils, one fixed and one which is moved along by the rotation of an accurately machined screw thread. One complete rotation of the screw typically moves the anvil by a distance of 0.5 mm. Such movements of the anvil are measured using a scale marked with divisions every 0.5 mm along the barrel of the instrument. A scale marked with 50 divisions is etched around the circumference of the spindle holder: each division therefore corresponds to an axial movement of 0.01 mm. Assuming that the user is able to judge the position of the spindle on this

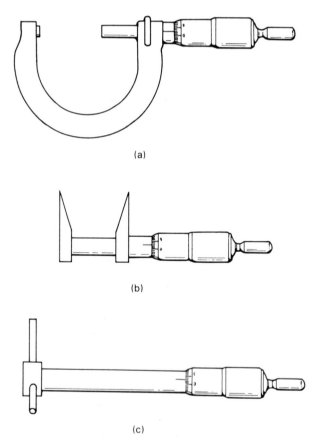

(a)

(b)

(c)

Figure 16.7 (a) Standard (external) micrometer; (b) internal micrometer; (b) bore micrometer

circumferential scale against the datum mark to within one-fifth of a division, a measurement resolution of 0.002 mm is possible. Such instruments are manufactured to high standards of accuracy, and typical specifications (see BS 870 (1950), ISO 3611 (1978)) might be: flatness of anvils, ±0.001 mm; parallelism of anvils, ±0.003 mm; screw traverse, ±0.003 mm; alignment error, 0.05 mm max.

The most common measurement ranges are either 0–25 mm or 25–50 mm, with inaccuracy levels down to ±0.003%. However, a whole family of micrometers is available, where each has a measurement span of 25 mm, but with the minimum distance measured varying from 0 mm up to 575 mm. Thus, the last instrument in this family measures the range from 575 to 600 mm. Some manufacturers also provide micrometers with two or more interchangeable anvils, which extends the span measurable with one instrument to between 50 and 100 mm according to the number of anvils supplied. Therefore, an instrument with four anvils might be used for instance to measure the range from 300 to 400 mm by making appropriate changes to the anvils.

An alternative form of micrometer (BS 959 (1950)) (see Figure 16.7(b)) is able to measure internal dimensions such as hole diameters. In the case of measuring holes, micrometers are inaccurate if there is any ovality in the hole, unless the diameter is measured at several points. An alternative solution to this problem is to use a special type of instrument known as a bore micrometer (Figure 16.7(c)). In this, three probes move out radially from the body of the instrument as the spindle is turned. These probes make contact with the sides of the hole at three equidistant points, thus averaging out any ovality.

Intelligent micrometers in the form of the electronic digital micrometer are now available. These have a self-calibration capability and a digital readout, with a measurement resolution of 0.001 mm (1 micron).

16.6 Gauge blocks (slip gauges) and length bars

Gauge blocks, also known as slip gauges, consist of rectangular blocks (see Figure 16.8) of hardened steel which have flat and parallel end faces. These faces are machined to very high standards of accuracy in terms of their surface finish and flatness. The purpose of gauge blocks is to provide a means of checking whether a particular dimension in a component is within the allowable tolerance rather than actually measuring what the dimension is. To do this, a number of gauge blocks are joined together to make up the required dimension to be checked. Apart from such inspection duties, higher-grade gauge blocks are also used to provide a reference standard in the calibration of other dimension-measuring instruments.

Gauge blocks are available in boxed sets containing a range of block sizes, which allows any dimension up to 200 mm to be constructed by joining together an appropriate number of blocks. Whilst 200 mm is the maximum dimension that should be set up with gauge blocks alone, they can be used in conjunction with length bars to set up much greater standard dimensions. Blocks are joined by 'wringing', a procedure

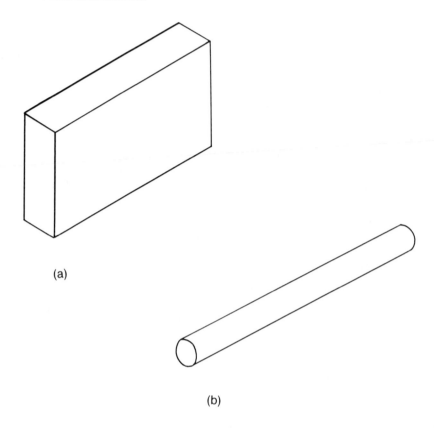

(a)

(b)

Figure 16.8 (a) Gauge block; (b) length bar

in which the two end faces are rotated slowly against each other. This removes the air film and allows adhesion to develop by intermolecular attraction. Adhesion is so good in fact that, if groups of blocks were not separated within a few hours, the molecular diffusion process would continue to the point where the blocks would be permanently welded together. The interblock gap resulting from wringing has been measured as typically $0.001 \, \mu m$, which is effectively zero. Thus, any number of blocks can be joined without creating any significant measurement error.

Gauge blocks are available in five grades of accuracy known as calibration, 00, 0, 1 and 2. Grades 1 and 2 are used for normal production inspection measurements. The length, end-face-flatness and end-face-parallelism tolerances allowed for gauge blocks, defined according to BS 4311 (1968)* for a 100 mm long block, are listed in Table 16.1.

The larger positive tolerance for grades 1 and 2 blocks is provided to allow for the relatively high level of wear which occurs immediately after a new block is put into

*ISO 3650 (1978) defines similar, though not identical, standards.

Table 16.1

	Length (μm)	Flatness (μm)	Parallelism (μm)
Calibration grade	±0.5	±0.05	±0.1
Grade 00	±0.15	±0.05	±0.1
Grade 0	±0.25	±0.1	±0.15
Grade 1	+0.6 / −0.3	±0.15	±0.25
Grade 2	+1.4 / −1.0	±0.25	±0.35

Table 16.2

Size or series	Increment	No. of pieces
1.0005	—	1
1.001–1.009	0.001	9
1.01–1.49	0.01	49
0.5–24.5	0.5	49
25–100	25	4
		112

use. It should be noted that the length tolerances of blocks wrung together are cumulative. Thus the length tolerance of five grade 2 blocks wrung together is $(+7\,\mu m, -5\,\mu m)$.

A typical set of gauge blocks allows any dimension between 3.0 and 200 mm to be built up in steps of 0.0005 mm. By careful choice of block combination, it is usually possible to construct any dimension using not more than five blocks. Consider the M112/1 set shown in Table 16.2, which consists of 112 blocks.

The procedure for choosing appropriate blocks to make up a particular length is to consider each decimal point in turn, as follows:

1. Choose the thinnest block to satisfy the last decimal place.
2. Choose the next block to satisfy the next-to-last decimal place.
3. Continue in this fashion until the full length required is constructed.

For example, to construct a length of 89.4365 mm, the following five blocks would be used:

$$
\begin{array}{r}
1.0005 \\
1.006 \\
1.43 \\
11.0 \\
\underline{75.0} \\
89.4365
\end{array}
$$

It is fairly common practice with blocks of grades 0, 1 and 2 to include an extra pair of 2 mm thick blocks in the set made from wear-resisting tungsten carbide. These are marked with a letter P and are designed to protect the other blocks from wear during use. Where such protector blocks are used, due allowance has to be made for their thickness (4 mm) in calculating the sizes of block needed to make up the required length.

A precaution to be followed when using gauge blocks is to avoid handling them more than is necessary. The length of a bar which was 100 mm long at 20 °C would increase to 100.02 mm at 37 °C (body temperature). Hence, after wringing bars together, they should be left to stabilize back to ambient room temperature before use. This might be several hours if the blocks have been handled to any significant extent.

Where a dimension greater than 200 mm is required, gauge blocks are used in conjunction with **length bars**. Length bars consist of straight, hardened, high-quality steel bars of a uniform 22 mm diameter and come in a range of lengths between 100 and 1200 mm. They are available in four grades of accuracy: reference, calibration, grade 1 and grade 2 (as defined by BS 5317 (1976)). Reference and calibration grades have accurately flat end faces, which allows a number of bars to be wrung together to obtain the required standard length. Bars of grades 1 and 2 have threaded ends, which allows them to be screwed together. Grade 2 bars are used for general measurement duties, with grade 1 bars being reserved for inspection duties. By combining length bars with gauge blocks, any dimension up to about 2 m can be set up with a resolution of 0.0005 mm.

The length tolerances of a 200 mm length bar are given by:

Grade 1: $^{+1.4}_{-0.6} \times (0.2 + 0.004L)\,\mu$m, i.e. $+1.4\,\mu$m or $-0.6\,\mu$m

Grade 2: $^{+1.4}_{-0.6} \times (0.4 + 0.006L)\,\mu$m, i.e. $+2.2\,\mu$m or $-1.0\,\mu$m.

where L is the length of the bar in millimetres.

As for gauge blocks, the larger positive tolerance for grades 1 and 2 is provided to allow for the high level of wear occurring immediately after new bars are put into use. These length tolerances are only valid if the bar is horizontal and at 20 °C. Horizontal alignment is obtained by mounting the bars at their 'Airy' points. The Airy points are at a distance of $0.2117L$ from the ends of the bar and are marked by circumferential lines.

16.7 **Height and depth measurement**

The height of objects and the depth of holes, slots, etc., are measured by the height gauge and depth gauge respectively. A dial gauge is often used in conjunction with these instruments to improve measurement accuracy. All these instruments are discussed in more detail below.

16.7.1 **Height and depth gauges**

The height gauge (see BS 1643 (1983)), shown in Figure 16.9(a), effectively consists of a vernier caliper mounted on a flat base. Measurement accuracy down to ±0.015% is possible. The depth gauge (Figure 16.9(b)) is a further variation on the standard vernier caliper principle and has the same measurement accuracy as the height gauge.

In practice, certain difficulties can arise in the use of these instruments where either the base of the instrument is not properly located on the measuring table or the point of contact between the moving anvil and the workpiece is unclear. In such cases, a dial gauge, which has a clearly defined point of contact with the measured object, is used in conjunction with the height or depth gauge to avoid these possible sources of error.

These instruments can also be obtained in intelligent versions which give a digital display and have self-calibration capabilities.

16.7.2 **Dial gauge**

The dial gauge (see BS 907 (1965), ISO/R 463 (1965)), shown in Figure 16.10, consists of a spring-loaded probe which drives a pointer around a circular scale via rack and pinion gearing. Typical measurement resolution is 0.01 mm. When used to measure the height of objects, it is clamped in a retort stand and a measurement taken of the height of the unknown component. Then it is put in contact with a height gauge (Figure 16.11) which is adjusted until the reading on the dial gauge is the same. At this stage, the height gauge is set to the height of the object. The dial gauge is also used in conjunction with the depth gauge in an identical manner. (Gauge blocks can be used instead of height/depth gauges in such measurement procedures if greater accuracy is required.)

16.8 **Angle measurement**

The angle between adjoining faces, etc., of a component is a further dimensional parameter where a measurement requirement sometimes exists. Three instruments are used to measure this (the protractor and two forms of spirit level) as discussed below.

16.8.1 **Protractors**

The simplest form of protractor is the type shown in Figure 16.12(a). This is of limited usefulness, but can be modified into a more useful form, as shown in Figure 16.12(b). This angle protractor consists of two straight edges, one of which is able to rotate with respect to the other. A circular scale attached to one of the straight edges rotates inside a fixed circular housing attached to the other straight edge. The relative angle between the two straight edges in contact with the component being measured is determined by the position of the moving scale with respect to a reference mark on

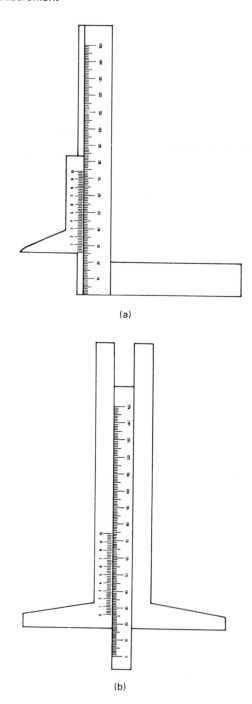

(a)

(b)

Figure 16.9 (a) Height gauge; (b) depth gauge

Figure 16.10 Dial gauge

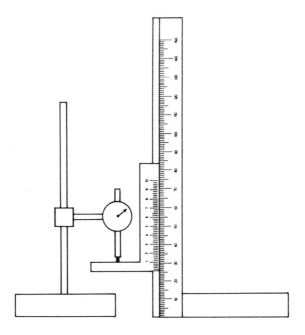

Figure 16.11 Use of dial gauge in conjunction with height gauge

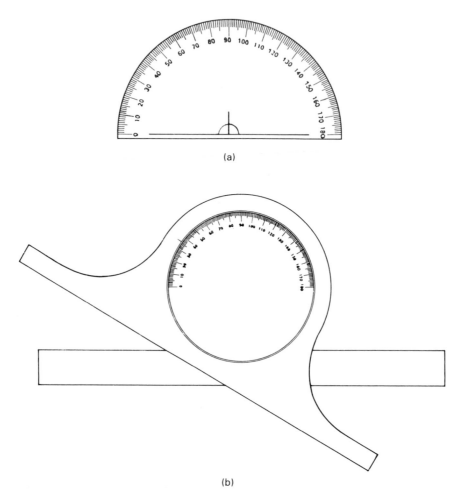

(a)

(b)

Figure 16.12 Angle measurement: (a) simple protractor; (b) angle protractor

the fixed housing. With this type of instrument, measurement inaccuracy is at least ±1%. An alternative form, the bevel protractor (see BS 1685 (1951)), is similar to the angle protractor but it has a vernier scale on the fixed housing. This allows the inaccuracy level to be reduced to ±10 minutes of arc.

16.8.2 Spirit level

The spirit level shown in Figure 16.13 is an alternative angle measuring instrument. It consists of a standard spirit level attached to a rotatable circular scale which is mounted inside an accurately machined square frame. When placed on the sloping surfaces of components, rotation of the scale to centralize the bubble in the spirit level

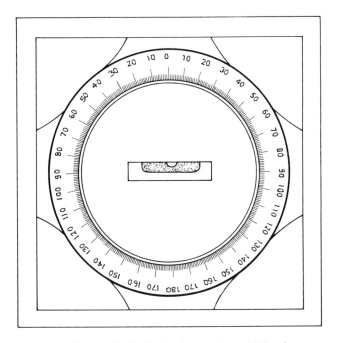

Figure 16.13 Angle measuring spirit level

allows the angle of slope to be measured. Again, measurement accuracies down to ±10 minutes of arc are possible if a vernier scale is incorporated in the instrument (see BS 958 (1968)).

16.8.3 **Electronic spirit level**

Electronic spirit levels contain a pendulum whose position is sensed electrically. With such instruments, measurement resolutions as good as 0.2 seconds of arc are possible.

16.9 **Flatness measurement**

The only dimensional parameter not so far discussed where a measurement requirement sometimes exists is the flatness of the surface of a component. This is measured by a variation gauge.

The variation gauge, shown in Figure 16.14, has four feet, three of which are fixed and one of which floats in a vertical direction. Motion of the floating foot is measured by a dial gauge which is calibrated such that its reading is zero when the floating foot is exactly level with the fixed feet. Thus, any non-zero reading on the dial gauge indicates non-flatness at the point of contact of the floating foot. By moving the

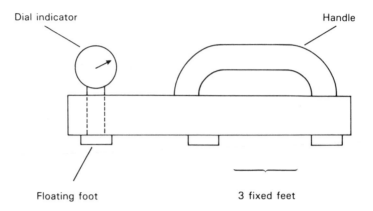

Figure 16.14 Variation gauge

variation gauge over the surface of a component and taking readings at various points, a contour map of the flatness of the surface can be obtained.

References and further reading

Anthony, D. M. (1986) *Engineering Metrology*, Pergamon: Oxford.
BS 817 (1988) Specifications for surface plates, British Standards Institution, London.
BS 870 (1950) Specifications for external micrometers, British Standards Institution, London.
BS 887 (1982) Specifications for precision vernier calipers, British Standards Institution, London.
BS 907 (1965) Specifications for dial gauges for linear measurements, British Standards Institution, London.
BS 958 (1968) Specifications for spirit levels for use in precision engineering, British Standards Institution, London.
BS 959 (1950) Specifications for internal micrometers (including stick micrometers), British Standards Institution, London.
BS 1643 (1983) Specifications for precision vernier height gauges, British Standards Institution, London.
BS 1685 (1951) Specifications for bevel protractors (mechanical and optical), British Standards Institution, London.
BS 4035 (1966) Specifications for linear measuring instruments for use in building and civil engineering constructional works. Steel measuring tapes, steel bands and retractable steel pocket rules, British Standards Institution, London.
BS 4311 (Parts 1 and 2) (1968) Gauge blocks and accessories, British Standards Institution, London.
BS 4372 (1968) Specifications for engineers' steel measuring rules, British Standards Institution, London.

BS 5317 (1976) Specification of metric length bars and their accessories, British Standards Institution, London.

Hume, K. J. (1970) *Engineering Metrology*, McDonald: London.

ISO 3611 (1978) Micrometer calipers for external measurement, International Organization for Standardization, Geneva.

ISO 3650 (1978) Gauge blocks, International Organization for Standardization, Geneva.

ISO 6906 (1984) Vernier calipers reading to 0.02 mm, International Organization for Standardization, Geneva.

ISO 8322 (1989) Building Construction – Measuring Instruments – Procedures for determining accuracy in use. Part 2 – Measuring Tapes, International Organization for Standardization, Geneva.

ISO/R 463 (1965) Dial gauges reading in 0.01 mm, 0.001 inch and 0.0001 inch, International Organization for Standardization, Geneva.

17 Translational displacement transducers

Translational displacement transducers are instruments which measure the motion of a body in a straight line between two points. Apart from their use as a primary transducer measuring the motion of a body, translational displacement transducers are also widely used as a secondary component in measurement systems, where some other physical quantity such as pressure, force, acceleration or temperature is translated into a translational motion by the primary measurement transducer.

Many different types of translational displacement transducer exist and these, along with their relative merits and characteristics, are discussed in the following sections of this chapter. The factors governing the choice of a suitable type of instrument in any particular measurement situation are considered in the final section at the end of the chapter.

17.1 The resistive potentiometer

The resistive potentiometer consists of a resistance element and a movable contact as shown in Figure 17.1. A voltage V_s is applied across the two ends A and B of the resistance element and an output voltage V_0 is measured between the point of contact C of the sliding element and the end of the resistance element A. A linear relationship exists between the output voltage V_0 and the distance AC, which can be expressed by:

$$\frac{V_0}{V_s} = \frac{AC}{AB} \tag{17.1}$$

The body whose motion is being measured is connected to the sliding element of the potentiometer, so that translational motion of the body causes a motion of equal magnitude of the slider along the resistance element and a corresponding change in the output voltage V_0.

Three different types of potentiometer exist, wire wound, carbon film and plastic film, so named according to the material used to construct the resistance element. Wire-wound potentiometers consist of a coil of resistance wire wound on a non-conducting former. As the slider moves along the potentiometer track, it makes contact with successive turns of the wire coil. This limits the resolution of the

340

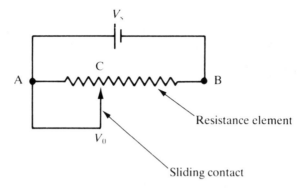

Figure 17.1 The resistive potentiometer

instrument to the distance from one coil to the next. Much better measurement resolution is obtained from potentiometers using either a carbon film or a conducting plastic film for the resistance element. The resolution of these is theoretically limited only by the grain size of the particles in the film, suggesting that measurement resolutions up to 10^{-4} ought to be attainable. In practice, the resolution is limited by mechanical difficulties in constructing the spring system which maintains the slider in contact with the resistance track, although these types are still considerably better than wire-wound types.

Operational problems of potentiometers all occur at the point of contact between the sliding element and the resistance track. The most common problem is dirt under the slider, which increases the resistance and thereby gives a false output voltage reading, or in the worst case causes a total loss of output. High-speed motion of the slider can also cause the contact to bounce, giving an intermittent output. Friction between the slider and the track can also be a problem in some measurement systems where the body whose motion is being measured is moved by only a small force of a similar magnitude to these friction forces.

The life expectancy of potentiometers is normally quoted as a number of reversals, i.e. as the number of times the slider can be moved backwards and forwards along the track. The figures quoted for wire-wound, carbon-film and plastic-film types are respectively 1 million, 5 million and 30 million. In terms of both life expectancy and measurement resolution, therefore, the carbon- and plastic-film types are clearly superior, although wire-wound types do have one advantage in respect of their lower temperature coefficient. This means that wire-wound types exhibit much less variation in their characteristics in the presence of varying ambient temperature conditions.

A typical accuracy which is quoted for translational motion resistive potentiometers is $\pm 1\%$ of full-scale reading. Manufacturers produce potentiometers to cover a large span of measurement ranges. At the bottom end of this span, instruments with a range of ± 2 mm are available, whilst at the top end, instruments with a range of ± 1 m are produced.

The resistance of the instrument measuring the output voltage at the potentiometer slider can affect the value of the output reading, as discussed in Chapter 3. As the slider moves along the potentiometer track, the ratio of the measured resistance to that of the measuring instrument varies, and thus the linear relationship between the measured displacement and the voltage output is distorted as well. This effect is minimized when the potentiometer resistance is small relative to that of the measuring instrument. This is achieved first by using a very high-impedance measuring instrument and secondly by keeping the potentiometer resistance as small as possible. Unfortunately, the latter is incompatible with achieving a high measurement sensitivity since this requires a high potentiometer resistance. A compromise between these two factors is therefore necessary. The alternative strategy of obtaining high measurement sensitivity by keeping the potentiometer resistance low and increasing the excitation voltage is not possible in practice because of the power rating limitation. This restricts the allowable power loss in the potentiometer to its heat dissipation capacity.

The process of choosing the best potentiometer from a range of instruments which are available, taking into account power rating and measurement linearity considerations, is illustrated in the example below.

Example 17.1

The output voltage from a translational motion potentiometer of stroke length 0.1 metre is to be measured by an instrument whose resistance is 10 kΩ. The maximum measurement error, which occurs when the slider is positioned two-thirds of the way along the element (i.e. when AC = 2AB/3 in Figure 17.1), must not exceed 1% of the full-scale reading. The highest possible measurement sensitivity is also required. A family of potentiometers having a power rating of 1 watt per 0.01 metre and resistances ranging from 100 Ω to 1 kΩ in 100 Ω steps are available. Choose the most suitable potentiometer from this range and calculate the sensitivity of measurement which it gives.

Solution

Referring to the labelling used in Figure 17.1, let the resistance of portion AC of the resistance element be R_i and that of the whole length AB of the element be R_t. Also, let the resistance of the measuring instrument be R_m and the output voltage measured by it be V_m. When the voltage measuring instrument is connected to the potentiometer, the net resistance across AC is the sum of two resistances in parallel (R_i and R_m) given by:

$$R_{AC} = \frac{R_i R_m}{R_i + R_m}$$

Let the excitation voltage applied across the ends AB of the potentiometer be V and the resultant current flowing between A and B be I. Then I and V are related by:

$$I = \frac{V}{R_{AC} + R_{CB}} = \frac{V}{[R_i R_m/(R_i + R_m)] + R_t - R_i}$$

V_m can now be calculated as:

$$V_m = IR_{AC} = \frac{VR_iR_m}{\{[R_iR_m/(R_i + R_m)] + R_t - R_i\}(R_i + R_m)}$$

If we express the voltage which exists across AC in the absence of the measuring instrument as V_0, then we can express the error due to the loading effect of the measuring instrument as:

$$\text{error} = V_0 - V_m$$

From Equation (17.1), $V_0 = (R_iV)/R_t$. Thus:

$$\text{error} = V_0 - V_m = V\left(\frac{R_i}{R_t} - \frac{R_iR_m}{\{[R_iR_m/(R_i + R_m)] + R_t - R_i\}(R_i + R_m)}\right)$$

$$= V\left(\frac{R_i^2(R_t - R_i)}{R_t(R_iR_t + R_mR_t - R_i^2)}\right) \tag{17.2}$$

Substituting $R_i = 2R_t/3$ into Equation (17.2) to find the maximum error:

$$\text{maximum error} = \frac{2R_t}{2R_t + 9R_m}$$

For a maximum error of 1%:

$$\frac{2R_t}{2R_t + 9R_m} = 0.01 \tag{17.3}$$

Substituting $R_m = 10\,000\,\Omega$ into the above expression (17.3) gives $R_t = 454\,\Omega$.

The nearest resistance values in the range of potentiometers available are $400\,\Omega$ and $500\,\Omega$. The value of $400\,\Omega$ has to be selected as this is the only one which gives a maximum measurement error of less than 1%.

The thermal rating of the potentiometers is quoted as 1 W per 0.01 m, i.e. 10 watts for the total length of 0.1 m. By Ohm's law:

$$\text{maximum supply voltage} = (\text{power} \times \text{resistance})^{1/2} = (10 \times 400)^{1/2} = 63.25\,\text{V}$$

Thus, the measurement sensitivity is 63.25/0.1 V/m = 632.5 V/m.

17.2 **Linear variable differential transformer (LVDT)**

The linear variable differential transformer, which is commonly known by the abbreviation LVDT, consists of a transformer with a single primary winding and two secondary windings connected in the series opposing manner shown in Figure 17.2. The object whose translational displacement is to be measured is physically attached to the central iron core of the transformer, so that all motions of the body are transferred to the core.

For an excitation voltage V_s given by $V_s = V_p \sin(\omega t)$, the e.m.f.s induced in the secondary windings V_a and V_b are given by:

$$V_a = K_a \sin(\omega t - \phi) \quad V_b = K_b \sin(\omega t - \phi)$$

The parameters K_a and K_b depend on the amount of coupling between the respective secondary and primary windings and hence on the position of the iron core. With the core in the central position, $K_a = K_b$, and we have

$$V_a = V_b = K \sin(\omega t - \phi)$$

Because of the series opposition mode of connection of the secondary windings, $V_0 = V_a - V_b$, and hence with the core in the central position, $V_0 = 0$.

Suppose now that the core is displaced upwards (i.e. towards winding A) by a distance x. If then $K_a = K_1$ and $K_b = K_2$, we have:

$$V_0 = (K_1 - K_2) \sin(\omega t - \phi)$$

If, alternatively, the core were displaced downwards from the null position (i.e.

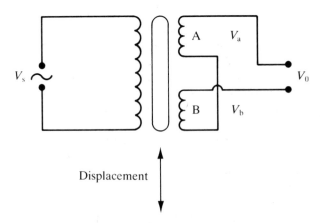

Figure 17.2 The linear variable differential transformer (LVDT)

towards winding B) by a distance x, the values of K_a and K_b would then be $K_a = K_2$ and $K_b = K_1$, and we would have:

$$V_0 = (K_2 - K_1) \sin(\omega t - \phi)$$
$$= (K_1 - K_2) \sin[\omega t + (\pi - \phi)]$$

Thus for equal magnitude displacements $+x$ and $-x$ of the core away from the central (null) position, the magnitude of the output voltage V_0 is the same in both cases. The only information about the direction of movement of the core is contained in the phase of the output voltage, which differs between the two cases by 180 degrees. If, therefore, measurements of the core position on both sides of the null position are required, it is necessary to measure the phase as well as the magnitude of the output voltage. The relationship between the magnitude of the output voltage and the core position is approximately linear over a reasonable range of movement of the core on either side of the null position and is expressed using a constant of proportionality C as:

$$V_0 = Cx$$

The only moving part in an LVDT is the central iron core. As the core is only moving in the air gap between the windings, there is no friction or wear during operation. For this reason, the instrument is a very popular one for measuring linear displacements and has a quoted life expectancy of 200 years. The typical accuracy is $\pm 0.5\%$ of full-scale reading and measurement resolution is almost infinite. Instruments are available to measure a wide span of measurements from $\pm 100\,\mu\text{m}$ to $\pm 100\,\text{mm}$. The instrument can be made suitable for operation in corrosive environments by enclosing the windings within a non-metallic barrier, which leaves the magnetic flux paths between the core and windings undisturbed. An epoxy resin is commonly used to encapsulate the coils for this purpose. One further operational advantage of the instrument is its insensitivity to mechanical shock and vibration.

Some problems which affect the accuracy of the LVDT are the presence of harmonics in the excitation voltage and stray capacitances, both of which cause a non-zero output of low magnitude when the core is in the null position. It is also impossible in practice to produce two identical secondary windings, and the small asymmetry which invariably exists between the secondary windings adds to this non-zero null output. The magnitude of this is always less than 1% of the full-scale output and in many measurement situations is of little consequence. If necessary, this non-zero null output can be eliminated by several means (see Doebelin 1975).

17.3 Variable inductance transducers

Variable inductance transducers, which are also sometimes referred to as variable reluctance or variable permeance transducers, exist in two separate forms. Both of these rely for their operation on the variation of the mutual inductance between two magnetically coupled parts. In this respect they are similar to the LVDT described in the last section. However, whereas the LVDT has three windings, both types of variable inductance transducer have only one winding.

The two types of variable inductance transducer are shown in Figure 17.3. In the first of these (Figure 17.3(a)), the single winding on the central limb of an E-shaped ferromagnetic body is excited with an alternating voltage. The displacement to be measured is applied to a ferromagnetic plate in close proximity to the E-piece. Movements of the plate alter the flux paths and hence cause a change in the current flowing in the winding. By Ohm's law, the current flowing in the winding is given by:

$$I = V/\omega L$$

For fixed values of ω and V, this equation becomes:

$$I = 1/KL$$

where K is a constant. The relationship between L and the displacement, d, applied to the plate is a non-linear one, and hence the output current–displacement characteristic has to be calibrated. A typical measurement range is 0–10 mm.

The second form of variable inductance transducer has a very similar size and physical appearance to the LVDT, but has a centre-tapped single winding, as shown in Figure 17.3(b). The two halves of the winding are connected to form two arms of a bridge circuit as shown in Figure 17.4, which is excited with an alternating voltage. With the core in the central position, the output from the bridge is zero. Displacements of the core either side of the null position cause a net output voltage which is approximately proportional to the displacement for small movements of the core. Instruments in this second form are available to cover a wide span of displacement measurements. At the lower end of this span, instruments with a range of 0–2 mm are available, whilst at the top end, instruments with a range of 0–5 m can be obtained.

Both forms of variable inductance transducer have infinite measurement resolution and a quoted life expectancy of 200 years. Their typical accuracy of ±0.5% of full-scale reading is better than that of the LVDT. In spite of this, their use is very much less common than that of the LVDT.

17.4 Variable capacitance transducers

Like variable inductance, the principle of variable capacitance is used in displacement measuring transducers in various ways. The three most common forms of variable capacitance transducer are shown in Figure 17.5. In Figure 17.5(a), the capacitor plates are formed by two concentric, hollow, metal cylinders. The displacement to be measured is applied to the inner cylinder which alters the capacitance. The second

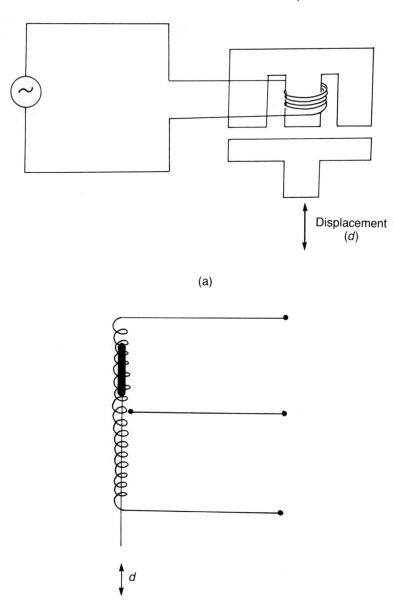

(a)

(b)

Figure 17.3 Variable inductance transducers

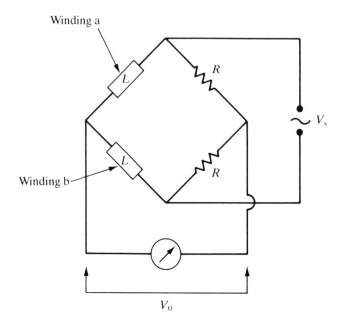

Figure 17.4 Connection of variable inductance transducer windings in bridge circuit

form, Figure 17.5(b), consists of two flat, parallel, metal plates, one of which is fixed and the other movable. Displacements to be measured are applied to the movable plate and the capacitance changes as this moves. Both of these first two forms use air as the dielectric medium between the plates. The final form, Figure 17.5(c), has two flat, parallel, metal plates with a sheet of solid dielectric material between them. The displacement to be measured causes a capacitance change by moving the dielectric sheet.

Accuracies up to ±0.01% are possible with these instruments with measurement resolutions of 1 micron (1 μm). Individual devices can be selected from manufacturers' ranges which measure displacements as small as 10^{-11} m or as large as 1 m. The fact that such instruments consist only of two simple conducting plates means that it is possible to fabricate devices which are tolerant of a wide range of environmental hazards such as extreme temperatures, radiation and corrosive atmospheres. As there are no contacting moving parts, there is no friction or wear in operation and the life expectancy quoted is 200 years. The major problem with variable capacitance transducers is their high impedance. This makes them very susceptible to noise and means that the length and position of connecting cables need to be chosen very carefully. In addition, very high-impedance instruments need to be used to measure the value of the capacitance. Because of these difficulties, use of these devices tends to be limited to those few applications where the high accuracy and measurement resolution of the instrument are required.

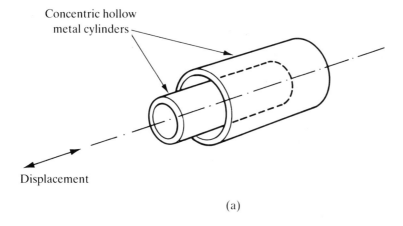

(a)

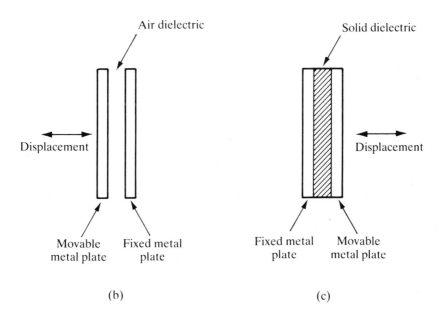

(b)   (c)

Figure 17.5 Variable capacitance transducers

17.5 **Strain gauges**

A strain gauge is a device which experiences a change in resistance when it is stretched or strained. It consists physically of a length of resistance wire formed into a zigzag pattern and mounted on to a flexible backing sheet, as shown in Figure 17.6. The wire is nominally of circular cross-section. As strain is applied to the gauge, the shape of the cross-section of the resistance wire distorts, changing the cross-sectional

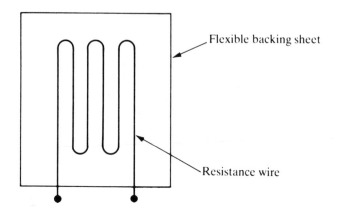

Figure 17.6 The strain gauge

area. As the resistance of the wire per unit length is inversely proportional to the cross-sectional area, there is a consequential change in resistance. The input–output relationship of a strain gauge is expressed by the gauge factor which is defined as the change in resistance (R) for a given value of strain (S), i.e.

$$\text{gauge factor} = \frac{\delta R}{\delta S}$$

Both metal alloy and semiconductor elements are used in strain gauges. Of the two, semiconductor elements have a much superior gauge factor (up to 50 times better than metal types) but are more expensive. Also, whilst metal alloy gauges have an almost zero temperature coefficient, semiconductor types have a relatively high temperature coefficient.

In use, strain gauges are bonded to the object whose displacement is to be measured. The process of bonding presents a certain amount of difficulty, particularly for semiconductor types. The resistance of the gauge is usually measured by a d.c. bridge circuit and the displacement is inferred from the bridge output measured. As the resistance change in a strain gauge is small, the bridge output voltage is also small and amplification has to be carried out. This adds to the cost of using strain gauges for displacement measurement.

A very common application of strain gauges is in measuring the very small displacements which occur in the diaphragms of pressure measuring instruments. The strain gauge has a typical range of measurement of 0–50 μm and so is very suitable for this task. Measurement accuracies of $\pm 0.15\%$ of full-scale reading are achievable and the quoted life expectancy is usually 3 million reversals.

The very small range of displacement measurement can be grossly extended by the scheme illustrated in Figure 17.7. In this, the displacement to be measured is applied to a wedge fixed between two beams carrying strain gauges. As the wedge is

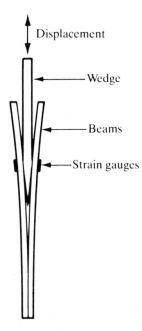

Figure 17.7 Strain gauges measuring large displacements

displaced downwards, the beams are forced apart and strained, causing an output reading on the strain gauges. Using this method, displacements up to about 50 mm can be measured.

17.6 **Nozzle flapper**

The nozzle flapper is a displacement transducer which translates displacements into a pressure change. A secondary pressure measuring device is therefore required within the instrument. The general form of a nozzle flapper is shown schematically in Figure 17.8. Fluid at a known supply pressure, P_s, flows through a fixed restriction and then through a variable restriction formed by the gap, x, between the end of the main vessel and the flapper plate. The body whose displacement is being measured is connected physically to the flapper plate. The output measurement of the instrument is the pressure P_0 in the chamber shown in Figure 17.8, and this is almost proportional to x over a limited range of movement of the flapper plate. The instrument typically has a first-order response characteristic. Air is very commonly used as the working fluid and this gives the instrument a time constant of about 0.1 seconds. The instrument has extremely high sensitivity but its range of measurement is quite small. A typical measurement range is ± 0.05 mm with a measurement

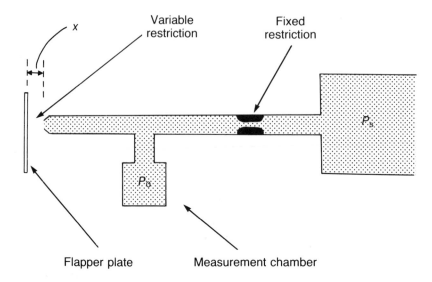

Figure 17.8 The nozzle flapper

resolution of ± 0.01 μm. One very common application of nozzle flappers is measuring the displacements within a load cell, which are typically very small.

17.7 **Piezoelectric transducers**

The piezoelectric transducer is effectively a force measuring device which is used in many instruments measuring force, or the force-related quantities of pressure and acceleration. It is included within this discussion of linear displacement transducers because its mode of operation is to generate an e.m.f. proportional to the distance by which it is compressed.

The device is manufactured from a crystal which can be either a natural material such as quartz or a synthetic material such as lithium sulphate. The crystal is mechanically stiff, i.e. a large force is required to compress it, and consequently piezoelectric transducers can only be used to measure the displacement of mechanical systems which are stiff enough themselves to be unaffected by the stiffness of the crystal. When the crystal is compressed, a charge is generated on the surface which is measured as the output voltage. As is normal with any induced charge, the charge leaks away over a period of time. Consequently, the output voltage–time characteristic is as shown in Figure 17.9. Because of this characteristic, piezoelectric transducers are not suitable for measuring static or slowly varying displacements, even though the time constant of the charge-decay process can be lengthened by adding a shunt capacitor across the device.

As a displacement measuring device, the piezoelectric transducer has a very high

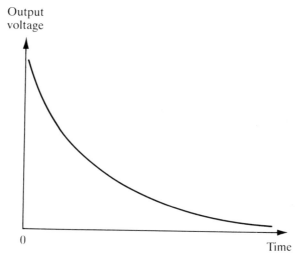

Figure 17.9 Voltage–time characteristic of piezoelectric transducer following step displacement

sensitivity, about 1000 times better than the strain gauge. Its typical accuracy is ±1% of full-scale reading and its life expectancy is 3 million reversals.

17.8 **Other methods of measuring small displacements**

Apart from the methods outlined above, several other techniques for measuring small translational displacements exist, as discussed below. Some of these involve special instruments which have a very limited sphere of application, for instance in measuring machine tool displacements. Others are very recent developments which may potentially gain wide use in the future but have few applications at present.

17.8.1 **Linear inductosyn**

The linear inductosyn is an extremely accurate instrument which is widely used for axis measurement and control within machine tools. Typical measurement resolution is 2.5 microns (2.5 μm).

The instrument consists of two magnetically coupled parts which are separated by an air gap, typically 0.125 mm wide, as shown in Figure 17.10. One part, the track, is attached to the axis along which displacements are to be measured. This would generally be the bed of a machine tool. The other part, the slider, is attached to the body which is to be measured or positioned. This would usually be a cutting tool.

The track, which may be several metres long, consists of a fine metal wire formed into the pattern of a continuous rectangular waveform and deposited on to a glass base. The typical pitch (cycle length), s, of the pattern is 2 mm, and this extends over

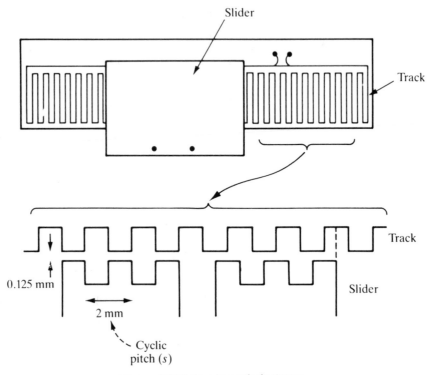

Figure 17.10 The linear inductosyn

the full length of the track. The slider is usually about 50 mm wide and carries two separate wires formed into continuous rectangular waveforms which are displaced with respect to each other by one-quarter of the cycle pitch, i.e. by 90 electrical degrees.

The wire waveform on the track is excited by an applied voltage given by:

$$V_s = V\sin(\omega t)$$

This excitation causes induced voltages in the slider windings. When the slider is positioned in the null position such that its first winding is aligned with the winding on the track, the output voltages on the two slider windings are given by:

$$V_1 = 0 \quad V_2 = V\sin(\omega t)$$

For any other position, the slider winding voltages are given by:

$$V_1 = V\sin(\omega t)\sin(2\pi x/s) \quad V_2 = V\sin(\omega t)\cos(2\pi x/s)$$

where x is the displacement of the slider away from the null position.

Consideration of these equations for the slider-winding outputs shows that the pattern of output voltages repeats every cycle pitch. The instrument can therefore only discriminate displacements of the slider within one cycle pitch of the windings. This means that the typical measurement range of an inductosyn is only 2 mm. This is of no use in normal applications and therefore an additional displacement transducer with coarser resolution but larger measurement range has to be used as well. This coarser measurement is commonly made by translating the linear displacements into rotary motion by suitable gearing which is then measured by a rotational displacement transducer such as a synchro or resolver.

One slight problem with the inductosyn is the relatively low level of coupling between the track and slider windings. Compensation for this is made by using a high-frequency excitation voltage (5–10 kHz is common).

17.8.2 Translation of linear displacements into rotary motion

In some applications, it is inconvenient to measure linear displacements directly, either because there is insufficient space to mount a suitable transducer or because it is inconvenient for other reasons. A suitable solution in such cases is to translate the translational motion into rotational motion by suitable gearing. Any of the rotational displacement transducers discussed in Chapter 19 can then be applied.

17.8.3 Integration of output from velocity transducers and accelerometers

If velocity transducers or accelerometers already exist in a system, displacement measurements can be obtained by integration of the output from these instruments. This, however, only gives information about the relative position with respect to some arbitrary starting point. It does not yield a measurement of the absolute position of a body in space unless all motions away from a fixed starting point are recorded.

17.8.4 Laser interferometer

This recently developed instrument is shown in Figure 17.11. In this particular design, a dual-frequency helium–neon (He–Ne) laser is used which gives an output pair of light waves at a nominal frequency of 5×10^{14} Hz. The two waves differ in frequency by 2×10^6 Hz and are of opposite polarization. This dual-frequency output waveform is split into measurement and reference beams by the first beam splitter.

The reference beam is sensed by the polarizer and photodetector, A, which converts both waves in the light to the same polarization. The two waves interfere constructively and destructively alternately, producing light–dark flicker at a frequency of 2×10^6 Hz. This excites a 2 MHz electrical signal in the photodetector.

The measurement beam is separated into the two component frequencies by a polarizing beam splitter. Light of the first frequency, f_1, is reflected by a fixed reflecting cube into a photodetector and polarizer, B. Light of the second frequency, f_2, is reflected by a movable reflecting cube and also enters B. The displacement to be

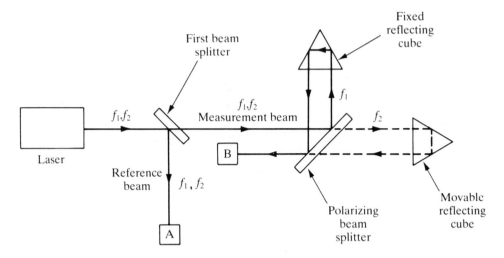

Figure 17.11 The laser interferometer

measured is applied to the movable cube. With the movable cube in the null position, the light waves entering B produce an electrical signal output at a frequency of 2 MHz, which is the same frequency as the reference signal output from A. Any displacement of the movable cube causes a Doppler shift in the frequency f_2 and changes the output from B. The frequency of the output signal from B varies between 0.5 and 3.5 MHz according to the speed and direction of movement of the movable cube.

The outputs from A and B are amplified and subtracted. The resultant signal is fed to a counter whose output indicates the magnitude of the displacement in the movable cube and whose rate of change indicates the velocity of motion.

This technique is beginning to be used in applications requiring high-accuracy measurement such as machine tool control. Such systems can measure to an accuracy of a few parts per million over ranges up to 2 m long. They are therefore an attractive alternative to the inductosyn, in having both high measurement resolution and a large measurement range within one instrument.

17.8.5 **Fotonic sensor**

The Fotonic sensor is one of many instruments developed recently which make use of fiber optic techniques. It consists of a light source, a light detector, a fiber optic light transmission system and a plate which moves with the body whose displacement is being measured, as shown in Figure 17.12. Light from the outward fiber optic cable travels across the air gap to the plate and some of it is reflected back into the return fiber optic cable. The amount of light reflected back from the plate is a function of the air gap length, x, and hence of the plate displacement. Measurement of the intensity of the light carried back along the return cable to the light detector allows the

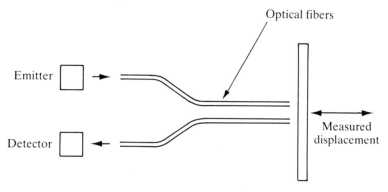

Figure 17.12 The Fotonic sensor

displacement of the plate to be calculated. Common applications of Fotonic sensors are measuring diaphragm displacements in pressure sensors and measuring the movement of bimetallic temperature sensors.

17.8.6 **Evanescent-field fiber optic sensor**

This sensor consists of a prism and a light source/detector system, as shown in Figure 17.13. The amount of light reflected into the detector depends on the proximity of a movable silver surface to the prism. Reflection varies from 96% when the surface is touching the prism to zero when it is 1 μm away. This provides a means of measuring very tiny displacements over the range between 0 and 1 μm (1 micron).

17.8.7 **Non-contacting optical sensor**

Figure 17.14 shows an optical technique which is used to measure small displacements. The motion to be measured is applied to a vane whose displacement

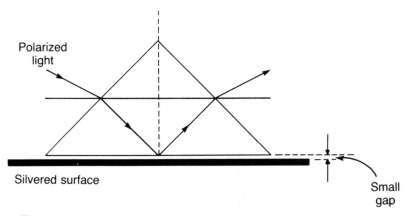

Figure 17.13 Evanescent-field fiber optic displacement sensor

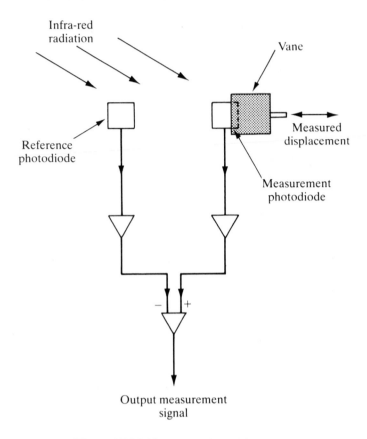

Figure 17.14 Non-contacting optical sensor

progressively shades one of a pair of monolithic photodiodes which are exposed to infrared radiation. A displacement measurement is obtained by comparing the output of the reference (unshaded) photodiode with that of the shaded one. The typical range of measurement is ±0.5 mm with an accuracy of ±0.1% of full scale. Such sensors are used in some intelligent pressure measuring instruments based on Bourdon tubes or diaphragms as described in section 14.4.

17.9 **Measurement of large displacements (range sensors)**

One final class of instruments which has not been mentioned so far consists of those designed to measure relatively large translational displacements. Most of these are known as range sensors and measure the motion of a body with respect to some fixed datum.

17.9.1 **Rotary potentiometer and spring-loaded drum**

One scheme for measuring large displacements which are beyond the measurement range of common displacement transducers is shown in Figure 17.15. This consists of a steel wire attached to the body whose displacement is being measured: the wire passes round a pulley and on to a spring-loaded drum whose rotation is measured by a rotary potentiometer. A multi-turn potentiometer is usually required for this to give an adequate measurement resolution.

17.9.2 **Range sensors**

Range sensors provide a well-used technique of measuring the translational displacement of a body with respect to some fixed boundary. The common feature of all range-sensing systems is an energy source, an energy detector and an electronic means of measuring the time of flight of the energy between the source and detector. The form of energy used is either ultrasonic or light. In some systems, both energy source and detector are fixed on the moving body and operation depends on the energy being reflected back from the fixed boundary as in Figure 17.16(a). In other systems, the energy source is attached to the moving body and the energy detector is located within the fixed boundary, as shown in Figure 17.16(b).

In ultrasonic systems, the energy is transmitted from the source in high-frequency bursts. A frequency of at least 20 kHz is usual and 40 kHz is common for measuring

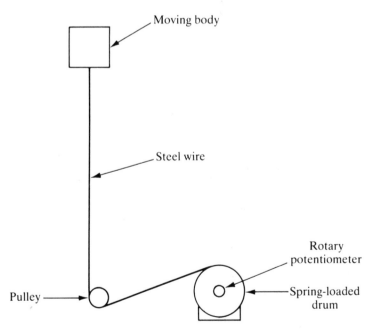

Figure 17.15 System for measuring large displacements

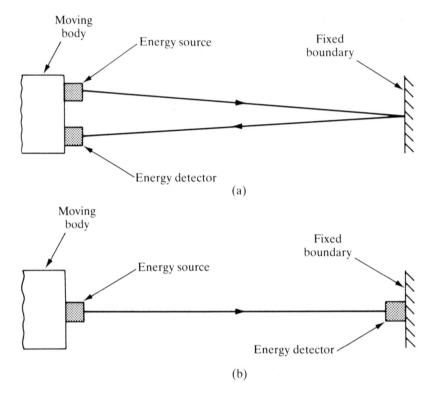

Figure 17.16 Range sensors

distances up to 5 m. By measuring the time of flight of the energy, the distance of the body from the fixed boundary can be calculated, using the fact that the speed of sound in air is 340 m/s. Because of difficulties in measuring the time of flight with sufficient accuracy, ultrasonic systems are not suitable for measuring distances of less than about 300 mm. Measurement resolution is limited by the wavelength of the ultrasonic energy and can be improved by operating at higher frequencies. At higher frequencies, however, attenuation of the magnitude of the ultrasonic wave as it passes through air becomes significant. Therefore, only low frequencies are suitable if large distances are to be measured. The typical accuracy of ultrasonic range-finding systems is ±0.5% of full scale.

Optical range-finding systems generally use a laser light source. The speed of light in air is about 3×10^8 m/s, so that light takes only a few nanoseconds to travel a metre. In consequence, such systems are only suitable for measuring very large displacements where the time of flight is long enough to be measured with reasonable accuracy.

17.10 **Proximity sensors**

For the sake of completeness, it is proper to conclude this chapter on translational displacement transducers with the consideration of proximity sensors. Proximity detectors provide information on the displacement of a body with respect to some boundary, but only in so far as to say whether the body is less than or greater than a certain distance away from the boundary. The output of a proximity sensor is thus binary in nature: the body is or is not close to the boundary.

Like range sensors, proximity detectors make use of an energy source and detector. The detector is a device whose output changes between two states when the magnitude of the incident reflected energy exceeds a certain threshold level. A common form of proximity sensor uses an infrared light-emitting diode (LED) source and a phototransistor. Light triggers the transistor into a conducting state when the LED is within a certain distance of a reflective boundary and the reflected light exceeds a threshold level. This system is physically small, occupying a volume of only a few cubic centimetres. If even this small volume is obtrusive, then fiber optic cables can be used to transmit light from a remotely mounted LED and phototransistor. The threshold displacement detected by optical proximity sensors can be varied between 0 and 2 m.

Another form of proximity sensor uses the principle of varying inductance. Such devices are particularly suitable for operation in aggressive environmental conditions and can be made vibration and shock resistant by vacuum encapsulation techniques. The sensor contains a high-frequency oscillator whose output is demodulated and fed via a trigger circuit to an amplifier output stage. The oscillator output radiates through the surface of the sensor and, when the sensor surface comes close to an electrically or magnetically conductive boundary, the output voltage is reduced because of the interference with the flux paths. At a certain point, the output voltage is reduced sufficiently for the trigger circuit to change state and reduce the amplifier output to zero. Inductive sensors can be adjusted to change state at displacements in the range of 1 to 20 mm.

A third form of proximity sensor uses the capacitive principle. These can operate in similar conditions to inductive types. The threshold level of displacement detected can be varied between 5 and 40 mm.

Fiber optic proximity sensors also exist where the amount of reflected light varies with the proximity of the fiber ends to a boundary, as shown in Figure 9.6.

17.11 **Selection of translational measurement transducers**

Choosing between the various translational motion transducers available for any particular application depends very much on the magnitude of the displacement to be measured.

Displacements larger than 5 m can only be measured by a range sensor, or possibly by the method of using a wire described in section 17.9 (Figure 17.15). Such methods are also used for displacements in the range between 2 and 5 m, except where the expense of a variable inductance transducer can be justified.

For measurements within the range of 2 mm to 2 m, the number of suitable instruments grows. Both the relatively cheap potentiometer and the LVDT, which is somewhat more expensive, are commonly used for such measurements. Variable inductance and variable capacitance transducers are also used in some applications. Additionally, strain gauges measuring the strain in two beams forced apart by a wedge (see section 17.5) can measure displacements up to 50 mm. If very high measurement resolution is required, either the linear inductosyn or the laser interferometer is used.

The requirement to measure displacements of less than 2 mm usually occurs as part of an instrument which is measuring some other physical quantity such as pressure, and several types of instrument have evolved to fulfil this task. The LVDT, strain gauges, the Fotonic sensor, variable capacitance transducers and the non-contacting optical transducer all find application in measuring diaphragm or Bourdon-tube displacements within pressure transducers. Load-cell displacements are also very small, and these are commonly measured by nozzle flapper devices.

Apart from the measurement range required, several other factors influence the choice between different instruments. Severe environmental operating conditions, for example hot, radioactive or corrosive atmospheres, often dictate the use of devices which can be easily protected from these conditions, such as the LVDT, variable inductance or variable capacitance instruments. Accessibility and maintenance requirements are also very important factors, as high maintenance costs can quickly destroy the initial advantage that a particular type of instrument may have in terms of low purchase cost.

References and further reading

Doebelin, E. O. (1975) *Measurement Systems*, McGraw-Hill: New York.

18 Measurement of translational velocity, acceleration, vibration and shock

18.1 Measurement of translational velocity

Translational velocity cannot be measured directly and therefore must be calculated indirectly by other means as set out below.

18.1.1 Differentiation of displacement measurements

Differentiation of position measurements obtained from any of the translational displacement transducers described in the previous chapter can be used to produce a translational velocity signal. Unfortunately, the process of differentiation always amplifies noise in a measurement system. Therefore, if this method has to be used, a low-noise instrument such as a d.c.-excited carbon-film potentiometer or laser interferometer should be chosen. In the case of potentiometers, a.c. excitation must be avoided because of the problem that harmonics in the power supply would cause.

18.1.2 Integration of the output of an accelerometer

Where an accelerometer is already included within a system, integration of its output can be performed to yield a velocity signal. The process of integration attenuates rather than amplifies measurement noise and this is therefore an acceptable technique.

18.1.3 Conversion to rotational velocity

Conversion from translational to rotational velocity is the final measurement technique open to the system designer and is the one most commonly used. This enables any of the rotational velocity measuring instruments described in Chapter 20 to be applied.

18.2 Measurement of acceleration

The only class of device available for measuring acceleration is the accelerometer. These are available in a wide variety of types and ranges designed to meet particular measurement requirements. They have a frequency response between zero and a high value and have a form of output which can be readily integrated to give displacement and velocity measurements. The frequency response of accelerometers can be improved by altering the level of damping in the instrument. Such adjustment must be done carefully, however, because frequency response improvements are only achieved at the expense of degrading the measurement sensitivity. Besides their use for general-purpose motion measurement, accelerometers are widely used to measure mechanical shocks and vibrations.

The structure of most forms of accelerometer is shown in Figure 18.1. The accelerometer is rigidly fastened to the body undergoing acceleration. Any acceleration of the body causes a force, F_a, on the mass, M, given by:

$$F_a = M\ddot{x}$$

This force is opposed by the restraining effect, F_s, of a spring with spring constant K, and the net result is that the mass is displaced by a distance x from its starting position such that:

$$F_s = Kx$$

In the steady state, when the mass inside is accelerating at the same rate as the case of the accelerometer, $F_a = F_s$ and so:

$$Kx = M\ddot{x} \quad \text{or} \quad \ddot{x} = (Kx)/M \tag{18.1}$$

This is the equation of motion of a second-order system, and, in the absence of damping, the output of the accelerometer would consist of non-decaying oscillations. A

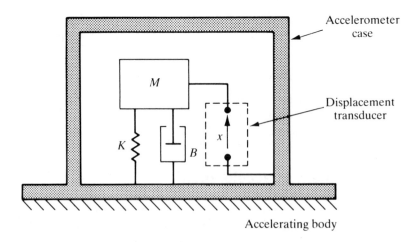

Figure 18.1 Structure of an accelerometer

damper is therefore included within the instrument, which produces a damping force, F_d, proportional to the velocity of the mass M given by:

$$F_d = B\dot{x}$$

This modifies the previous equation of motion (18.1) to the following:

$$Kx + B\dot{x} = M\ddot{x} \tag{18.2}$$

One important characteristic of accelerometers is their sensitivity to accelerations at right angles to the sensing axis (the direction along which the instrument is designed to measure acceleration). This is defined as the cross-sensitivity and is specified in terms of the output, expressed as a percentage of the full-scale output, when an acceleration of some specified magnitude (e.g. $30\,g$) is applied at 90 degrees to the sensing axis.

The acceleration reading is obtained from the instrument by measurement of the displacement of the mass within the accelerometer. Many different displacement measuring techniques are used in the various types of accelerometer which are commercially available. Different types of accelerometer also vary in terms of the type of spring element and form of damping used.

Resistive potentiometers are one such displacement measuring instrument used in accelerometers. These are used mainly for measuring slowly varying accelerations and low-frequency vibrations in the range of 0–$50\,g$. The measurement resolution obtainable is about 1 in 400 and typical values of cross-sensitivity are $\pm 1\%$. Accuracy is about $\pm 1\%$ and life expectancy is quoted at 2 million reversals. A typical size and weight are $125\ \text{cm}^3$ and 500 g.

Strain gauges are also used in accelerometers for measuring accelerations up to $200\,g$. These serve as the spring element as well as measuring mass displacement, thus simplifying the instrument's construction. Their typical characteristics are a resolution of 1 in 1000, an accuracy of $\pm 1\%$ and a cross-sensitivity of 2%. They have a major advantage over potentiometer-based accelerometers in terms of their much smaller size and weight ($3\ \text{cm}^3$ and 25 g).

Another displacement transducer found in accelerometers is the LVDT. This class can measure accelerations up to $700\,g$ with an accuracy of $\pm 1\%$ of full scale. They are of a similar physical size to potentiometer-based instruments but are lighter in weight (100 g).

Accelerometers based on variable inductance displacement measuring devices have extremely good characteristics and are suitable for measuring accelerations up to $40\,g$. Typical specifications of such instruments are an accuracy of $\pm 0.25\%$ of full scale, a resolution of 1 in 10 000 and a cross-sensitivity of 0.5%. Their physical size and weight are similar to potentiometer-based devices. Such instruments are often found in oil exploration applications measuring vibrations reflected from underground rock strata.

The other common displacement transducer used in accelerometers is the piezoelectric type. The major advantage of using piezoelectric crystals is that they also act as the spring and damper within the instrument. In consequence, the device is quite small ($15 \, cm^3$) and very light (50 g), but because of the nature of piezoelectric crystal operation, such instruments are not suitable for measuring constant or slowly time-varying accelerations. As the electrical impedance of a piezoelectric crystal is itself high, the output voltage must be measured with a very high-impedance instrument to avoid loading effects. Many recent piezoelectric-crystal-based accelerometers incorporate a high-impedance charge amplifier within the body of the instrument. This simplifies the signal conditioning requirements external to the accelerometer but can lead to problems in certain operational environments because these internal electronics are exposed to the same environmental hazards as the rest of the accelerometer. Typical measurement resolution of this class of accelerometer is 0.1% of full scale with an accuracy of $\pm 1\%$. Individual instruments are available to cover a wide range of measurements from $0.03 \, g$ full scale up to $1000 \, g$ full scale.

Recently, very small microsensors have become available for measuring acceleration. These consist of a small mass subject to acceleration which is mounted on a thin silicon membrane. Displacements are measured either by piezoresistors deposited on the membrane or by etching a variable capacitor plate into the membrane.

Two forms of fiber-optic-based accelerometer also exist. One form measures the effect on light transmission intensity caused by a mass subject to acceleration resting on a multimode fiber. The other form measures the change in phase of light transmitted through a monomode fiber which has a mass subject to acceleration resting on it.

In choosing between the different types of accelerometer available for any particular situation, several factors need to be considered. Of major importance is the mass of the instrument. This should be very much less than that of the body whose motion is being measured, in order to avoid loading effects which affect the accuracy of the readings obtained. In this respect, strain-gauge-based instruments are best.

18.3 Vibration measurement

Vibrations are very commonly encountered in machinery operation, and therefore measurement of the accelerations associated with such vibrations is extremely important in industrial environments. The peak accelerations involved in such vibrations can be of $100 \, g$ or more in magnitude, whilst both the frequency of oscillation and the magnitude of displacements from the equilibrium position in vibrations have a tendency to vary randomly.

Vibrations normally consist of linear harmonic motion which can be expressed mathematically as:

$$X = X_0 \sin(\omega t) \tag{18.3}$$

where X is the displacement from the equilibrium position at any general point in time,

X_0 is the peak displacement from the equilibrium position, and ω is the angular frequency of the oscillations.

By differentiating Equation (18.3) with respect to time, an expression for the velocity v of the vibrating body at any general point in time is obtained as:

$$v = -\omega X_0 \cos(\omega t) \tag{18.4}$$

Differentiating Equation (18.4) again with respect to time, we obtain an expression for the acceleration, α, of the body at any general point in time as:

$$\alpha = -\omega^2 X_0 \sin(\omega t) \tag{18.5}$$

Inspection of Equation (18.5) shows that the peak acceleration is given by:

$$\alpha_{peak} = \omega^2 X_0 \tag{18.6}$$

This square-law relationship between peak acceleration and oscillation frequency is the reason why high values of acceleration occur during relatively low-frequency oscillations. For example, an oscillation at 10 Hz produces peak accelerations of $2\,g$.

Example 18.1
A pipe carrying a fluid vibrates at a frequency of 50 Hz with displacements of 8 mm from the equilibrium position. Calculate the peak acceleration.

Solution
From Equation (18.6):

$$\alpha_{peak} = \omega^2 X_0 = (2\pi 50)^2 \times (0.008) = 789.6 \text{ m/s}^2$$

Using the fact that the acceleration due to gravity, g, is 9.81 m/s^2, this answer can be expressed alternatively as:

$$\alpha_{peak} = 789.6/9.81 = 80.5\,g$$

It is apparent that the intensity of vibration can be measured in terms of displacement, velocity or acceleration. Acceleration is clearly the best parameter to measure at high frequencies. However, because displacements are large at low frequencies according to Equation (18.3), it would seem that measuring either displacement or velocity would be best at low frequencies. The amplitude of vibrations can be measured by various forms of displacement transducer. Fiber-optic-based devices are particularly attractive and can give measurement resolution as high as $1\,\mu$m. Unfortunately, there are considerable practical difficulties in mounting and calibrating displacement and velocity transducers and therefore they are rarely used. Thus, vibration is usually measured by accelerometers at all frequencies. The most common type of transducer used is the piezo-accelerometer, which has typical inaccuracy levels of $\pm 2\%$.

The frequency response of accelerometers is particularly important in vibration measurement in view of the inherently high-frequency characteristics of the measurement situation. The bandwidth of both potentiometer-based accelerometers and accelerometers using variable-inductance-type displacement transducers goes up only to 25 Hz. Accelerometers including either the LVDT or strain gauges can measure frequencies up to 150 Hz and the latest instruments using piezoresistive strain gauges have bandwidths up to 2 kHz. Finally, the inclusion of piezoelectric crystal displacement transducers yields an instrument with a bandwidth which can be as high as 7 kHz.

When measuring vibration, consideration must be given to the fact that attaching an accelerometer to the vibrating body will significantly affect the vibration characteristics if the body has a low mass. The effect of such 'loading' of the measured system can be quantified by the following equation:

$$a_1 = a_b \left(\frac{m_b}{m_b + m_a} \right)$$

where a_1 is the acceleration of the body with the accelerometer attached, a_b is the acceleration of the body without the accelerometer, m_a is the mass of the accelerometer and m_b is the mass of the body.

Such considerations emphasize the advantage of piezo-accelerometers, as these have a lower mass than other forms of accelerometer and so contribute least to this system-loading effect.

As well as an accelerometer, a vibration measurement system requires other elements, as shown in Figure 18.2, to translate the accelerometer output into a recorded signal. The three other necessary elements are a signal conditioning element, a signal analyzer and a signal recorder. The signal conditioning element amplifies the relatively weak output signal from the accelerometer and also transforms the high output impedance of the accelerometer to a lower impedance value. The signal analyzer then converts the signal into the form required for output. The output parameter may be displacement, velocity or acceleration and this may be expressed as the peak value, r.m.s. value or average absolute value. The final element of the measurement system is the signal recorder.

All elements of the measurement system, and especially the signal recorder, must be chosen very carefully to avoid distortion of the vibration waveform. The bandwidth should be such that it is at least a factor of ten better than the bandwidth of the vibration frequency components at both ends. Thus its lowest frequency limit should be less than or equal to 0.1 times the fundamental frequency of vibration and its upper frequency limit should be greater than or equal to ten times the highest significant vibration frequency component.

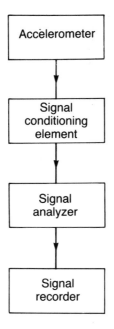

Figure 18.2 Vibration measurement system

18.4 **Shock measurement**

Shock describes a type of motion where a moving body is brought suddenly to rest, often because of a collision. This is very common in industrial situations and usually involves a body being dropped and hitting the floor.

Shocks characteristically involve large-magnitude decelerations (e.g. 500 g) which last for a very short time (e.g. 5 ms). An instrument having a very high-frequency response is required for shock measurement, and for this reason, piezoelectric-crystal-based accelerometers are commonly used. Again, other elements for analyzing and recording the signal are required as shown in Figure 18.2 and described in the previous section. A storage oscilloscope is a suitable instrument for recording the output signal, as this allows the time duration as well as the acceleration levels in the shock to be measured. Alternatively, if a permanent record is required, the screen of a standard oscilloscope can be photographed. A further option is to record the output on magnetic tape, which facilitates computerized signal analysis.

Example 18.2
A body is dropped from a height of 10 m and suffers a shock when it hits the ground. If the duration of the shock is 5 ms, calculate the magnitude of the shock in terms of g.

Solution

The equation of motion for a body falling under gravity gives the following expression for the terminal velocity, v:

$$v = (2gx)^{1/2}$$

where x is the height through which the body falls. Having calculated v, the average deceleration during the collision can be calculated as:

$$\alpha = v/t$$

where t is the time duration of the shock. Substituting the appropriate numerical values into these expressions:

$$v = (2 \times 9.81 \times 10)^{1/2} = 14.0 \text{ m/s} \quad \alpha = 14.0/0.005 = 2801 \text{ m/s}^2 = 286 \, g$$

19 Rotational displacement transducers

Rotational displacement transducers measure the angular motion of a body about some rotation axis. They are important not only for measuring the rotation of bodies such as shafts, but also as part of systems which measure translational displacement by converting the translational motion to a rotary form. The various devices available for measuring rotational displacements are presented below, and the arguments for choosing a particular form in any given measurement situation are considered at the end of the chapter.

19.1 Circular and helical potentiometers

This class of instrument is the cheapest available for measuring rotational displacements, with the circular type being the cheaper of the two. Their structure is shown in Figure 19.1. The mode of operation is exactly the same as for translational motion potentiometers, except that the track is bent round into a circular shape. The output voltage measured at the sliding contact is proportional to the angular displacement of the sliding contact from the starting position, thus satisfying the usual requirement for a linear measured quantity–output reading relationship.

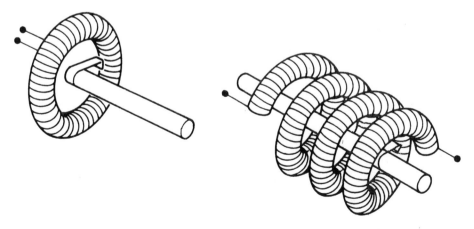

Figure 19.1 Rotary motion potentiometers: (a) circular; (b) helical

The circular potentiometer has a track consisting of no more than one full turn, and hence has a maximum angular measurement range of 0–360 degrees, although potentiometers are often produced with measurement ranges as small as 0–10 degrees. The measurement resolution afforded by this arrangement is very limited, and is the reason for the existence of multi-turn devices where the resistance element is formed into the shape of a helix. Such devices are produced with up to 60 turns. Their accuracy varies from ±1% of full scale for circular potentiometers up to ±0.002% of full scale for the best helical potentiometers. As with linear track potentiometers, all rotational potentiometers can give performance problems through dirt on the track causing loss of contact and they have a limited life because of wear between the sliding surfaces.

19.2 **Rotational differential transformer**

Special forms of differential transformer are available which measure rotational rather than translational motion. The method of construction and connection of the windings is exactly the same as for the linear variable differential transformer (LVDT), except that a specially shaped core is used which varies the mutual inductance between the windings as it rotates, as shown in Figure 19.2. Apart from the difficulty of avoiding some asymmetry between the secondary windings, great care has to be taken in these instruments to machine the core to exactly the right shape. In consequence, the best accuracy obtainable is only ±1%. This accuracy is only obtained for limited excursions of the core of ±40 degrees away from the null position. For angular displacements of ±60 degrees, the accuracy drops to ±3%, and the instrument is unsuitable for

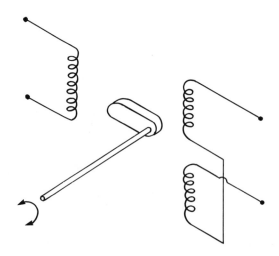

Figure 19.2 Rotary differential transformer

measuring displacements greater than this. Like its linear equivalent, the instrument suffers no wear in operation and therefore has a very long life with almost no maintenance requirements. It can also be modified for operation in harsh environments by enclosing the windings inside a protective enclosure.

19.3 Incremental shaft encoders

Incremental shaft encoders are one of a class of encoder devices which give an output in digital form. They measure the instantaneous angular position of a shaft relative to some arbitrary datum, but are unable to give any indication about the absolute position of a shaft. The principle of operation is to generate pulses as the shaft whose displacement is being measured rotates. These pulses are counted and the total angular rotation inferred from the pulse count. The pulses are generated either by optical or by magnetic means and are detected by suitable sensors. Of the two, the optical system is considerably cheaper and therefore much more common. Such instruments are very convenient for computer control applications, as the measurement is already in the required digital form and therefore the usual analog-to-digital signal conversion process is avoided.

An example of an optical incremental shaft encoder is shown in Figure 19.3. It can be seen that the instrument consists of a pair of disks, one of which is fixed and one of which rotates with the body whose angular displacement is being measured. Each disk is basically opaque but has a pattern of windows cut into it. The fixed disk has only one

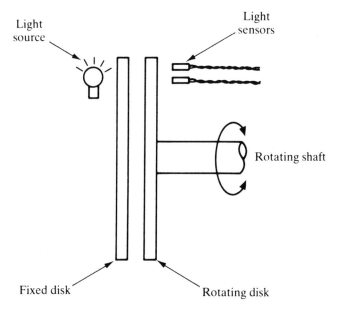

Figure 19.3 The optical incremental shaft encoder

window and the light source is aligned with this so that the light shines through all the time. The second disk has two tracks of windows cut into it which are spaced equidistantly around the disk, as shown in Figure 19.4. Two light detectors are positioned beyond the second disk so that one is aligned with each track of windows. As the second disk rotates, light alternately enters and does not enter the detectors, as windows and then opaque regions of the disk pass in front of them. These pulses of light are fed to a counter, with the final count after motion has ceased corresponding to the angular position of the moving body relative to the starting position. The primary information about the magnitude of rotation is obtained by the detector aligned with the outer track of windows. The pulse count obtained from this gives no information about the direction of rotation, however. Direction information is provided by the second, inner track of windows, which have an angular displacement with respect to the outer set of windows of half a window-width. The pulses from the detector aligned with the inner track of windows therefore lag or lead the primary set of pulses according to the direction of rotation.

The maximum measurement resolution obtainable is limited by the number of windows which can be machined on to a disk. The maximum number of windows per

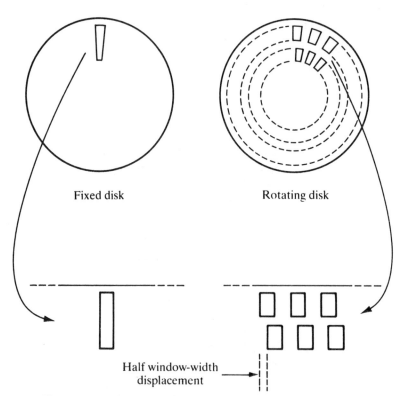

Fixed disk Rotating disk

Half window-width displacement

Figure 19.4 Window arrangement in incremental shaft encoder

track for a 150 mm diameter disk is 5000, which gives a basic angular measurement resolution of 1 in 5000. By using more sophisticated circuits which increment the count on both the rising and falling edges of the pulses through the outer track of windows, it is possible to double the resolution to a maximum of 1 in 10 000. At the expense of even greater complexity in the counting circuit, it is possible also to include the pulses from the inner track of windows in the count, so giving a maximum measurement resolution of 1 in 20 000.

Optical incremental shaft encoders are a popular instrument for measuring relative angular displacements and are generally reliable in operation. Problems with noise in the system giving false counts can sometimes cause difficulties, although this can usually be eliminated by squaring the output from the light detectors.

Such instruments are found in many applications where rotational motion has to be measured. Incremental shaft encoders are also commonly used in circumstances where a translational displacement has been transformed to a rotational one by suitable gearing. One example of this practice is in measuring the translational motions in numerically controlled (NC) drilling machines. Typical gearing used for this would give one revolution per millimetre of translational displacement. By using an incremental shaft encoder with 1000 windows per track in such an arrangement, a measurement resolution of 1 micron (1 μm) is obtained.

19.4 Coded disk shaft encoders

Unlike the incremental shaft encoder which gives a digital output in the form of pulses that have to be counted, the digital shaft encoder has an output in the form of a binary number of several digits which provides an absolute measurement of shaft position. This is particularly useful for computer control applications but these devices have a significantly higher cost than incremental encoders. Three different forms of digital shaft encoder exist using respectively optical, electrical and magnetic energy systems.

19.4.1 Optical digital shaft encoder

The optical digital shaft encoder is the cheapest form available of this class of instrument and is the one used most commonly. It is found in a variety of applications, and one where it is particularly popular is in measuring the position of rotational joints in robot manipulators.

The instrument is similar in physical appearance to the incremental shaft encoder, in having a pair of disks (one movable and one fixed) with a light source on one side and light detectors on the other side, as shown in Figure 19.5. The fixed disk has a single window as in the incremental shaft encoder, but the principal way in which the device differs from the incremental shaft encoder is in the design of the windows on the movable disk, as shown in Figure 19.6. These are cut in four or more tracks instead of two and are arranged in sectors as well as tracks. An energy detector is aligned with each track, and these give an output of 1 when energy is detected and an output of 0

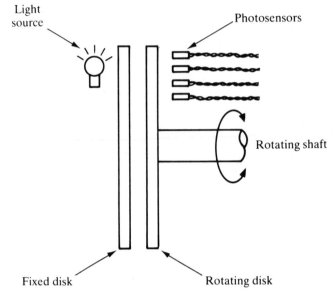

Figure 19.5 The coded disk shaft encoder

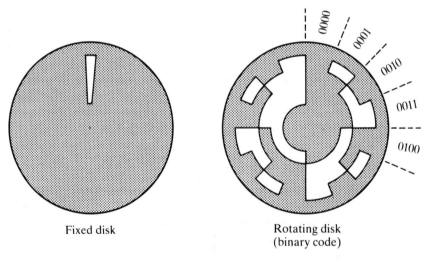

Fixed disk

Rotating disk
(binary code)

Figure 19.6 Window arrangement for coded disk shaft encoder

otherwise. The measurement resolution obtainable depends on the number of tracks used. For a four-track version, the resolution is 1 in 16, with progressively higher measurement resolution being attained as the number of tracks is increased. These binary outputs from the detectors are combined together to give a binary number of several digits. The number of digits corresponds to the number of tracks on the disk, which in the example shown in Figure 19.6 is four. The pattern of windows in each sector is cut such that, as that particular sector passes across the window in the fixed disk, the four energy detector outputs combine to give a unique binary number. In the binary-coded example shown in Figure 19.6, the binary number output increments by one as each sector in the rotating disk passes in turn across the window in the fixed disk. Thus the output from sector 1 is 0001, from sector 2 is 0010, from sector 3 is 0011, etc.

Whilst this arrangement is perfectly adequate in theory, serious problems can arise in practice because of the manufacturing difficulty of machining the windows of the movable disk such that the edges of the windows in each track are aligned with each other exactly. Any misalignment means that, as the disk turns across the boundary between one sector and the next, the outputs from each track will switch at slightly different instants of time, and therefore the binary number output will be incorrect over small angular ranges corresponding to the sector boundaries. The worst error can occur at the boundary between sectors 7 and 8, where the output is switching from 0111 to 1000. If the energy sensor corresponding to the first digit switches before the others, then for a very small angular range of movement the output will be 1111, indicating that sector 15 is aligned with the fixed disk rather than sector 7 or 8. This represents an error of 100% in the indicated angular position.

There are two ways used in practice to overcome this difficulty, which both involve an alteration to the manner in which windows are machined on the movable disk, as shown in Figure 19.7. The first of these methods adds an extra outer track on the disk, known as an anti-ambiguity track, which consists of small windows which span a small angular range on either side on each sector boundary of the main track system. When energy sensors associated with this extra track sense energy, this is used to signify that the disk is aligned on a sector boundary and the output is unreliable.

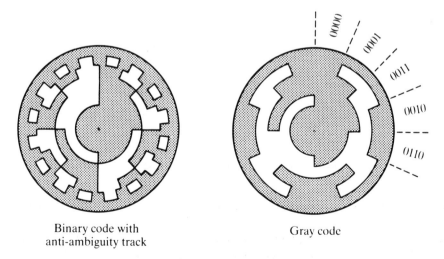

Binary code with anti-ambiguity track

Gray code

Figure 19.7 Modified window arrangements for the rotating disk

The second method is somewhat simpler and cheaper because it avoids the expense of machining the extra anti-ambiguity track. It does this by using a special code known as the Gray code to cut the tracks in each sector on the movable disk. The Gray code is a special binary representation where only one binary digit changes in moving from one decimal number representation to the next, i.e. from one sector to the next in the digital shaft encoder. The code is illustrated in Table 19.1.

It is possible to manufacture optical digital shaft encoders with up to 21 tracks. Such devices have a measurement resolution of 1 part in 10^6 (about 1 second of arc) but are very expensive. There is a high cost involved in the special photolithography techniques used to cut the windows and very high-quality mounts and bearings are needed.

19.4.2 Contacting (electrical) digital shaft encoder

The contacting digital shaft encoder consists of only one disk which rotates with the body whose displacement is being measured. The disk has conducting and non-conducting segments rather than the transparent and opaque areas found on the movable disk of the optical form of instrument, but these are arranged in an identical pattern of sectors and tracks. The disk is charged to a low potential by an electrical brush in contact with one side of the disk, and a set of brushes on the other side of the disk measures the potential in each track. The output of each detector brush is interpreted as a binary value of 1 or 0 according to whether the track in that particular segment is conducting or not and hence whether a voltage is sensed or not. As for the

Table 19.1 The Gray code

Decimal	Binary	Gray
0	0000	0000
1	0001	0001
2	0010	0011
3	0011	0010
4	0100	0110
5	0101	0111
6	0110	0101
7	0111	0100
8	1000	1100
9	1001	1101
10	1010	1111
11	1011	1110
12	1100	1010
13	1101	1011
14	1110	1001
15	1111	1000

case of the optical form of instrument, these outputs are combined together to give a multi-bit binary number.

Contacting digital shaft encoders have a similar cost to the equivalent optical instruments and have operational advantages in severe environmental conditions of high temperature or mechanical shock. They suffer from the usual problem of output ambiguity at the sector boundaries but this problem is overcome by the same methods as used in optical instruments.

A serious problem in the application of contacting digital shaft encoders arises from their use of brushes. These introduce friction into the measurement system, and the combination of dirt and brush wear causes contact problems. Consequently, problems of intermittent output can occur, and such instruments generally have limited reliability and a high maintenance cost. Measurement resolution is also limited because of the lower limit on the minimum physical size of the contact brushes. The maximum number of tracks possible is ten, which limits the resolution to 1 part in 1000. Thus contacting digital shaft encoders are only used where the environmental conditions are too severe for optical instruments.

19.4.3 **Magnetic digital shaft encoder**

Magnetic digital shaft encoders consist of a single rotatable disk, as in the contacting form of encoder discussed in the previous section. The pattern of sectors and tracks consists of magnetically conducting and non-conducting segments, and the sensors aligned with each track consist of small toroidal magnets. Each of these sensors has a coil wound on it which has a high or low voltage induced in it according to the magnetic field close to it. This field is dependent on the magnetic conductivity of that segment of the disk which is closest to the toroid.

These instruments have no moving parts in contact and therefore have a similar reliability to optical devices. Their major advantage over optical equivalents is an ability to operate in very harsh environmental conditions. Unfortunately, the process of manufacturing and accurately aligning the toroidal magnet sensors required makes such instruments very expensive. Their use is therefore limited to a few applications where both high measurement resolution and operation in harsh environments are required.

19.5 **The resolver**

The resolver, also known as a synchro-resolver, is an electromechanical device which gives an analog output by transformer action. Physically, resolvers resemble a small a.c. motor and have a diameter ranging from 10 mm to 100 mm. They have no contacting moving surfaces, and are therefore frictionless and reliable in operation, and have a long life. The best devices give measurement resolutions of 0.1%.

Resolvers have two stator windings, which are mounted at right angles to one another, and a rotor, which can have either one or two windings. As the angular position of the rotor changes, the output voltage changes. The simpler configuration of a resolver with only one winding on the rotor is illustrated in Figure 19.8. This exists in two separate forms which are distinguished according to whether the output voltage changes in amplitude or in phase as the rotor rotates relative to the stator winding.

19.5.1 **Varying-amplitude output resolver**

The stator of this type of resolver is excited with a single-phase sinusoidal voltage of frequency ω, where the amplitudes in the two windings are given by:

$$V_1 = V \sin(\beta) \quad V_2 = V \cos(\beta)$$

where $V = V_s \sin(\omega t)$. The effect of this is to give a field at an angle of $\beta + \pi/2$ relative to stator winding 1. (A full proof of this can be found in Healey (1975).)

Suppose that the angle of the rotor winding relative to that of the stator winding is given by θ. Then the magnetic coupling between the windings is a maximum for $\theta = \beta + \pi/2$ and a minimum for $\theta = \beta$. The rotor output voltage (see Healey (1975) for proof) is of fixed frequency and varying amplitude given by:

$$V_0 = KV_s \sin(\beta - \theta) \sin(\omega t)$$

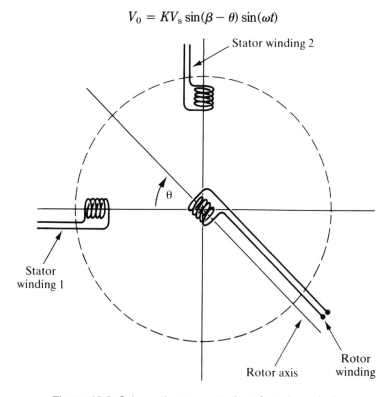

Figure 19.8 Schematic representation of resolver windings

This relationship between shaft angle position and output voltage is non-linear, but approximate linearity is obtained for small angular motions where $|\beta - \theta| < 15$ degrees.

An intelligent version of this type of resolver is now available which uses a microprocessor to process the sine and cosine outputs, giving a measurement resolution of 2 minutes of arc (Analogue Devices 1988).

19.5.2 **Varying phase output resolver**

This is a less common form of resolver but is used in a few applications. The stator windings are excited with a two-phase sinusoidal voltage of frequency ω, and the instantaneous voltage amplitudes in the two windings are given by:

$$V_1 = V_s \sin(\omega t) \quad V_2 = V_s \sin(\omega t + \pi/2) = V_s \cos(\omega t)$$

The net output voltage in the rotor winding is the sum of the voltages induced due to each stator winding. This is given by:

$$V_0 = KV_s \sin(\omega t) \cos(\theta) + KV_s \cos(\omega t) \cos(\pi/2 - \theta)$$

$$= KV_s[\sin(\omega t) \cos(\theta) + \cos(\omega t) \sin(\theta)]$$

$$= KV_s \sin(\omega t + \theta)$$

This represents a linear relationship between shaft angle and the phase shift of the rotor output relative to the stator excitation voltage. The measurement accuracy of shaft rotation, which can be up to $\pm 0.1\%$, depends on the accuracy with which the phase shift can be measured. This is improved by increasing the excitation frequency, ω, but this also increases magnetizing losses. Consequently, a compromise excitation frequency of about 400 Hz is used. The conversion of the phase shift into a varying-amplitude output signal requires relatively complex and expensive electronics. It is mainly for this reason of cost that this type of resolver is not very common.

19.6 **The synchro**

Like the resolver, the synchro is a motor-like, electromechanical device with an analog output. Apart from having three stator windings instead of two, the instrument is similar in appearance and operation to the resolver and has the same range of physical dimensions. Also, like the resolver, the rotor can have either one or two windings, and usually has a dumb-bell shape.

Synchros have been in use for many years for the measurement of angular position, especially in military applications, and achieve similar levels of accuracy and measurement resolution to digital encoders. One common application is axis measurement in machine tools, where the translational motion of the tool is translated into a rotational displacement by suitable gearing. Synchros are tolerant of high temperatures, high humidity, shock and vibration and are therefore suitable for operation in such harsh environmental conditions. Some maintenance problems are associated with the slip ring and brush system used to supply power to the rotor. However, the only major source of error in the instrument is asymmetry in the windings, and measurement accuracies of $\pm0.5\%$ are easily achievable.

Figure 19.9 shows the simpler form of synchro with a single rotor winding. If an a.c. excitation voltage is applied to the rotor via slip rings and brushes, this sets up a certain pattern of fluxes and induced voltages in the stator windings by transformer action.

For a rotor excitation voltage, V_r, given by:

$$V_r = V\sin(\omega t)$$

the voltages induced in the three stator windings are:

$$V_1 = V\sin(\omega t)\sin(\beta)$$
$$V_2 = V\sin(\omega t)\sin(\beta + 2\pi/3)$$
$$V_3 = V\sin(\omega t)\sin(\beta - 2\pi/3)$$

where β is the angle between the rotor and stator windings.

If the rotor is turned at constant velocity through one full revolution, the voltage waveform induced in each stator winding is as shown in Figure 19.10. This has the form of a carrier-modulated waveform in which the carrier frequency corresponds to the excitation frequency, ω. It follows that if the rotor is stopped at any particular angle, β', the peak-to-peak amplitude of the stator voltage is a function of β'. If therefore the stator winding voltage is measured, generally as its root-mean-square (r.m.s.) value, this indicates the magnitude of the rotor rotation away from the null position. The direction of rotation is determined by the phase difference between the stator voltages, which is indicated by their relative instantaneous magnitudes.

Although a single synchro thus provides a means of measuring angular displacements, it is much more common to find a pair of them used for this purpose. When used in pairs, one member of the pair is known as the synchro transmitter and the other as the synchro transformer, and the two sets of stator windings are connected together, as shown in Figure 19.11. Each synchro is of the form shown in Figure 19.9, but the rotor of the transformer is fixed for displacement measuring applications. A sinusoidal excitation voltage is applied to the rotor of the transmitter, setting up a pattern of fluxes and induced voltages in the transmitter stator windings. These voltages are transmitted to the transformer stator windings where a similar flux pattern is established. This in turn causes a sinusoidal voltage to be induced in the fixed transformer rotor winding. For an excitation voltage, $V\sin(\omega t)$, applied to the transmitter rotor, the voltage measured in the transformer rotor is given by:

$$V_0 = V\sin(\omega t)\sin(\theta)$$

where θ is the relative angle between the two rotor windings.

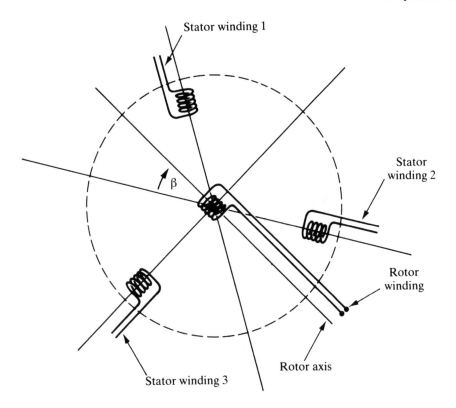

Figure 19.9 Schematic representation of synchro windings

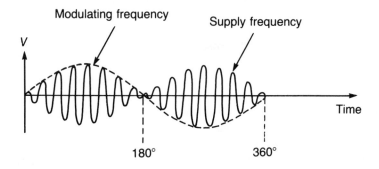

Figure 19.10 Synchro stator voltage waveform

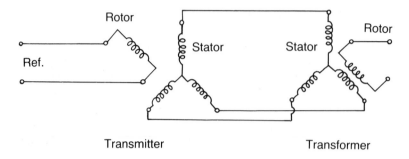

Figure 19.11 Synchro transmitter–transformer pair

Apart from their use as a displacement transducer, such synchro pairs are commonly used to transmit angular displacement information over some distance, for instance to transmit gyro compass measurements in an aircraft to remote meters. They are also used for load positioning, allowing a load connected to the transformer rotor shaft to be controlled remotely by turning the transmitter rotor. For these applications, the transformer rotor is free to rotate and is also damped to prevent oscillatory motions. In the simplest arrangement, a common sinusoidal excitation voltage is applied to both rotors. If the transmitter rotor is turned, this causes an imbalance in the magnetic flux patterns and results in a torque on the transformer rotor which tends to bring it into line with the transmitter rotor. This torque is typically small for small displacements, and so this technique is only useful if the load torque on the transformer shaft is very small. In other circumstances, it is necessary to incorporate the synchro pair within a servomechanism, where the output voltage induced in the transformer rotor winding is amplified and applied to a servomotor which drives the transformer rotor shaft until it is aligned with the transmitter shaft.

19.7 **The induction potentiometer**

These instruments belong to the same class as resolvers and synchros but have only one rotor winding and one stator winding. They are of a similar size and appearance to other devices in the class. A single-phase sinusoidal excitation is applied to the rotor winding and this causes an output voltage in the stator winding through the mutual inductance linking the two windings. The magnitude of this induced stator voltage varies with the rotation of the rotor. The variation of the output with rotation is naturally sinusoidal if the coils are wound such that their field is concentrated at one point, and only small excursions can be made away from the null position if the output relationship is to remain approximately linear. However, if the rotor and stator windings are distributed around the circumference in a special way, an approximately linear relationship for angular displacements of up to ±90 degrees can be obtained.

19.8 **The rotary inductosyn**

This instrument is similar in operation to the linear inductosyn, except that it measures rotary displacements and has tracks which are arranged radially on two circular disks, as shown in Figure 19.12. Typical diameters of the instrument vary between 75 mm and 300 mm. The larger versions give a measurement resolution of up to 0.05 seconds of arc. Like its linear equivalent, however, the rotary inductosyn has a very small measurement range, and a lower-resolution rotary displacement transducer with a larger measurement range must be used in conjunction with it.

19.9 **Gyroscopes**

Gyroscopes measure both absolute angular displacement and absolute angular velocity. The predominance of mechanical, spinning-wheel gyroscopes in the market place is now being challenged by recently introduced optical gyroscopes.

19.9.1 **Mechanical gyroscopes**

Mechanical gyroscopes consist essentially of a large, motor-driven wheel whose angular momentum is such that the axis of rotation tends to remain fixed in space, thus acting as a reference point. The gyro frame is attached to the body whose motion is to be measured. The output is measured in terms of the angle between the frame and the axis of the spinning wheel. Two different forms of mechanical gyroscope are used for measuring angular displacement, the free gyro and the rate-integrating gyro. A third type of mechanical gyroscope, the rate gyro, measures angular velocity and is described in Chapter 20.

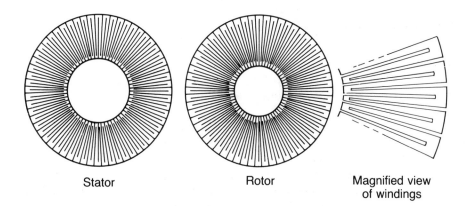

Stator　　　　　　　　Rotor　　　　　　　Magnified view
of windings

Figure 19.12 The rotary inductosyn

Free gyroscope

The free gyroscope is illustrated in Figure 19.13. This measures the absolute angular rotation of the body to which its frame is attached about two perpendicular axes. Two alternative methods of driving the wheel are used in different versions of the instrument. One of these is to enclose the wheel in stator-like coils which are excited with a sinusoidal voltage. A voltage is applied to the wheel via slip rings at both ends of the spindle carrying the wheel. The wheel behaves as a rotor and motion is produced by motor action. The other, less common, method is to fix vanes on the wheel which is then driven by directing a jet of air on to the vanes.

The instrument can measure angular displacements of up to 10 degrees with a high

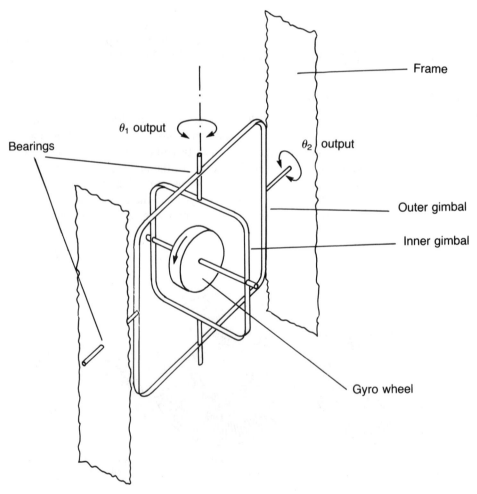

Figure 19.13 The free gyroscope

accuracy. For greater angular displacements, the interaction between the measurements on the two perpendicular axes starts to cause a serious loss of accuracy. The physical size of the coils in the motor-action-driven system also limits the measurement range to 10 degrees. For these reasons, this type of gyroscope is only suitable for measuring rotational displacements of up to 10 degrees.

A further operational problem of free gyroscopes is the presence of angular drift (precession) due to bearing friction torque. This has a typical magnitude of 0.5 degrees per minute and means that the instrument can only be used over short time intervals of, say, 5 minutes. This time duration can be extended if the angular momentum of the spinning wheel is increased.

A major application of gyroscopes is in inertial navigation systems. Only two free gyros mounted along orthogonal axes are needed to monitor motion in three dimensions, because each instrument measures displacement about two axes. The limited angular range of measurement is not usually a problem in such applications, as control action prevents the error in the direction of motion about any axis ever exceeding 1 or 2 degrees. Precession is a much greater problem, however, and for this reason the rate-integrating gyro is used much more commonly.

Rate-integrating gyroscope

The rate-integrating gyroscope, or 'integrating gyro' as it is commonly known, is illustrated in Figure 19.14. It measures angular displacements about a single axis only, and therefore three instruments are required in a typical inertial navigation system. The major advantage of the instrument over the free gyro is the almost total absence of precession, with typical specifications quoting drifts of only 0.01 degrees per hour. The instrument has a first-order type of response given by:

$$\frac{\theta_0}{\alpha_i}(D) = \frac{K}{\tau D + 1} \tag{19.1}$$

where $K = H/\beta$, $\tau = M/\beta$, α_i is the input angle, θ_0 is the output angle, D is the D operator, H is the angular momentum, M is the moment of inertia of the system about the measurement axis and β is the damping coefficient. (The mathematical derivation of this equation is given in Doebelin (1975).)

Inspection of Equation (19.1) shows that to obtain a high value of measurement sensitivity, K, a high value of H and a low value of β are required. A large H is normally obtained by driving the wheel with a hysteresis-type motor revolving at high speeds of up to 24 000 rpm. The damping coefficient β can only be reduced so far, however, because a small value of β results in a large value for the system time constant, τ, and an unacceptably low speed of system response. Therefore, the value of β has to be chosen as a compromise between these constraints.

Besides their use as a fixed reference in inertial guidance systems, integrating gyros are also commonly used within aircraft autopilot systems and in military applications such as stabilizing weapon systems in tanks.

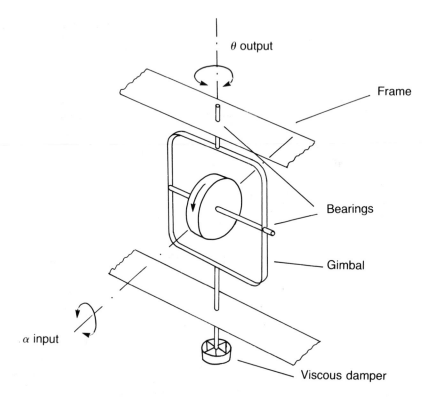

Figure 19.14 The rate-integrating gyroscope

19.9.2 **Optical gyroscopes**

Optical gyroscopes have been developed only recently and come in two forms, the ring laser gyroscope and the fiber optic gyroscope.

The **ring laser gyroscope** consists of a glass ceramic chamber containing a helium–neon gas mixture in which two laser beams are generated by a single-anode/twin-cathode system, as shown in Figure 19.15. Three mirrors, supported by the ceramic block and mounted in a triangular arrangement, direct the pair of laser beams around the cavity in opposite directions. Any rotation of the ring affects the coherence of the two beams, raising one in frequency and lowering the other. The clockwise and anticlockwise beams are directed into a photodetector which measures the beat frequency according to the frequency difference which is proportional to the angle of rotation. A more detailed description of the mode of operation can be found elsewhere (Nuttall 1987).

The advantages of the ring laser gyroscope are considerable. The measurement accuracy obtained is substantially better than that afforded by mechanical gyros in a similar price range. The device is also considerably smaller physically, which is of considerable benefit in many applications.

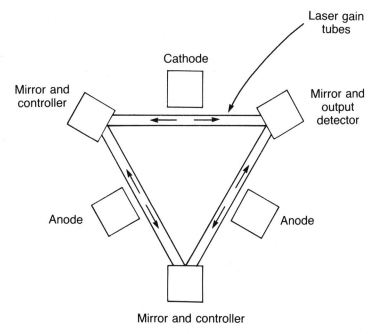

Figure 19.15 Ring laser gyroscope

The **fiber optic gyroscope** measures angular velocity and is described in Chapter 20.

19.10 **Choice between different instruments**

Choosing between the various rotational displacement transducers which might be used in any particular measurement situation depends first of all upon whether absolute measurement of angular position is required or whether the measurement of rotation relative to some arbitrary starting point is acceptable. Other factors affecting the choice between instruments are the required measurement range, the resolution of the transducer and the measurement accuracy afforded.

Where only measurement of relative angular position is required, the incremental encoder is a very suitable instrument. The best commercial instruments of this type can measure rotations to a resolution of 1 part in 20 000 of a full revolution, and the measurement range is an infinite number of revolutions. Instruments with such a high measurement resolution are very expensive, but much cheaper versions are available according to what lower level of measurement resolution is acceptable.

All the other instruments presented in this chapter provide an absolute measurement of angular position. The required measurement range is a dominant factor in the choice between them. If this exceeds one full revolution, then the only instrument available is the helical potentiometer. Such devices can measure rotations of up to 60 full turns, but they are expensive because the procedure involved in manufacturing a helical resistance element to a reasonable standard of accuracy is difficult.

For measurements of less than one full revolution, the range of available instruments widens. The cheapest one available is the circular potentiometer, but much better measurement accuracy and resolution is obtained from coded disk encoders. The cheapest of these is the optical form, but certain operating environments necessitate the use of the alternative contacting (electrical) and magnetic versions. All types of coded disk encoder are particularly attractive in computer control schemes, as the output of encoders is in digital form. A varying-phase output resolver is yet another instrument which can measure angular displacements up to one full revolution in magnitude. Unfortunately, these instruments are expensive because of the complicated electronics incorporated to measure the phase variation and convert it to a varying-amplitude output signal, and hence their use is not common.

An even greater range of instruments becomes available as the required measurement range is reduced further. These include the synchro (± 90 degrees), the varying-amplitude output resolver (± 90 degrees), the induction potentiometer (± 90 degrees) and the differential transformer (± 40 degrees). All these instruments have a high reliability and a long service life.

Finally, two further instruments are available for satisfying special measurement requirements, the rotary inductosyn and the gyroscope. The rotary inductosyn is used in applications where very high measurement resolution is required, although the measurement range afforded is extremely small and a coarser-resolution instrument must be used in parallel to extend the measurement range. Gyroscopes, in both mechanical and optical forms, are used to measure small angular displacements up to ± 10 degrees in magnitude in inertial navigation systems and similar applications.

References and further reading

Analogue Devices (1988) 'Resolver to digital converter', *Measurement and Control*, **21**, 291.

Doebelin, E. O. (1975) *Measurement Systems*, McGraw-Hill: New York.

Healey, M. (1975) *Principles of Automatic Control*, Hodder and Stoughton: London.

Nuttall, J. D. (1987) 'Optical gyroscopes', *Electronics and Power*, **33**, 703–7.

<u>20</u> Rotational velocity transducers

The main application of rotational velocity transducers is in speed control systems. They also provide the usual means of measuring translational velocities, which are transformed into rotational motions for measurement purposes by suitable gearing. Many different instruments and techniques are available for measuring rotational velocity as presented below. The factors governing the choice between them for particular applications are discussed at the end of the chapter.

20.1 Differentiation of angular displacement measurements

Angular velocity measurements can be obtained by differentiating the output signal from angular displacement transducers. Unfortunately, the process of differentiation amplifies any noise in the measurement signal, and therefore this technique has rarely been used in the past. The technique has become more feasible with the advent of intelligent instruments, and one such instrument which processes the output of a resolver claims a velocity measurement accuracy of ±1% (Analogue Devices 1988).

20.2 Integration of the output from an accelerometer

In measurement systems which already contain an angular acceleration transducer, such as a gyro accelerometer, it is possible to obtain a velocity measurement by integrating the acceleration measurement signal. This produces a signal of acceptable quality, as the process of integration attenuates any measurement noise. However, the method is of limited value in many measurement situations because the measurement obtained is the average velocity over a period of time, rather than a profile of the instantaneous velocities as motion takes place along a particular path.

20.3 Direct current tachometric generator

The d.c. tachometric generator, or d.c. tachometer as it is generally known, has an output which is approximately proportional to its speed of rotation. Its basic structure is identical to that found in a standard d.c. generator used for producing power, and is shown in Figure 20.1. Both permanent-magnet types and separately excited field types are used. However, certain aspects of the design are optimized to improve its accuracy as a speed measuring instrument. One significant design modification is to reduce the weight of the rotor by constructing the windings on a hollow fiberglass shell. The effect of this is to minimize any loading effect of the instrument on the system being measured.

The d.c. output voltage from the instrument is of a relatively high magnitude, giving a high measurement sensitivity which is typically 5 volts per 1000 rpm. The direction of rotation is determined by the polarity of the output voltage. A common range of measurement is 0–5000 rpm. Maximum non-linearity is usually about ±1% of full-scale reading.

One problem with these devices which can cause further problems under some circumstances is the presence of an a.c. ripple in the output signal. The magnitude of this ripple can be up to 2% of the output d.c. level.

20.4 **Alternating current tachometric generator**

The a.c. tachometric generator, or a.c. tachometer as it is generally known, has an output approximately proportional to rotational speed, as in the d.c. tachogenerator. Its mechanical structure takes the form of a two-phase induction motor, with two stator windings and (usually) a drag-cup rotor, as shown in Figure 20.2. One of the stator windings is excited with an a.c. voltage and the measurement signal is taken from the output voltage induced in the second winding. The magnitude of this output voltage is zero when the rotor is stationary, and otherwise proportional to the angular velocity of the rotor. The direction of rotation is determined by the phase of the output voltage, which switches by 180 degrees as the direction reverses. Therefore, both the phase and the magnitude of the output voltage have to be measured. A

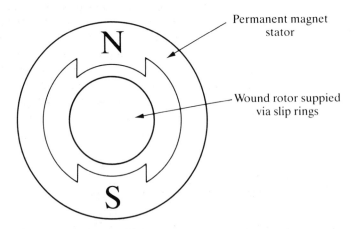

Figure 20.1 The d.c. tachometer

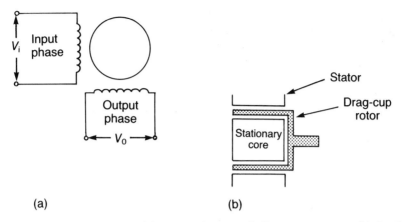

Figure 20.2 The a.c. tachometer: (a) rotor and stator winding arrangement; (b) detail of drag-cup rotor

typical range of measurement is 0–4000 rpm with an accuracy of ±0.05% of full-scale reading.

Whilst the form of a.c. tachometer described above is the commonest one, a second form also exists. This has a squirrel-cage rather than a drag-cup rotor and so is cheaper. Its structure and mode of operation are otherwise identical, but the measurement accuracy is reduced.

20.5 Drag-cup tachometric generator

Like the last two devices, this instrument also has a more common name, the 'drag-cup tachometer'. The instrument has a central spindle carrying a permanent magnet which rotates inside a non-magnetic drag cup consisting of a cylindrical sleeve of electrically conductive material, as shown in Figure 20.3. As the spindle and magnet rotate, a voltage is induced which causes circulating eddy currents in the cup. These currents interact with the magnetic field from the permanent magnet and produce a torque. In response, the drag cup turns until the induced torque is balanced by the torque due to the restraining springs connected to the cup. When equilibrium is reached, the angular displacement of the cup is proportional to the rotational velocity of the central spindle. The instrument has a typical measurement accuracy of ±0.5% and is commonly used in the speedometers of motor vehicles and as a speed indicator for aeroengines.

20.6 Variable reluctance velocity transducer

The form of a variable reluctance transducer is shown in Figure 20.4. It can be seen that this consists of two parts, a rotating disk connected to the moving body being

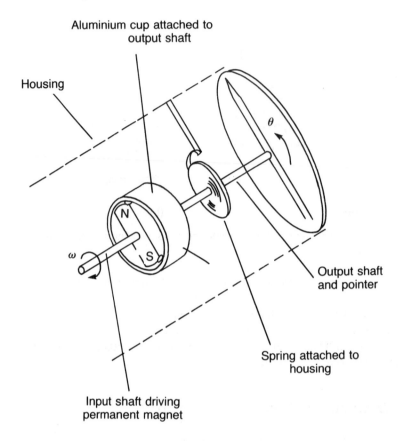

Aluminium cup attached to
output shaft

Housing

θ

ω

Output shaft
and pointer

Spring attached to
housing

Input shaft driving
permanent magnet

Figure 20.3 The drag-cup tachometer

measured and a pick-up unit. The rotating disk is constructed from a bonded-fiber material into which soft iron poles are inserted at regular intervals around its periphery. The pick-up unit consists of a permanent magnet with a shaped pole piece which carries a wound coil. The distance between the pick-up and the outer perimeter of the disk is about 0.5 mm.

As the disk rotates, the soft iron inserts on the disk move in turn past the pick-up unit. As each iron insert moves towards the pole piece, the reluctance of the magnetic circuit increases and hence the flux in the pole piece also increases. Similarly, the flux in the pole piece decreases as each iron insert moves away from the pick-up unit, and the pattern of flux changes with time as shown in Figure 20.5. The changing magnetic flux inside the pick-up coil causes a voltage to be induced in the coil whose magnitude is proportional to the rate of change of flux. This voltage is positive whilst the flux is increasing and negative whilst it is decreasing and therefore its variation with time is as shown in Figure 20.6.

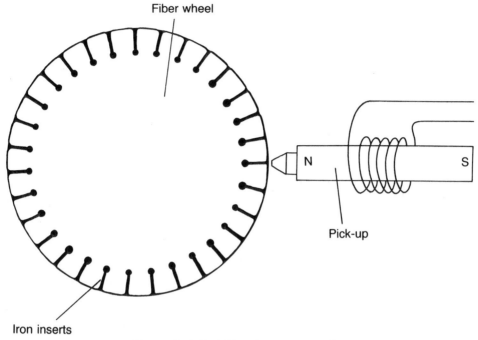

Fiber wheel

N

S

Pick-up

Iron inserts

Figure 20.4 Variable reluctance transducer

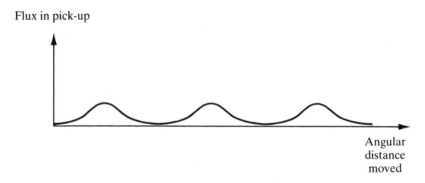

Flux in pick-up

Angular
distance
moved

Figure 20.5 Pattern of flux change with rotation of a variable reluctance transducer

The form of this output can be regarded as a sequence of positive and negative pulses whose frequency is proportional to the rotational velocity of the disk. This can be converted into an analog, varying-amplitude, d.c. voltage output by means of a frequency-to-voltage converter circuit connected to the output terminals of the pick-up. However, greater measurement accuracy can be obtained by converting the output waveform into sharp pulses which are counted by an electronic counter. It is

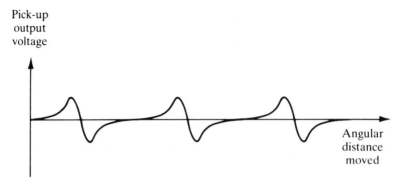

Figure 20.6 Pattern of induced voltage change with rotation of a variable reluctance transducer

normal procedure to produce the pulses at each instant that the induced voltage in the coil changes sign as it passes through zero. This is achieved by electronic means.

The maximum angular velocity which the instrument can measure is limited because of the finite width of the induced pulses. As the velocity increases, the distance between the pulses is reduced, and at a certain velocity, the pulses start to overlap. At this point, the pulse counter ceases to be able to distinguish the separate pulses.

A simpler and cheaper form of variable reluctance transducer uses a ferromagnetic gear wheel in place of a fiber disk. The motion of the tip of each gear tooth towards and away from the pick-up unit causes a similar variation in the flux pattern to that produced by the iron inserts in the fiber disk. The pulses produced by these means are less sharp, however, and consequently the maximum angular velocity measurable is lower.

In either form of variable reluctance transducer, simply outputting the total pulse count measured over a certain length of time only gives information about the average velocity over that period. Measurement of the actual velocities at the instants of time that each output pulse occurs can be achieved by the scheme shown in Figure 20.7. In this circuit, the pulses from the transducer gate the train of pulses from a 1 MHz clock into a counter. Control logic resets the counter and updates the digital output value after receipt of each pulse from the transducer. The measurement resolution of this system is highest when the speed of rotation is low.

20.7 **Photoelectric pulse-counting methods**

An alternative to the variable reluctance transducer, but which uses very similar principles to it, is the method where pulses are produced by photoelectric techniques and counted. These pulses are generated by one of the two alternative methods illustrated in Figure 20.8. In Figure 20.8(a), the pulses are produced as the windows

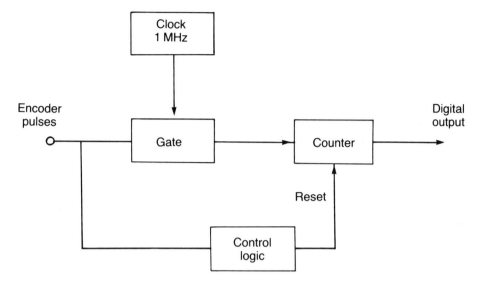

Figure 20.7 Scheme to measure instantaneous angular velocities

in a slotted disk pass in sequence between a light source and a detector. The alternative form, Figure 20.8(b), has both light source and detector mounted on the same side of a reflective disk which has black sectors painted on to it at regular angular intervals. In either case, the pulses are counted by an electronic counter. The frequency of the pulses is proportional to the angular velocity of the body connected to the rotating disk. Pulses generated in this manner are narrower than those generated by magnetic means and so the instrument is capable of measuring higher velocities.

As for the variable reluctance transducer, a simple pulse count only yields information about the average velocity, but instantaneous velocities can be measured using the scheme illustrated in Figure 20.7.

20.8 **Stroboscopic methods**

The stroboscopic technique of rotational velocity measurement operates on a similar physical principle to the variable reluctance and photoelectric pulse-counting methods, except that the pulses involved consist of flashes of light generated electronically and whose frequency is adjustable so that it can be matched with the frequency of occurrence of some feature on the rotating body being measured. This feature can either be some naturally occurring one such as the spokes of a wheel or gear teeth, or it can be an artificially created pattern of black and white stripes. In either case, the rotating body appears stationary when the frequencies of the light pulses and body features are in synchronism. Flashing rates up to 25 000 per minute are available from

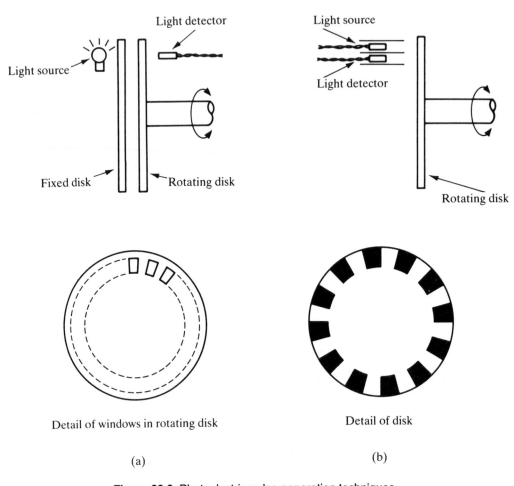

Detail of windows in rotating disk

Detail of disk

(a)

(b)

Figure 20.8 Photoelectric pulse generation techniques

commercial stroboscopes according to the range of velocity measurement required, and the typical measurement accuracy obtained is ±1% of the reading.

Measurement of the flashing rate at which the rotating body appears stationary does not automatically indicate the rotational velocity, because synchronism also occurs when the flashing rate is some integral submultiple of the rotational speed. The practical procedure followed is therefore to adjust the flashing rate until synchronism is obtained at the largest flashing rate possible, R_1. The flashing rate is then carefully decreased until synchronism is again achieved at the next lowest flashing rate, R_2. The rotational velocity is then given by:

$$v = \frac{R_1 R_2}{R_1 - R_2}$$

20.9 **Mechanical flyball**

The mechanical flyball is a velocity measuring instrument which was first developed many years ago and is still used extensively in speed-governing systems for engines, turbines, etc. As shown in Figure 20.9, it consists of a pair of spherical balls pivoted on the rotating shaft. These balls move outwards under the influence of centrifugal forces as the rotational velocity of the shaft increases and lift a pointer against the resistance of a spring. The pointer can be arranged to give a visual indication of speed by causing it to move in front of a calibrated scale, or its motion can be converted by a translational displacement transducer into an electrical signal.

In equilibrium, the centrifugal force, F_c, is balanced by the spring force, F_s, where:

$$F_c = K_c \omega^2 \quad F_s = K_s x$$

and K_c and K_s are constants, ω is the rotational velocity and x is the displacement of the pointer. Thus:

$$K_c \omega^2 = K_s x \quad \text{or} \quad \omega = (K_s x / K_c)^{1/2}$$

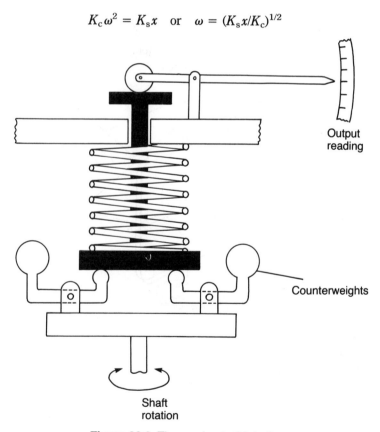

Output reading

Counterweights

Shaft rotation

Figure 20.9 The mechanical flyball

This is inconvenient because it involves a non-linear relationship between the pointer displacement and the rotational velocity. If this is not acceptable, a linear relationship can be obtained by using a spring with a non-linear characteristic such that $F_s = K'_s x^2$. This gives:

$$\omega = (K'_s/K_c)^{1/2} x$$

20.10 The rate gyroscope

The rate gyro, illustrated in Figure 20.10, has an almost identical construction to the rate-integrating gyro described in the previous chapter, and differs only by including a spring system which acts as an additional restraint on the rotational motion of the frame. The instrument measures the absolute angular velocity of a body, and is widely used in generating stabilizing signals within vehicle navigation systems. The typical measurement resolution given by the instrument is 0.01 degrees per second.

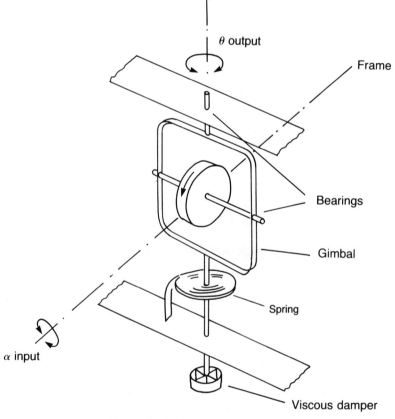

Figure 20.10 The rate gyroscope

The angular velocity, α, of the body is related to the angular deflection of the gyroscope, θ, by the equation:

$$\frac{\theta}{\alpha} (D) = \frac{HD}{MD^2 + \beta D + K} \qquad (20.1)$$

where H is the angular momentum of the spinning wheel, M is the moment of inertia of the system, β is the viscous damping coefficient, K is the spring constant, and D is the D operator. (The derivation of this equation is given in Doebelin (1975).)

The relationship (20.1) is a second-order differential equation and therefore we must expect the device to have a response typical of second-order instruments, as discussed in Chapter 2. The instrument must therefore be designed carefully so that the output response is neither oscillatory nor too slow in reaching a final reading. To assist in the design process, it is useful to re-express Equation (20.1) in the following form:

$$\frac{\theta}{\alpha} (D) = \frac{K'}{D^2/\omega^2 + 2\varepsilon D/\omega + 1} \qquad (20.2)$$

where $K' = H/K$, $\omega = (K/M)^{1/2}$ and $\varepsilon = \beta/2(KM)^{1/2}$.

The static sensitivity of the instrument, K', is made as large as possible by using a high-speed motor to spin the wheel and so make H high. Reducing the spring constant, K, further improves the sensitivity but this cannot be reduced too far as it makes the resonant frequency, ω, of the instrument too small. The value of β is chosen such that the damping ratio, ε, is close to 0.7.

20.11 Fiber optic gyroscope

This is a relatively new instrument which makes use of fiber optic technology. Incident light from a source is separated by a beam splitter into a pair of beams a and b, as shown in Figure 20.11. These travel in opposite directions around an optic fiber coil (which may be several hundred metres long) and emerge from the coil as the beams marked a' and b'. The beams a' and b' are directed by the beam splitter into an interferometer. Any motion of the coil causes a phase shift between a' and b' which is detected by the interferometer. The instrument has great promise but is currently (1993) still at the development stage. Further details can be found in Nuttall (1987).

20.12 Choice between different instruments

Choosing between different rotational velocity transducers is influenced strongly by whether an analog or digital form of output is required.

Probably the most common form of analog output device used is the d.c.

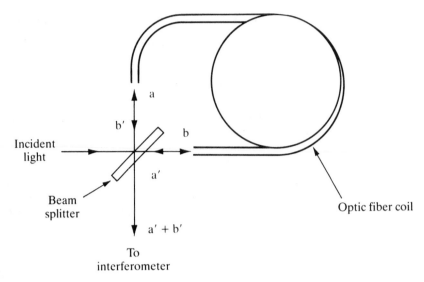

a

b′

b

Incident
light

a′

Beam
splitter

Optic fiber coil

a′ + b′

To
interferometer

Figure 20.11 The fiber optic gyroscope

tachometer. This is a relatively simple device which measures speeds up to about 5000 rpm with an accuracy of ±1%. Where better accuracy is required within a similar range of speed measurement, a.c. tachometers are used. The squirrel-cage rotor type has an accuracy of ±0.25% and drag-cup rotor types have accuracies up to ±0.05%.

Two other devices with an analog output are used in some applications, the drag-cup tachometer and the mechanical flyball. The drag-cup tachometer has an accuracy of only ±5% but is cheap and therefore very suitable for use in vehicle speedometers. The mechanical flyball has a linear displacement type of output which is commonly connected mechanically by a system of links to the throttle of an engine, thereby fulfilling a speed-governing function.

Digital-output-type instruments are represented by the variable reluctance transducer, devices using electronic light-pulse-counting methods, and the stroboscope. The first two of these are used to measure angular speeds up to about 10 000 rpm and the last one is used to measure speeds up to 25 000 rpm.

References and further reading

Analogue Devices (1988) 'Resolver to digital converter', *Measurement and Control*, **21**, 291.
Doebelin, E. O. (1975) *Measurement Systems*, McGraw-Hill: New York.
Nuttall, J. D (1987) 'Optical gyroscopes', *Electronics and Power*, **33**, 703–7.

21 Mass, force and torque measurement

21.1 Mass measurement

Mass describes the quantity of matter which a body contains. One way in which it can be measured is to compare the gravitational force on the body with the gravitational force on another body of known mass, using a beam balance, weigh beam, pendulum scale or electromagnetic balance, in a procedure known as 'weighing'. Alternative mass measurement techniques are to use either a spring balance or a load cell. Of these, the electronic load cell has definite advantages and is the preferred instrument nowadays in more and more industrial applications.

21.1.1 Beam balance (equal-arm balance)

In the beam balance, shown in Figure 21.1, standard masses are added to a pan on one side of a pivoted beam until the magnitude of the gravitational force on them balances the magnitude of the gravitational force on the unknown mass acting at the

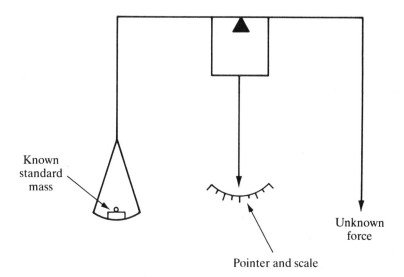

Known standard mass

Pointer and scale

Unknown force

Figure 21.1 The beam balance

other end of the beam. This equilibrium position is indicated by a pointer which moves against a calibrated scale.

Instruments of this type are capable of measuring a wide span of masses. Those at the top end of the range can typically measure masses up to 1000 g whereas those at the bottom end of the range can measure masses of less than 0.01 g. Measurement resolution can be as good as 1 part in 10^7 of full-scale reading if the instrument is designed and manufactured very carefully. The lowest measurement inaccuracy figure attainable is ±0.002%.

One serious disadvantage of this type of instrument is its lack of ruggedness. Continuous use and the inevitable shock loading which will occur from time to time both cause damage to the knife edges, leading to problems in measurement accuracy and resolution, as discussed in section 21.5. A further problem in industrial use is the relatively long time needed to make each measurement. For these reasons, the beam balance is normally reserved as a calibration standard and is not used in day-to-day production environments.

21.1.2 **Weigh beam**

The weigh beam, sketched in two alternative forms in Figure 21.2, operates on similar principles to the beam balance but is much more rugged. In the first form, standard masses are added to balance the unknown mass and fine adjustment is provided by a known mass which is moved along a notched, graduated bar until the pointer is brought to the null balance point. The alternative form has two or more graduated bars (three bars are shown in Figure 21.2). Each bar carries a different standard mass and these are moved to appropriate positions on the notched bar to balance the unknown mass. Versions of these instruments are used to measure masses up to 50 tonnes.

21.1.3 **Pendulum scale**

The pendulum scale, sketched in Figure 21.3, is another instrument which works on the mass-balance principle. The unknown mass is put on a platform which is attached by steel tapes to a pair of cams. Downward motion of the platform, and hence rotation of the cams, under the influence of the gravitational force on the mass, is opposed by the gravitational force acting on two pendulum-type masses attached to the cams. The amount of rotation of the cams when the equilibrium position is reached is determined by the deflection of a pointer against a scale. The shape of the cams is such that this output deflection is linearly proportional to the applied mass.

This instrument is particularly useful in some applications because it is a relatively simple matter to replace the pointer and scale system by a rotational displacement transducer which gives an electrical output. Various versions of the instrument can measure masses in the range between 1 kg and 500 t, with a typical measurement inaccuracy of ±0.1%.

One potential source of difficulty with the instrument is oscillation of the weigh

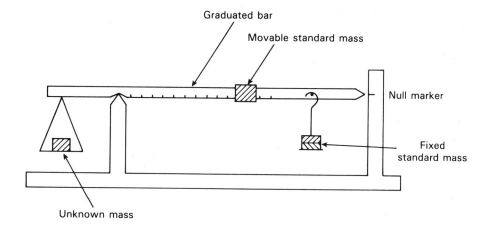

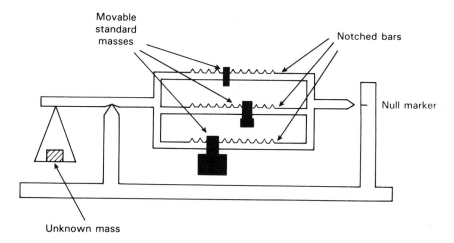

Figure 21.2 Two alternative forms of weigh beam

platform when the mass is applied. Where necessary, in instruments measuring larger masses, dashpots are incorporated into the cam system to damp out such oscillations. A further possible problem can arise, mainly when measuring large masses, if the mass is not placed centrally on the platform. This can be avoided by designing a second platform to hold the mass which is hung from the first platform by knife edges. This lessens the criticality of mass placement.

21.1.4 **Electromagnetic balance**

The electromagnetic balance uses the torque developed by a current-carrying coil suspended in a permanent magnetic field to balance the unknown mass against the

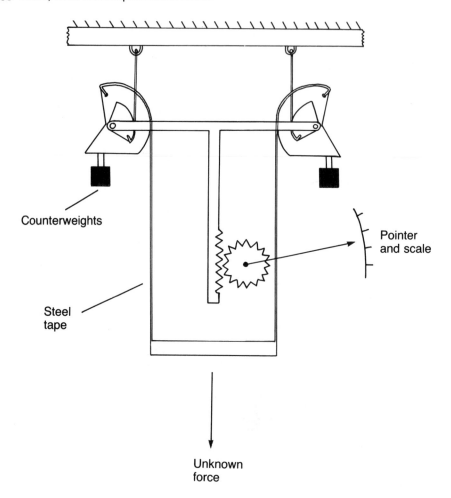

Counterweights

Steel
tape

Pointer
and scale

Unknown
force

Figure 21.3 The pendulum scale

known gravitational force produced on a standard mass, as shown in Figure 21.4. A light source and detector system is used to determine the null balance point. The voltage output from the light detector is amplified and applied to the coil, thus creating a servo system where the deflection of the coil in equilibrium is proportional to the applied force. Its advantages over beam balances, weigh beams and pendulum scales include its smaller size, its insensitivity to environmental changes (modifying inputs) and its electrical form of output.

21.1.5 Spring balance

The spring balance is a well-known instrument for measuring mass. The mass is hung on the end of a spring and the deflection of the spring due to the downward

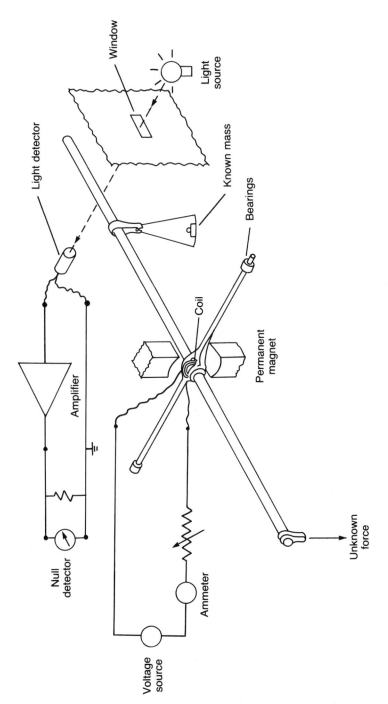

Figure 21.4 The electromagnetic balance

gravitational force on the mass is measured against a scale. Because the characteristics of the spring are very susceptible to environmental changes, measurement accuracy is usually relatively poor. If, however, compensation is made for the changes in spring characteristics, then a measurement inaccuracy of less than ±0.2% is achievable. According to the design of the instrument, masses between 0.5 kg and 10 t can be measured.

21.1.6 **Electronic load cell (electronic balance)**

Electronic load cells, also known as electronic balances, have significant advantages over most other forms of mass measuring instrument and so are the preferred type for most industrial applications. These advantages include relatively low cost, wide measurement range, tolerance of dusty and corrosive environments, remote measurement capability, tolerance of shock loading and ease of installation.

The electronic load cell uses the physical principle that a force applied to an elastic element produces a measurable deflection. The elastic elements used are specially shaped and designed, some examples of which are shown in Figure 21.5. The design aims are to obtain a linear output relationship between the applied force and the measured deflection and to make the instrument insensitive to forces which are not applied directly along the sensing axis. Various types of displacement transducer are used to measure the deflection of the elastic elements.

One problem which can affect the performance of this class of instruments is the phenomenon of creep. Creep describes the permanent deformation which an elastic element undergoes after it has been under load for a period of time. This can lead to significant measurement errors in the form of a bias on all readings if the instrument is not recalibrated from time to time. However, careful design and choice of materials can largely eliminate the problem.

The best accuracy in elastic force transducers is obtained by those instruments which use strain gauges to measure displacements of the elastic element, with an inaccuracy figure of less than ±0.05% of full-scale reading being obtainable. Typical resistance values of the strain gauges used are 120 Ω, 350 Ω and 480 Ω. Instruments of this type are used to measure masses over a very wide range between 0 and 3×10^6 kg. The measurement capability of an individual instrument designed to measure masses at the bottom end of this range would typically be 0.1–5 kg, whereas instruments designed for the top of the range would have a typical measurement span of 10–3000 t.

Elastic force transducers based on differential transformers (LVDTs) to measure deflections are used to measure masses up to 25 t. Apart from having a lower maximum measuring capability, they are also inferior to strain-gauge-based instruments in terms of their inaccuracy figure of ±0.2%. Their major advantage is their longevity and almost total lack of maintenance requirements.

The final type of displacement transducer used in this class of instrument is the piezoelectric device. Such instruments are used to measure masses in the range of 0 to 10^6 kg. Piezoelectric crystals replace the specially designed elastic member

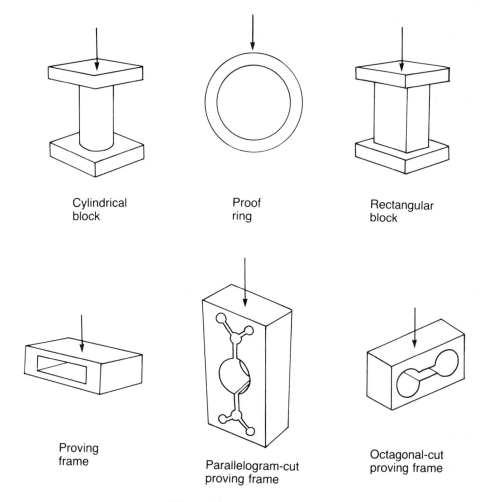

Cylindrical
block

Proof
ring

Rectangular
block

Proving
frame

Parallelogram-cut
proving frame

Octagonal-cut
proving frame

Figure 21.5 Elastic elements

normally used in this class of instrument, allowing the device to be physically small. As discussed previously, such devices can only measure dynamically changing forces because the output reading results from an induced electrical charge whose magnitude leaks away with time. The fact that the elastic element consists of the piezoelectric crystal means that it is very difficult to design such instruments to be insensitive to forces applied at an angle to the sensing axis. Therefore, special precautions have to be taken in applying these devices. Although they are relatively cheap, their lowest inaccuracy is ±1% of full-scale reading, and they also have a high temperature coefficient.

Either compression or tension load cells are used, depending on whether the mass is placed on top of a platform resting on the cell, which is therefore compressed, or whether the mass is hung from a cell, which is therefore put in tension. Compression-type, load-cell-based electronic balances normally contain several load cells, as illustrated in Figure 21.6. Such arrangements are dictated by the need for stability in the mechanical construction. Commonly, either three or four load cells are used in the balance, with the output mass measurement being formed from the sum of the outputs of each cell. Where appropriate, the upper platform can be replaced by a tank for weighing liquids, powders, etc.

21.1.7 Pneumatic/hydraulic load cells

Alternative forms of load cell also exist which work on either pneumatic or hydraulic principles and convert mass measurement into a pressure measurement problem. A pneumatic load cell is shown schematically in Figure 21.7. Application of a mass to the cell causes deflection of a diaphragm acting as a variable restriction in a nozzle flapper mechanism. The output pressure measured in the cell is approximately proportional to the magnitude of the gravitational force on the applied mass. The instrument requires a flow of air at its input of around $0.25 \, \text{m}^3/\text{h}$ at a pressure of 4 bar. Standard cells are available to measure a wide range of masses. For measuring small masses, instruments are available with a full-scale reading of 25 kg, whilst at the top of the range, instruments with a full-scale reading of 25 t are obtainable. Inaccuracy is typically $\pm 0.5\%$ of full scale in pneumatic load cells.

The alternative, hydraulic load cell is shown in Figure 21.8. In this, the gravitational force due to the unknown mass is applied, via a diaphragm, to oil contained within an enclosed chamber. The corresponding increase in oil pressure is measured by a suitable pressure transducer. These instruments are designed for measuring much larger masses than pneumatic cells, with a load capacity of 500 t being common. Special units can be obtained to measure masses as large as 50 000 t. Besides their much greater measuring range, hydraulic load cells are much more accurate than

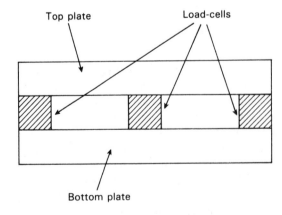

Figure 21.6 Load-cell-based electronic balance

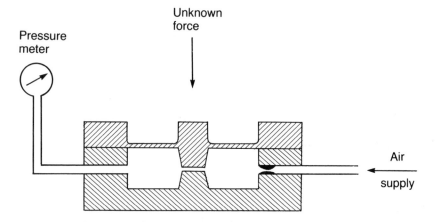

Figure 21.7 The pneumatic load cell

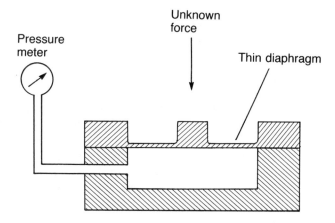

Figure 21.8 The hydraulic load cell

pneumatic cells, with an inaccuracy figure of ±0.05% of full scale being typical, although to obtain this level of accuracy, correction for the local value of g (acceleration due to gravity) is necessary. A measurement resolution of 0.02% is attainable.

21.1.8 Intelligent load cells

Intelligent load cells are formed by adding a microprocessor to a standard cell. This brings no improvement in accuracy because the load cell is already a very accurate device. What it does produce is an intelligent weighing system which can compute total cost from the measured weight, using stored cost per unit weight information, and provide an output in the form of a digital display. Cost per weight figures can be pre-stored for a large number of substances, making the instrument very flexible in its application.

21.2 **Force measurement**

If a force of magnitude F is applied to a body of mass M the body will accelerate at a rate A according to the equation:

$$F = MA$$

The standard unit of force is the newton, this being the force which will produce an acceleration of 1 metre per second squared in the direction of the force when it is applied to a mass of 1 kilogram. One way of measuring an unknown force is therefore to measure the acceleration when it is applied to a body of known mass.

An alternative technique is to measure the variation in the resonant frequency of a vibrating wire as it is tensioned by an applied force.

21.2.1 **Use of accelerometers**

The technique of applying a force to a known mass and measuring the acceleration produced can be carried out using any type of accelerometer. Unfortunately, the method is of very limited practical value because, in most cases, forces are not free entities but are part of a system (from which they cannot be decoupled) in which they are acting on some body which is not free to accelerate. However, the technique can be of use in measuring some transient forces, and also for calibrating the forces produced by thrust motors in space vehicles.

21.2.2 **Vibrating-wire sensor**

This instrument, illustrated in Figure 21.9, consists of a wire which is kept vibrating at its resonant frequency by a variable frequency oscillator. The resonant frequency of a wire under tension is given by:

$$f = \frac{0.5}{L} \left(\frac{M}{T} \right)^{1/2}$$

where M is the mass per unit length of the wire, L is the length of the wire, and T is the tension due to the applied force, F.

Thus, measurement of the output frequency of the oscillator allows the force applied to the wire to be calculated.

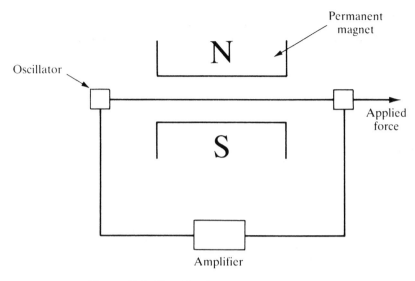

Figure 21.9 The vibrating-wire sensor

21.3 **Torque measurement**

Measurement of applied torques is of fundamental importance in all rotating bodies to ensure that the design of the rotating element is adequate to prevent failure under shear stresses. Torque measurement is also a necessary part of measuring the power transmitted by rotating shafts.

The three traditional methods of measuring torque consist of (a) measuring the reaction force in cradled shaft bearings, (b) the 'Prony brake' method and (c) measuring the strain produced in a rotating body due to an applied torque. However, recent developments in electronics and fiber optic technology now offer an alternative method also described below.

21.3.1 **Reaction forces in shaft bearings**

Any system involving torque transmission through a shaft contains both a power source and a power absorber where the power is dissipated. The magnitude of the transmitted torque can be measured by cradling either the power source or the power absorber end of the shaft in bearings and then measuring the reaction force F and the arm length L, as shown in Figure 21.10. The torque is then calculated as the simple product, FL. Pendulum scales are very commonly used for measuring the reaction force. Inherent errors in the method are bearing friction and windage torques.

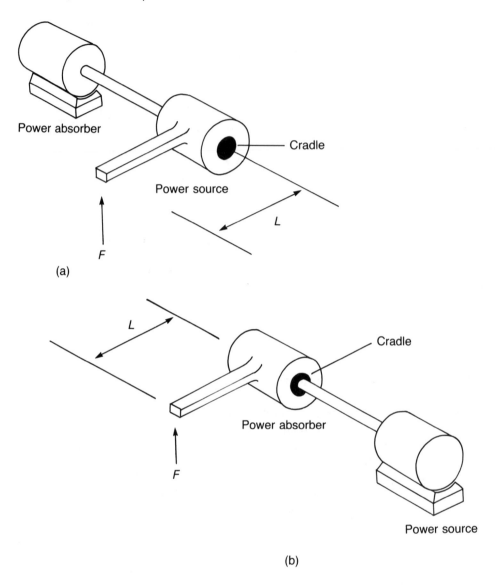

Figure 21.10 Measuring reaction forces in cradled shaft bearings: (a) cradled power source; (b) cradled power absorber

21.3.2 **Prony brake**

The principle of the Prony brake is illustrated in Figure 21.11. It is used to measure the torque in a rotating shaft and consists of a rope wound round the shaft, one end of the rope being attached to a spring balance and the other end carrying a load in the

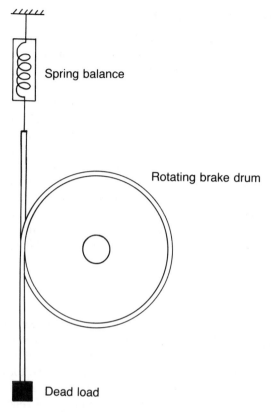

Spring balance

Rotating brake drum

Dead load

Figure 21.11 The Prony brake

form of a standard mass, m. If the measured force in the spring balance is F_s, then the effective force, F_e, exerted by the rope on the shaft is given by:

$$F_e = mg - F_s$$

If the radius of the shaft is R_s and that of the rope is R_r, then the effective radius, R_e, of the rope and drum with respect to the axis of rotation of the shaft is given by:

$$R_e = R_s + R_r$$

The torque in the shaft, T, can then be calculated as:

$$T = F_e R_e$$

Whilst this is a well-known method of measuring shaft torque, a lot of heat is generated because of friction between the rope and shaft, and water cooling is usually necessary.

21.3.3 **Measurement of induced strain**

Measuring the strain induced in a shaft due to an applied torque has been the most common method used for torque measurement in recent years. It is a very attractive method because it does not disturb the measured system by introducing friction torques in the same way as do the last two methods described. The method involves bonding four strain gauges on to the shaft as shown in Figure 21.12, where the strain gauges are arranged in a d.c. bridge circuit. The output from the bridge circuit is a function of the strain in the shaft and hence of the torque applied. It is very important that the positioning of the strain gauges on the shaft is precise, and the difficulty in achieving this makes the instrument relatively expensive.

The technique is ideal for measuring the stalled torque in a shaft before rotation commences. However, a problem is encountered in the case of rotating shafts because a suitable method then has to be found for making the electrical connections to the strain gauges. One solution to this problem found in many commercial instruments is to use a system of slip rings and brushes, although this increases the cost of the instrument still further.

21.3.4 **Optical torque measurement**

Optical techniques for torque measurement have become available recently with the development of laser diodes and fiber optic light transmission systems. One such system is shown in Figure 21.13. Two black-and-white-striped wheels are mounted at either end of the rotating shaft and are in alignment when no torque is applied to the shaft. Light from a laser diode light source is directed by a pair of fiber optic cables on to the wheels. The rotation of the wheels causes pulses of reflected light and these are transmitted back to a receiver by a second pair of fiber optic cables. Under zero-torque conditions, the two pulse trains of reflected light are in phase with each other.

If torque is now applied to the shaft, the reflected light is modulated. Measurement

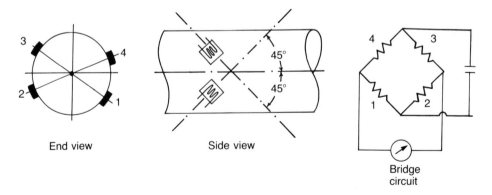

Figure 21.12 Position of torque measuring strain gauges on shaft

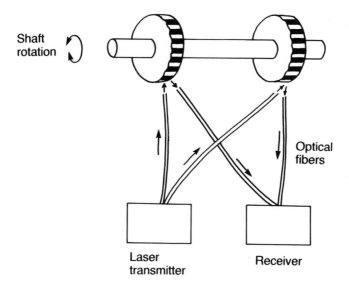

Figure 21.13 Optical torque measurment

by the receiver of the phase difference between the reflected pulse trains therefore allows the magnitude of torque in the shaft to be calculated. The cost of such instruments is relatively low, and an additional advantage in many applications is their small physical size.

22 Measurement of miscellaneous quantities

22.1 Viscosity measurement

Viscosity measurement is important in many process industries. In the food industry, the viscosity of raw materials such as dough, batter and ice cream has a direct effect on the quality of the product. Similarly, in other industries such as the ceramics industry, the quality of raw materials affects the quality of the final product. Viscosity control is also very important in assembly operations which involve the application of mastics and glue flowing through tubes. Clearly, successful assembly requires such materials to flow through tubes at the correct rate and therefore it is essential that their viscosity is correct.

Viscosity describes the way in which a fluid flows when it is subject to an applied force. Consider an elemental cubic volume of fluid and a shear force F applied to one of its faces of area A. If this face moves a distance L and at a velocity V relative to the opposite face of the cube under the action of F, the shear stress s and shear rate r are given by:

$$s = F/A \quad r = V/L$$

The coefficient of viscosity C_v is the ratio of shear stress to shear rate, i.e.:

$$C_v = s/r$$

C_v is often described simply as the 'viscosity'. A further term, kinematic viscosity, is also sometimes used, given by:

$$K_v = C_v/\rho$$

where K_v is the kinematic viscosity and ρ is the fluid density. To avoid confusion, C_v is often known as the dynamic viscosity to distinguish it from K_v. C_v is measured in units of poise or $N \, s/m^2$ and K_v is measured in units of stokes or m^2/s.

Viscosity was originally defined by Newton, who assumed that it was constant with respect to shear rate. However, it has since been shown that the viscosity of many fluids varies significantly at high shear rates and the viscosity of some varies even at low shear rates. The worst non-Newtonian characteristics tend to occur with emulsions, pastes and slurries. For non-Newtonian fluids, a subdivision into further classes can also be made according to the manner in which the viscosity varies with shear rate, as shown in Figure 22.1.

The relationship between the input variables and output measurement for instruments which measure viscosity normally assumes that the measured fluid has Newtonian characteristics. For non-Newtonian fluids, a correction must be made for shear rate variations (see Miller 1975, pp. 62–106). If such a correction is not made, the measurement obtained is known as the **apparent viscosity** and this can differ from the true viscosity by a large factor. The true viscosity is often called the **absolute viscosity** to avoid ambiguity. Viscosity also varies with fluid temperature and density.

Instruments for measuring viscosity work on one of three physical principles:

1. The rate of flow of the liquid through a tube.
2. The rate of fall of a body through the liquid.
3. The viscous friction force exerted on a rotating body.

22.1.1 Capillary and tube viscometers

These are the most accurate types of viscometer, with measurement inaccuracy levels typically down to ±0.3%. Liquid is allowed to flow, under gravity from a reservoir, through a tube of known cross-section. In different instruments, the tube can vary from capillary sized to a large diameter. The pressure difference across the

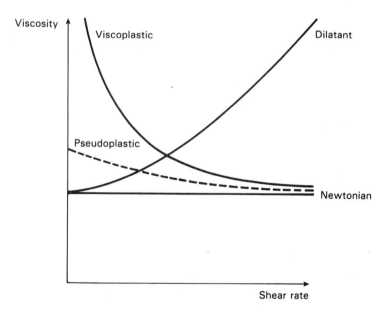

Figure 22.1 Different viscosity/shear rate relationships

ends of the tube and the time for a given quantity of liquid to flow are measured, and then the liquid viscosity for Newtonian fluids can be calculated as (in units of poise):

$$C_v = \frac{1.25\pi R^4 PT}{LV} \tag{22.1}$$

where R is the radius (m) of the tube, L is its length (m), P is the pressure difference (N/m^2) across the ends and V is the volume of liquid flowing in time T (m^3/s). For non-Newtonian fluids, corrections must be made for shear rate variations (Miller 1975).

For any given viscometer, R, L and V are constant and Equation (22.1) can be written as:

$$C_v = KPT \tag{22.2}$$

where K is known as the **viscometer constant**.

22.1.2 Falling-body viscometer

The falling-body viscometer is particularly recommended for the measurement of high-viscosity fluids and can give measurement uncertainty levels down to $\pm 1\%$. It involves measuring the time taken for a spherical body to fall a given distance through the liquid. The viscosity for Newtonian fluids is then given by Stoke's formula as (in units of poise):

$$C_v = \frac{R^2 g(\rho_s - \rho_l)}{450V}$$

where R is the radius (m) of the sphere, g is the acceleration due to gravity (m/s^2), ρ_s and ρ_l are the specific gravities (g/m^3) of the sphere and liquid respectively, and V is the velocity (m/s) of the sphere.

For non-Newtonian fluids, correction for the variation in shear rate is very difficult.

22.1.3 Rotational viscometers

Rotational viscometers are relatively easy to use but their measurement inaccuracy is at least $\pm 10\%$. All types have some form of element rotating inside the liquid at a constant rate. One common version has two coaxial cylinders with the fluid to be measured contained between them. One cylinder is driven at a constant angular velocity by a motor and the other is suspended by torsion wire. After the driven cylinder starts from rest, the suspended cylinder rotates until an equilibrium position is reached where the force due to the torsion wire is just balanced by the viscous force transmitted through the liquid. The viscosity (in poise) for Newtonian fluids is then given by:

$$C_v = 2.5G\left(\frac{1/R_1^2 - 1/R_2^2}{\pi h \omega}\right)$$

where G is the couple (N m) formed by the force exerted by the torsion wire and its deflection, R_1 and R_2 are the radii (m) of the inner and outer cylinders respectively, h is the length of the cylinder (m) and ω is the angular velocity (rad/s) of the rotating cylinder. Again, corrections have to be made for non-Newtonian fluids.

22.2 **Moisture measurement**

There are many industrial requirements for the measurement of the moisture content of solids, liquids and gases. The physical properties and storage stability of most solid materials is affected by their water content. There is also a statutory requirement to limit the moisture content in the case of many materials sold by weight. In consequence, the requirement for moisture measurement pervades a large number of industries involved in the manufacture of foodstuffs, pharmaceuticals, cement, plastics, textiles and paper.

Measurement of the water content in liquids is commonly needed for fiscal purposes, but is also often necessary to satisfy statutory requirements. The petrochemical industry has wide-ranging needs for moisture measurement in oil, etc. The food industry also needs to measure the water content of products such as beer and milk.

In the case of moisture in gases, the most common measurement is the amount of moisture in air. This is usually known as the humidity level. Humidity measurement and control is an essential requirement in many buildings, greenhouses and vehicles.

As there are several ways in which humidity can be defined, three separate terms have evolved so that ambiguity can be avoided. **Absolute humidity** is the mass of water in a unit volume of moist air; **specific humidity** is the mass of water in a unit mass of moist air; and **relative humidity** is the ratio of the actual water vapour pressure in air to the saturation vapour pressure, usually expressed as a percentage.

22.2.1 **Industrial moisture measurement techniques**

Industrial methods for measuring moisture are based on the variation of some physical property of the material with moisture content. Many different properties can be used and therefore the range of available techniques, as listed below, is large.

Electrical methods

Measuring the amount of absorption of **microwave energy** beamed through the material is the most common technique for measuring moisture content and is described in detail in Anderson (1989) and Thompson (1989). Microwaves at wavelengths between 1 mm and 1 m are absorbed to a much greater extent by water than most other materials. Wavelengths of 30 mm or 100 mm are commonly used because 'off-the-shelf' equipment to produce these is readily available from instrument suppliers. The technique is suitable for moisture measurement in solids, liquids and gases at moisture-content levels up to 45% and measurement uncertainties down to ±0.3% are possible.

The **capacitance moisture meter** uses the principle that the dielectric constant of materials varies according to their water content. Capacitance measurement is therefore related to moisture content. The instrument is useful for measuring moisture-content levels up to 30% in both solids and liquids, and measurement uncertainty down to ±0.3% has been claimed for the technique (Slight 1989). Drawbacks of the technique include (a) limited measurement resolution owing to the difficulty of measuring small changes in a relatively large standing capacitance value and (b) difficulties when the sample has a high electrical conductivity. An alternative capacitance charge-transfer technique has been reported (Gimson 1989) which overcomes these problems by measuring the charge-carrying capacity of the material. In this technique, wet and dry samples of the material are charged to a fixed voltage and then simultaneously discharged into charge measuring circuits.

The **electrical conductivity** of most materials varies with moisture content and this therefore provides another means of measurement. Techniques using electrical conductivity variation are cheap and can measure moisture levels up to 25%. However, the presence of other conductive substances in the material, such as salts or acids, affects the measurement.

A further technique is to measure the frequency change in a **quartz crystal** which occurs as it takes in moisture.

Neutron moderation

Neutron moderation measures moisture content using a radioactive source and a neutron counter. Fast neutrons emitted from the source are slowed down by hydrogen nuclei in the water, forming a cloud whose density is related to the moisture content. Measurements take a long time because the output density reading may take up to a few minutes to reach steady state, according to the nature of the materials involved. Also, the method cannot be used with any materials which contain hydrogen molecules, such as oils and fats, as these slow down the neutrons as well. Specific humidities up to 15% (±1% error) can be measured.

Low-resolution nuclear magnetic resonance (NMR)

Low-resolution nuclear magnetic resonance involves subjecting the sample to both a unidirectional and an alternating radio frequency (RF) magnetic field. The amplitude of the unidirectional field is varied cyclically, which causes resonance once per cycle in the coil producing the RF field. Under resonance conditions, protons are released from the hydrogen content of the water in the sample. These protons cause a measurable moderation of the amplitude of the RF oscillator waveform which is related to the moisture content of the sample. The technique is described more fully in Young (1989).

Materials with their own natural hydrogen content cannot normally be measured.

However, pulsed NMR techniques have been developed which overcome this problem by taking advantage of the different relaxation times of hydrogen nuclei in water and oil. In such pulsed techniques, the dependence on the relaxation time limits the maximum fluid flow rate for which moisture can be measured.

Optical methods

The **refractometer** is a well-established instrument which is used for measuring the water content of liquids. It measures the refractive index of the liquid which changes according to the moisture content.

Moisture-related **energy absorption** of near-infrared light can be used for measuring the moisture content of solids, liquids and gases. At a wavelength of 1.94 μm, energy absorption due to moisture is high, whereas at 1.7 μm, it is zero. Therefore, measuring absorption at both 1.94 μm and 1.7 μm allows absorption due to components in the material other than water to be compensated for, and the resulting measurement is directly related to energy content. The latest instruments use multiple-frequency infrared energy and have an even greater capability for eliminating the effect of components in the material other than water which absorb energy. Such multi-frequency instruments also cope much better with variations in particle size in the measured material.

In alternative versions of this technique, energy is either transmitted through the material or reflected from its surface. In either case, materials which are either very dark or highly reflective give poor results. The technique is particularly attractive, where applicable, because it is a non-contact method which can be used to monitor moisture content continuously at moisture levels up to 50%, with inaccuracy as low as $\pm0.1\%$ in the measured moisture level. A deeper treatment can be found in Benson (1989).

Ultrasonic methods

The presence of water changes the speed of propagation of ultrasonic waves through liquids. The moisture content of liquids can therefore be determined by measuring the transmission speed of ultrasound. This has the inherent advantage of being a non-invasive technique but temperature compensation is essential because the velocity of ultrasound is particularly affected by temperature changes. The method is best suited to measurement of high moisture levels in liquids which are not aerated or of high viscosity. Typical measurement uncertainty is $\pm1\%$ but measurement resolution is very high, with changes in moisture level as small as 0.05% being detectable. Further details can be found in Wiltshire (1989).

Mechanical properties

Density changes in many liquids and slurries can be measured and related to moisture content, with good measurement resolution up to a moisture level of 0.2%. Moisture content can also be estimated by measuring the moisture-level-dependent viscosity of liquids, pastes and slurries.

22.2.2 **Laboratory techniques for moisture measurement**

Laboratory techniques for measuring moisture content generally take much longer to obtain a measurement than the industrial techniques described above. However, the measurement accuracy obtained is usually much better.

Water separation

Various laboratory techniques are available which enable the moisture content of liquids to be measured accurately by separating the water from a sample of the host liquid. Separation is effected by titration (Karl Fischer technique), distillation (Dean and Stark technique) or a centrifuge. Any of these methods can measure the water content in a liquid with measurement uncertainty levels down to ±0.03%.

Gravimetric methods

The moisture content of solids can be measured accurately by weighing the moist sample, drying it and then weighing again. Great care must be taken in applying this procedure, as many samples rapidly take up moisture again if they are removed from the drier and exposed to the atmosphere before being weighed. Normal procedure is to put the sample in an open container, dry it in an oven and then screw an airtight top on to the container before it is removed from the oven.

Phase-change methods

The boiling and freezing points of materials are altered by the presence of moisture, and therefore the moisture level can be determined by measuring the phase-change temperature. This technique is used for measuring the moisture content of many food products and of some oil and alcohol products.

Equilibrium relative humidity measurement

This technique involves placing a humidity sensor in close proximity to the sample in an airtight container. The water vapour pressure close to the sample is related to the moisture content of the sample. The moisture level can therefore be determined from the humidity measurement.

22.2.3 **Humidity measurement**

The three major instruments used for measuring humidity in industry are the electrical hygrometer, the psychrometer and the dew point meter. The dew point meter is the most accurate of these and is commonly used as a calibration standard. The various types of hygrometer are described more fully in Miller (1975, pp. 180–209).

The electrical hygrometer

The electrical hygrometer measures the change in capacitance or conductivity of a hygroscopic material as its moisture level changes. Conductivity types use two noble-metal electrodes either side of an insulator coated in a hygroscopic salt such as calcium chloride. Capacitance types have two plates either side of a hygroscopic dielectric such as aluminium oxide.

These instruments are suitable for measuring moisture levels between 15% and 95%, with a typical measurement uncertainty of ±3%. Atmospheric contaminants and operation in saturation conditions both cause characteristics to drift, and therefore the recalibration frequency has to be determined according to the conditions of use.

The psychrometer (wet and dry bulb hygrometer)

The psychrometer, also known as the wet and dry bulb hygrometer, has two temperature sensors, one exposed to the atmosphere and one enclosed in a wet wick. Air is blown across the sensors which causes evaporation and a reduction in temperature in the wet sensor. The temperature difference between the sensors is related to the humidity level. The lowest measurement uncertainty attainable is ±4%.

Dew point meter

The elements of the dew point meter, also known as the dew point hygrometer, are shown in Figure 22.2. The sample is introduced into a vessel with an electrically cooled mirror surface. The mirror surface is cooled until a light source–light detector system detects the formation of dew on the mirror, and the condensation temperature is measured by a sensor bonded to the mirror surface. The dew point is the temperature at which the sample becomes saturated with water. Therefore, this temperature is related to the moisture level in the sample. A microscope is also provided in the instrument so that the thickness and nature of the condensate can be observed. The instrument is described in greater detail in Pragnell (1989).

Even small levels of contaminants on the mirror surface can cause large changes in the dew point and therefore the instrument must be kept very clean. When necessary, the mirror should be cleaned with deionized or distilled water applied with a lint-free swab. Any contamination can be detected by a skilled operator, as this makes the condensate look 'blotchy' when viewed through the microscope. The microscope also shows up other potential problems such as large ice crystals in the condensate which cause temperature gradients between the condensate and the temperature sensor. When used carefully, the instrument is very accurate and is often used as a reference standard.

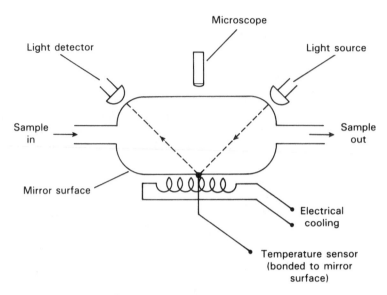

Microscope

Light detector

Light source

Sample in

Sample out

Mirror surface

Electrical cooling

Temperature sensor (bonded to mirror surface)

Figure 22.2 Dew point meter

22.3 **Volume measurement**

Volume measurement is required in its own right as well as being a necessary component in some techniques for the measurement and calibration of other quantities such as volume flow rate and viscosity. The volume of vessels of a regular shape, where the cross-section is circular or oblong, can be readily calculated from the dimensions of the vessel, using instruments as described in Chapter 13. Otherwise, for vessels of irregular shape, either gravimetric techniques or a set of calibrated volumetric measures are required.

In the gravimetric technique, the dry vessel is weighed and then completely filled with water and weighed again. The volume is then simply calculated from this weight difference and the density of water.

The alternative technique involves transferring the liquid from the vessel into an appropriate number of volumetric measures taken from a standard-capacity, calibrated set. Each vessel in the set has a mark which shows the volume of liquid contained when the vessel is filled up to the mark. Special care is needed to ensure that the meniscus of the water is in the correct position with respect to the reference mark on the vessel when it is deemed to be full. Normal practice is to set the water level such that the reference mark forms a smooth tangent with the convex side of the meniscus. This is made easier to achieve if the meniscus is viewed against a white background and the vessel is shaded from stray illumination.

The measurement uncertainty using calibrated volumetric measures depends on the number of measures used for any particular measurement. The total error is a

Table 22.1 Measurement uncertainty of calibrated volumetric measures

Capacity	Volumetric uncertainty
1 ml	±4%
10 ml	±0.8%
100 ml	±0.2%
1 l	±0.1%
10 l	±0.05%
100 l	±0.02%
1000 l	±0.02%

multiple of the individual error of each measure, typical values of which are shown in Table 22.1.

22.4 **Sound measurement**

Noise can arise from many sources in both industrial and non-industrial environments. Even low levels of noise can cause great annoyance to those people subjected to it and high levels can actually cause hearing damage. Apart from annoyance and possible hearing loss, noise in the workplace also causes loss of output where the persons subjected to it are involved in tasks requiring high concentration. Extreme noise can even cause material failures through fatigue stresses set up by noise-induced vibration.

Various legislation exists to control the creation of noise. Court orders can be made against houses or factories in a neighbourhood which create noise exceeding a certain acceptable level. In extreme cases, where hearing damage may be possible, health and safety legislation comes into effect. Such legislation clearly requires the existence of accurate methods of quantifying sound levels.

Sound is measured in terms of the **sound pressure level**, S_p, which is defined as:

$$S_p = 20 \log_{10}\left(\frac{P}{0.0002}\right) \text{decibels (dB)}$$

where P is the r.m.s. sound pressure in microbars.

The quietest sound which the average human ear can detect is a tone at a frequency of 1 kHz and a sound pressure level of 0 dB ($2 \times 10^{-4} \mu$bar). At the upper end, sound pressure levels of 144 dB (3.45 mbar) cause physical pain.

Sound is usually measured with a sound meter. This essentially processes the signal collected by a microphone, as shown in Figure 22.3. The microphone is a diaphragm-type pressure measuring device which converts sound pressure into a displacement. The displacement is applied to a displacement transducer (normally a

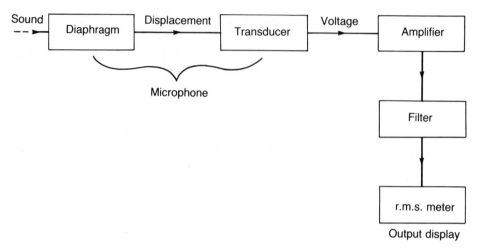

Figure 22.3 Sound meter

capacitive, inductive or piezoelectric type) which produces a low-magnitude voltage output. This is amplified, filtered and finally gives an output display on an r.m.s. meter. The filtering process has a frequency response approximating that of the human ear so that the sound meter 'hears' sounds in the same way as a human ear. In other words, the meter selectively attenuates frequencies according to the sensitivity of the human ear at each frequency, so that the sound-level measurement output accurately reflects the sound level heard by humans. If sound-level meters are being used to measure sound to predict vibration levels in machinery, then they are used without filters so that the actual rather than the human-perceived sound level is measured.

22.5 **pH measurement**

pH is a parameter which quantifies the level of acidity or alkalinity in a chemical solution. It defines the concentration of hydrogen atoms in the solution in grams/litre and is expressed as:

$$pH = \log_{10}[1/H^+]$$

where H^+ is the hydrogen ion concentration in the solution.

The value of pH can range from 0, which describes extreme acidity, to 14, which describes extreme alkalinity. Pure water has a pH of 7.

pH measurement is required in many process industries, especially those involving food and drink production. The most universally known method of measuring pH is to use litmus paper or some similar chemical indicator which changes colour according to the pH value. Unfortunately, this method gives only a very approximate indication of pH unless used under highly controlled laboratory conditions. Much research is going on into on-line pH sensors and the various activities are described later. However, at the present time, the device known as the glass electrode is by far the most common on-line sensor.

22.5.1 **The glass electrode**

The glass electrode consists of a glass probe containing two electrodes, a measuring one and a reference one, separated by a solid glass partition. Neither of the electrodes is in fact glass. The reference electrode is a screened electrode, immersed in a buffer solution, which provides a stable reference e.m.f. which is usually 0 V. The tip of the measuring electrode is surrounded by a pH-sensitive glass membrane at the end of the probe which permits the diffusion of ions according to the hydrogen ion concentration in the fluid outside the probe. The measuring electrode therefore generates an e.m.f. proportional to pH which is amplified and fed to a display meter. The characteristics of the glass electrode are very dependent on ambient temperature, with both zero drift and sensitivity drift occurring. Thus temperature compensation is essential. This is normally achieved through calibrating the system output before use by immersing the probe in solutions at reference pH values. Whilst the system is theoretically capable of measuring the full range of pH values between 0 and 14, the upper limit in practice is generally a pH value of about 12 because electrode contamination at very high-alkaline concentrations becomes a serious problem and also glass starts to dissolve at such high pH values. Glass also dissolves in acid solutions containing fluoride, and this represents a further limitation of use. If required, the latter problems can be overcome to some extent by using special types of glass.

Great care is necessary in the use of the glass electrode type of pH probe. First, the measuring probe has a very high resistance (typically $10^8 \, \Omega$) and a very low output. Hence, the output signal from the probes must be electrically screened to prevent any stray pick-up and electrical insulation of the assembly must be very high. The assembly must also be very efficiently sealed to prevent the ingress of moisture.

A second problem with the glass electrode is the deterioration in accuracy which occurs as the glass membrane becomes coated with the various substances it is exposed to in the measured solution. Cleaning at prescribed intervals is therefore necessary and this must be carried out carefully, using the correct procedures, to avoid damaging the delicate glass membrane at the end of the probe. The best cleaning procedure varies according to the nature of the contamination. In some cases careful brushing or wiping is adequate, whereas in other cases spraying with chemical solvents is necessary. Ultrasonic cleaning is often a useful technique, though it tends to be expensive. Steam cleaning should not be attempted, as this damages the pH-sensitive membrane. Mention must also be made about storage. The glass electrode must not be allowed to dry out during storage, as this would cause serious damage to the pH-sensitive layer.

Finally, caution must be taken with the response time of the instrument. The glass electrode has a relatively large time constant of 1 to 2 minutes, and so it must be left to settle for a long time before the reading is taken. If this causes serious difficulties, special forms of low-resistivity glass electrode are now available which have smaller time constants.

22.5.2 Other methods of pH measurement

Whilst the glass electrode predominates at present in pH measurement, several other devices and techniques exist. Most of these are still under development and unproven in long-term use, but a few are in practical use, especially for special measurement situations.

One alternative which is in current use is the antimony electrode. This is of a similar construction to the glass electrode but uses antimony instead of glass. The device is more robust than the glass electrode and can be cleaned by rubbing it with emery cloth. However, its time constant is very large and its output response is grossly non-linear, limiting its application to environments where the glass electrode is unsuitable. Such applications include acidic environments containing fluoride and environments containing very abrasive particles. The normal measurement range is pH 1 to 11.

A fiber optic pH sensor is another viable device. This was described in Figure 9.6(d). The pH level is indicated by the intensity of light reflected from the tip of a probe coated in a chemical indicator whose colour changes with pH.

Various techniques being researched are described in Colville (1985). One of these is the use of fiber optic cable to transmit colour changes resulting from the interaction of the sampled measured fluid with a chemical pH indicator. The principle of fluorescence can also be used by introducing fluoresceinamine into the measured fluid and measuring the pH-sensitive intensity of fluorescence.

22.6 Gas measurement

Gas sensing and analysis are required in many applications. A primary role of gas sensing is in hazard monitoring to predict the onset of conditions where flammable gases are reaching dangerous concentrations. Danger is quantified in terms of the **lower explosive level** which is usually reached when the concentration of gas in air is in the range between 1% and 5%.

Gas sensing also provides a fire detection and prevention function. When materials burn, a variety of gaseous products result. Most sensors which are used for fire detection measure the concentration of carbon monoxide, as this is the most common combustion product.

Early fire detection enables fire extinguishing systems to be triggered, preventing serious damage from occurring in most cases. However, fire prevention is even better than early fire detection and current research is looking at developing sensors to measure the gaseous products, generally various types of hydrocarbons, which are generated when materials become hot but before they actually burn. The major difficulty in this development is that such early products arising out of heating occur only in very small concentrations, and demand very sensitive sensors.

Health and safety legislation creates a further requirement for gas sensors. Certain gases, such as carbon monoxide, hydrogen sulphide, chlorine and nitrous oxide, cause fatalities above a certain concentration and sensors must provide warning of impending danger. For other gases, health problems are caused by prolonged exposure and so the sensors in this case must integrate gas concentration over time to determine whether the allowable exposure limit over a given period of time has been exceeded.

Concern about general environmental pollution is also making the development of gas sensors necessary in many new areas. Legislation is growing rapidly to control the emission of everything which is proven or suspected to cause health problems or environmental damage. The present list of controlled emissions includes nitrous oxide, oxides of sulphur, carbon monoxide and dioxide, CFCs, ammonia and hydrocarbons. Sensors are required both at the source of these pollutants, where concentrations are high, and to monitor the much lower concentrations in the general environment. The measurement of oxygen concentration is often also of great importance in pollution control, as the products of combustion processes are greatly affected by the air/fuel ratio.

Sensors associated with pollution monitoring and control often have to satisfy quite stringent specifications, particularly where the sensors are located at the source of the pollutant. Robustness is usually essential, as the sensors are subjected to bombard-ment from a variety of particulate matter, and they must also endure conditions of high humidity and temperature. They are also frequently located in inaccessible locations, such as in chimneys and flues, which means that they must have stable characteristics over long periods of time without calibration checks being necessary. The need for such high-specification sensors makes such pollutant monitoring potentially very expensive if there are several problem gases involved. However, because the concentration of all output gases tends to vary to a similar extent according to the condition of filters, etc., it is frequently only necessary to measure the concentration of one gas, from which the concentration of other gases can be predicted reliably. This greatly reduces the cost involved in such monitoring.

A number of devices which sense, measure the concentration of or analyze gases exist. In terms of frequency of usage, they vary from those which have been in use for a number of years to those which have appeared recently, and finally to those which are still under research and development. In the following list of devices, their status in terms of current usage will be indicated. Fuller information can be found in Jones (1989).

22.6.1 Catalytic (calorimetric) sensors

Catalytic sensors, otherwise known as calorimetric sensors, are in widespread use for measuring the concentration of flammable gases. Their principle of operation is to measure the heat evolved during the catalytic oxidation of reducing gases. They are cheap and robust but are unsuitable for measuring either very low or very high gas concentrations. The catalysts which have been commonly used in these devices in the past are adversely affected by many common industrial substances such as lead, phosphorus, silicon and sulphur, and this catalyst poisoning has previously prevented this type of device being used in many applications. However, new types of poison-resistant catalyst are now becoming available which greatly extend the applicability of this type of device.

22.6.2 Paper-tape sensors

By moving a paper tape impregnated with a reagent sensitive to a specific gas (e.g. lead acetate tape to detect hydrogen sulphide) through an air stream, the time history of the concentration of gas is indicated by the degree of colour change in the tape. This is used as a low-accuracy, but reliable and cheap, means of detecting the presence of hydrogen sulphide and ammonia.

22.6.3 Liquid-electrolyte electrochemical cells

These consist of two electrodes separated by an electrolyte, to which the measured air supply is directed through a permeable membrane, as shown in Figure 22.4. The gas in the air to which the cell is sensitive reacts at the electrodes to form ions in the solution which produces a voltage output from the cell.

Electrochemical cells have stable characteristics and give a good measurement sensitivity. However, they are expensive and their durability is relatively poor, with life being generally limited to about 1 or 2 years at most. A further restriction is that they cannot be used above temperatures of about 50 °C, as their performance deteriorates rapidly at high temperatures because of interference from other atmospheric substances.

The main use of such cells is in measuring toxic gases to satisfy health and safety legislation. Versions of the cell for this purpose are currently available to measure carbon monoxide, chlorine, nitrous oxide, hydrogen sulphide and ammonia. Cells to measure other gases are currently under development.

In addition, electrochemical cells are also used to a limited extent to monitor carbon monoxide emissions in flue gases for environmental control purposes. Pre-cooling of the emitted gases is a necessary condition for this application.

22.6.4 Solid-state electrochemical cells (zirconia sensor)

At present these cells are used only for measuring oxygen concentration, but ways of extending their use to other gases are currently being investigated. The oxygen measurement cell consists of two chambers separated by a zirconia wall. One chamber contains gas with a known oxygen concentration and the other contains the air being measured. Ions are conducted across the zirconia wall according to the difference in oxygen concentration across it and this produces an output e.m.f. The device is rugged but requires high temperatures to operate efficiently. It is, however, well

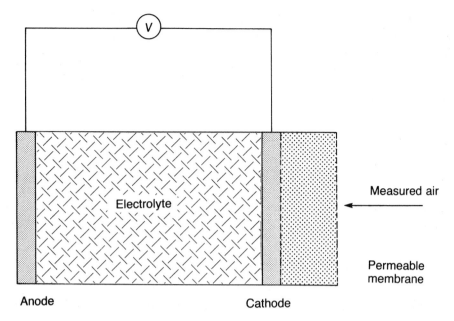

Figure 22.4 Liquid-electrolyte electrochemical cell

proven and a standard choice for oxygen measurement. In industrial uses, it is often located in chimney stacks, where quite expensive mounting and protection systems are needed. However, very low-cost versions (around £200) are now used in some vehicle exhaust systems as part of the engine management system.

22.6.5 **Catalytic gate FETs**

These consist of field-effect transistors (FETs) with a catalytic, palladium gate which is sensitive to hydrogen ions in the environment. The gate voltage, and hence the characteristics of the device, change according to the hydrogen concentration. They can be made to be sensitive to gases such as hydrogen sulphide, ammonia and hydrocarbons as well as hydrogen. They are cheap and find application in workplace monitoring, to satisfy of health and safety legislation, and in fire detection (mainly detecting hydrocarbon products).

22.6.6 **Semiconductor (metal oxide) sensors**

In these devices, use is made of the fact that the surface conductivity of semiconductor metal oxides (generally, tin or zinc oxides) changes according to the concentration of certain gases with which they are in contact. Unfortunately, they have a similar response for the range of gases to which they are sensitive. Hence, they show that a gas is present but not which one. Such sensors are cheap, robust, very durable and sensitive to very low gas concentrations. However, because their discrimination between gases is low and their accuracy in quantitative measurement is poor, they are mainly used only for qualitative indication of the presence of a gas. In this role, they are particularly useful for fire prevention in detecting the presence of combustion products which occur in low concentrations when the temperature starts to rise due to a fault.

22.6.7 Organic sensors

These work on similar principles to metal oxide semiconductors but use an organic surface layer which is designed to respond selectively to only one gas. At present, these devices are still the subject of ongoing research but industrial exploitation is anticipated in the near future. They promise to be cheap and have high stability and sensitivity.

22.6.8 Piezoelectric devices

In these devices, piezoelectric crystals are coated with an absorbent layer. As this layer absorbs gases, the crystal undergoes a change in resonant frequency which can be measured. There is no discrimination in this effect between different gases but the technique potentially offers a high-sensitivity mechanism for detecting the presence of a gas. At present, the problems of finding a suitable type of coating material where absorption is reversible have not been generally solved, and the device only finds limited application for measuring moisture concentrations.

22.6.9 Infrared absorption

This technique uses infrared light at a particular wavelength which is directed across a chamber between a source and a detector. The amount of light absorption is a function of the unknown gas concentration in the chamber. The instrument normally has a second chamber containing gas at a known concentration across which infrared light at the same wavelength is directed to provide a reference. Sensitivity to carbon monoxide, carbon dioxide, ammonia or hydrocarbons can be provided according to the wavelength used. Microcomputers are now routinely incorporated in the instrument to reduce its sensitivity to gases other than the one being sensed and so improve measurement accuracy. The instrument finds widespread use in chimney/flue emission monitoring and in general process measurements.

22.6.10 Mass spectrometers

The mass spectrometer is a laboratory device for analyzing gases. It first reduces a gas sample to a very low pressure. The sample is then ionized, accelerated and separated into its constituent components according to the respective charge-to-mass ratios. Almost any mixture of gases can be analyzed and the individual components quantified, but the instrument is very expensive and requires a skilled user. Mass spectrometers have existed for over half a century but recent advances in electronic data processing techniques have greatly improved their performance.

22.6.11 Gas chromatography

This is also a laboratory instrument in which a gaseous sample is passed down a packed column. The column separates the gas into its components, which are washed out of the column in turn and measured by a detector. Like the mass spectrometer, the instrument is versatile but expensive and it requires skilled use.

References and further reading

Anderson, J. G. (1989) 'Paper moisture measurement using microwaves', *Measurement and Control*, **22**, 82–4.

Benson, I. B. (1989) 'Industrial applications of near infrared reflectance for the measurement of moisture', *Measurement and Control*, **22**, 45–9.

Colville, R. F. (1985) 'On-line pH measurement and control', *Measurement and Control*, **18**, 396–9.

Gimson, C. (1989) 'Using the capacitance charge transfer principle for water content measurement', *Measurement and Control*, **22**, 79–81.

Jones, T. A. (1989) 'Trends in the development of gas sensors', *Measurement and Control*, **22**, 176–82.

Miller, J. T. (ed.) (1975) *The Instrument Manual*, United Trade Press: London.

Pragnell, R. F. (1989) 'The modern condensation dewpoint hygrometer', *Measurement and Control*, **22**, 74–7.

Slight, H. A. (1989) 'Further thoughts on moisture measurement', *Measurement and Control*, **22**, 85–6.

Thompson, F. (1989) 'Moisture measurement using microwaves', *Measurement and Control*, **22**, 210–15.

Wiltshire, M. P. (1989) 'Ultrasonic moisture measurement', *Measurement and Control*, **22**, 51–3.

Young, L. (1989) 'Moisture measurement using low resolution nuclear magnetic resonance', *Measurement and Control*, **22**, 54–5.

Appendix 1
Imperial–metric–SI conversion tables

Length

SI units: mm, m, km
Imperial units: in, ft, mile

	mm	m	km	in	ft	mile
mm	1	10^{-3}	10^{-6}	0.039 3701	3.281×10^{-3}	–
m	1000	1	10^{-3}	39.3701	3.280 84	6.214×10^{-4}
km	10^6	10^3	1	39 370.1	3280.84	0.621 371
in	25.4	0.0254	–	1	0.083 333	–
ft	304.8	0.3048	3.048×10^{-4}	12	1	1.894×10^{-4}
mile	–	1609.34	1.609 34	63 360	5280	1

Area

SI units: mm^2, m^2, km^2
Imperial units: in^2, ft^2, $mile^2$

	mm^2	m^2	km^2	in^2	ft^2	$mile^2$
mm^2	1	10^{-6}	–	1.550×10^{-3}	1.076×10^{-5}	–
m^2	10^6	1	10^{-6}	1550	10.764	–
km^2	–	10^6	1	–	1076×10^7	0.3861
in^2	645.16	6.452×10^{-4}	–	1	6.944×10^{-3}	–
ft^2	92 903	0.092 90	–	144	1	–
$mile^2$	–	2.590×10^6	2.590	–	2.788×10^7	1

Second moment of area

SI units: mm^4, m^4
Imperial units: in^4, ft^4

	mm^4	m^4	in^4	ft^4
mm^4	1	10^{-12}	2.4025×10^{-6}	1.159×10^{-10}
m^4	10^{12}	1	2.4025×10^6	115.86
in^4	416 231	4.1623×10^{-7}	1	4.8225×10^{-5}
ft^4	8.631×10^9	8.631×10^{-3}	20 736	1

Volume

SI units: mm^3, m^3
Metric units: ml, l
Imperial units: in^3, ft^3, UK gallon

	mm^3	ml	l	m^3	in^3	ft^3	UK gallon
mm^3	1	10^{-3}	10^{-6}	10^{-9}	6.10×10^{-5}	–	–
ml	10^3	1	10^{-3}	10^{-6}	0.061 024	3.53×10^{-5}	2.2×10^{-4}
l	10^6	10^3	1	10^{-3}	61.024	0.035 32	0.22
m^3	10^9	10^6	10^3	1	61 024	35.31	220
in^3	16 387	16.39	0.0164	1.64×10^{-5}	1	5.79×10^{-4}	3.61×10^{-3}
ft^3	–	2.83×10^4	28.32	0.028 32	1728	1	6.229
UK gallon	–	4546	4.546	4.55×10^{-3}	277.4	0.1605	1

Note: Additional unit: 1 US gallon = 0.8327 UK gallon.

Density

SI unit: kg/m^3
Metric unit: g/cm^3
Imperial units: lb/ft^3, lb/in^3

	kg/m^3	g/cm^3	lb/ft^3	lb/in^3
kg/m^3	1	10^{-3}	0.062 428	3.605×10^{-5}
g/cm^3	1000	1	62.428	0.036 127
lb/ft^3	16.019	0.016 019	1	5.787×10^{-4}
lb/in^3	27 680	27.680	1728	1

Mass

SI units: g, kg, t
Imperial units: lb, cwt, ton

	g	kg	t	lb	cwt	ton
g	1	10^{-3}	10^{-6}	2.205×10^{-3}	1.968×10^{-5}	9.842×10^{-7}
kg	10^3	1	10^{-3}	2.204 62	0.019 684	9.842×10^{-4}
t	10^6	10^3	1	2204.62	19.6841	0.984 207
lb	453.592	0.453 59	4.536×10^{-4}	1	8.929×10^{-3}	4.464×10^{-4}
cwt	50 802.3	50.8023	0.050 802	112	1	0.05
ton	1.016×10^6	1016.05	1.016 05	2240	20	1

Force

SI units: N, kN
Metric unit: kg_f
Imperial units: pdl (poundal), lb_f, UK ton_f

	N	kg_f	kN	pdl	lb_f	UK ton_f
N	1	0.1020	10^{-3}	7.233	0.2248	1.004×10^{-4}
kg_f	9.807	1	9.807×10^{-3}	70.93	2.2046	9.842×10^{-4}
kN	1000	102.0	1	7233	224.8	0.1004
pdl	0.1383	0.0141	1.383×10^{-4}	1	0.0311	1.388×10^{-5}
lb_f	4.448	0.4536	4.448×10^{-3}	32.174	1	4.464×10^{-4}
UK ton_f	9964	1016	9.964	72 070	2240	1

Note: Additional unit: 1 dyne = 10^{-5} N = 7.233×10^{-5} pdl.

Torque (moment of force)

SI unit: N m
Metric unit: kg_f m
Imperial units: pdl ft, lb_f ft

	N m	kg_f m	pdl ft	lb_f ft
N m	1	0.1020	23.73	0.7376
kg_f m	9.807	1	232.7	7.233
pdl ft	0.042 14	4.297×10^{-3}	1	0.031 08
lb_f ft	1.356	0.1383	32.17	1

Inertia

SI unit: $N\,m^2$
Imperial unit: $lb_f\,ft^2$

$$1\,lb_f\,ft^2 = 0.4132\,N\,m^2$$
$$1\,N\,m^2 = 2.420\,lb_f\,ft^2$$

Pressure

SI units: mbar, bar, N/m^2 (Pascal)
Imperial units: lb/in^2, in Hg, atm

	mbar	bar	N/m^2	lb/in^2	in Hg	atm
mbar	1	10^{-3}	100	0.014 50	0.029 53	9.869×10^{-4}
bar	1000	1	10^5	14.50	29.53	0.9869
N/m^2	0.01	10^{-5}	1	1.450×10^{-4}	2.953×10^{-4}	9.869×10^{-6}
lb/in^2	68.95	0.068 95	6895	1	2.036	0.068 05
in Hg	33.86	0.033 86	3386	0.4912	1	0.033 42
atm	1013	1.013	1.013×10^5	14.70	29.92	1

Additional conversion factors

1 inch water $= 0.073\,56$ in Hg $= 2.491$ mbar
1 torr $= 1.333$ mbar
1 pascal $= 1\,N/m^2$

Energy, work, heat

SI unit: J
Metric units: $kg_f\,m$, kW h
Imperial units: $ft\,lb_f$, cal, Btu

	J	$kg_f\,m$	kW h	$ft\,lb_f$	cal	Btu
J	1	0.1020	2.778×10^{-7}	0.7376	0.2388	9.478×10^{-4}
$kg_f\,m$	9.8066	1	2.724×10^{-6}	7.233	2.342	9.294×10^{-3}
kW h	3.600×10^6	367 098	1	2.655×10^6	859 845	3412.1
$ft\,lb_f$	1.3558	0.1383	3.766×10^{-7}	1	0.3238	1.285×10^{-3}
cal	4.1868	0.4270	1.163×10^{-6}	3.0880	1	3.968×10^{-3}
Btu	1055.1	107.59	2.931×10^{-4}	778.17	252.00	1

Additional conversion factors

1 therm $= 10^5$ Btu $= 1.0551 \times 10^8$ J
1 thermie $= 4.186 \times 10^6$ J
1 hp h $= 0.7457$ kW h $= 2.6845 \times 10^6$ J
1 ft pdl $= 0.042\,14$ J
1 erg $= 10^{-7}$ J

Power

SI units W, kW
Imperial units: HP, ft lb$_f$/s

	W	kW	HP	ft lb$_f$/s
W	1	10^{-3}	1.341×10^{-3}	0.735 64
kW	10^3	1	1.341 02	735.64
HP	745.7	0.7457	1	548.57
ft lb$_f$/s	1.359 35	1.359×10^{-3}	1.823×10^{-3}	1

Velocity

SI units: mm/s, m/s
Metric unit: km/h
Imperial units: ft/s, mile/h

	mm/s	m/s	km/h	ft/s	mile/h
mm/s	1	10^{-3}	3.6×10^{-3}	3.281×10^{-3}	2.237×10^{-3}
m/s	1000	1	3.6	3.280 84	2.236 94
km/h	277.778	0.277 778	1	0.911 344	0.621 371
ft/s	304.8	0.3048	1.097 28	1	0.681 818
mile/h	447.04	0.447 04	1.609 344	1.466 67	1

Acceleration

SI unit: m/s^2
Other metric unit: cm/s^2
Imperial unit: ft/s^2
Other unit: g

	m/s^2	cm/s^2	ft/s^2	g
m/s^2	1	100	3.281	0.102
cm/s^2	0.01	1	0.0328	0.001 02
ft/s^2	0.3048	30.48	1	0.031 09
g	9.81	981	32.2	1

Mass flow rate

SI unit: g/s
Metric units: kg/h, tonne/d
Imperial units: lb/s, lb/h, ton/d

	g/s	kg/h	tonne/d	lb/s	lb/h	ton/d
g/s	1	3.6	0.086 40	2.205×10^{-3}	7.937	0.085 03
kg/h	0.2778	1	0.024 00	6.124×10^{-4}	2.205	0.023 62
tonne/d	11.57	41.67	1	0.025 51	91.86	0.9842
lb/s	453.6	1633	39.19	1	3600	38.57
lb/h	0.1260	0.4536	0.010 89	2.788×10^{-4}	1	0.010 71
ton/d	11.76	42.34	1.016	0.025 93	93.33	1

Volume flow rate

SI unit: m^3/s
Metric units: l/h, ml/s
Imperial units: gal/h, ft^3/s, ft^3/h

	l/h	ml/s	m^3/s	gal/h	ft^3/s	ft^3/h
l/h	1	0.2778	2.778×10^{-7}	0.2200	9.810×10^{-6}	0.035 316
ml/s	3.6	1	10^{-6}	0.7919	3.532×10^{-5}	0.127 14
m^3/s	3.6×10^6	10^6	1	7.919×10^5	35.31	1.271×10^5
gal/h	4.546	1.263	1.263×10^{-6}	1	4.460×10^{-5}	0.160 56
ft^3/s	1.019×10^5	2.832×10^4	0.028 32	2.242×10^4	1	3600
ft^3/h	28.316	7.8653	7.865×10^{-6}	6.2282	2.778×10^{-4}	1

Specific energy (heat per unit volume)

SI units: J/m^3, kJ/m^3, MJ/m^3
Imperial units: $kcal/m^3$, Btu/ft^3, therm/UK gal

	J/m^3	kJ/m^3	MJ/m^3	$kcal/m^3$	Btu/ft^3	therm/UK gal
J/m^3	1	10^{-3}	10^{-6}	1.388×10^{-4}	2.684×10^{-5}	–
kJ/m^3	1000	1	10^{-3}	0.2388	0.026 84	–
MJ/m^3	10^6	1000	1	238.8	26.84	4.309×10^{-5}
$kcal/m^3$	4187	4.187	4.187×10^{-3}	1	0.1124	1.804×10^{-7}
Btu/ft^3	3.726×10^4	37.26	0.037 26	8.899	1	1.605×10^{-6}
therm/gal	–	–	2.321×10^4	5.543×10^6	6.229×10^5	1

Dynamic viscosity

SI unit: $N\,s/m^2$
Metric unit: cP (centipoise), P (poise) [1 P = 100 g/m s]
Imperial unit: $lb_m/ft\ h$

	$lb_m/ft\ h$	P	cP	$N\,s/m^2$
$lb_m/ft\ h$	1	4.133×10^{-3}	0.4134	4.134×10^{-4}
P	241.9	1	100	0.1
cP	2.419	0.01	1	10^{-3}
$N\,s/m^2$	2419	10	1000	1

Note: Additional unit: 1 pascal second = $1\ N\,s/m^2$.

Kinematic viscosity

SI unit: m^2/s
Metric unit: cSt (centistokes), St (stokes)
Imperial unit: ft^2/s

	ft^2/s	m^2/s	cSt	St
ft^2/s	1	0.0929	9.29×10^4	929
m^2/s	10.764	1	10^6	10^4
cSt	1.0764×10^{-5}	10^{-6}	1	0.01
St	1.0764×10^{-3}	10^{-4}	100	1

Appendix 2
Colour codes for resistors and capacitors

The British Standards Institution has defined two standards for the marking of values on components: a colour-code system for resistors and a character/number system for both resistors and capacitors (BS 1852 (1975) Resistor and capacitor marking).

A2.1 Colour-code system for resistors

The value and tolerance of resistors is defined by a set of either four or five coloured bands. These are displaced towards one end of the resistor with band 1 defined as that closest to the end of the resistor.

A2.1.1 Four-band system

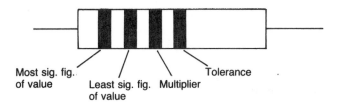

Most sig. fig. of value — Least sig. fig. of value — Multiplier — Tolerance

Codes for bands 1–3				Code for band 4	
Black	0	Green	5	Brown	∓1%
Brown	1	Blue	6	Red	∓2%
Red	2	Purple	7	Gold	∓5%
Orange	3	Grey	8	Silver	∓10%
Yellow	4	White	9		

A2.1.2 **Five-band system (not often used)**

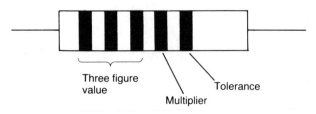

Three figure value

Multiplier

Tolerance

Codes for bands 1–4				Code for band 5	
Black	0	Green	5	Brown	∓1%
Brown	1	Blue	6	Red	∓2%
Red	2	Purple	7	Gold	∓5%
Orange	3	Grey	8	Silver	∓10%
Yellow	4	White	9		

Example 1

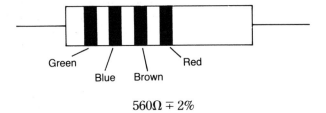

Green
Blue   Brown
Red

$$560\Omega \mp 2\%$$

Example 2 (not common)

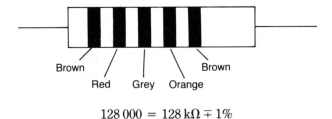

Brown
Red   Grey   Orange
Brown

$$128\,000 = 128\,k\Omega \mp 1\%$$

A2.2 **Character/number system for resistors and capacitors**

This code uses two, three or four figures plus one letter to indicate value. The tolerance is indicated by a letter following the value code. The date of manufacture is sometimes indicated by a separate code.

A2.2.1 **Resistors**

The value is defined by two, three or four figures.

The decimal point is represented by a letter which also acts as a multiplier: R, K, M, G, T define respectively $\times 1$, $\times 10^3$, $\times 10^6$, $\times 10^9$, $\times 10^{12}$. For example:

    6M8   means 6.8×10^6, i.e. $6.8\,\text{M}\Omega$
   59R04 means $59.04\,\Omega$

The code letters for tolerance are as follows:

B	$\mp 0.1\%$	F	$\mp 1\%$	K	$\mp 10\%$
C	$\mp 0.25\%$	G	$\mp 2\%$	M	$\mp 20\%$
D	$\mp 0.5\%$	J	$\mp 5\%$	N	$\mp 30\%$

The date of manufacture is indicated by either a two- or a four-character code.

Two-character systems

Code	Meaning	Code	Meaning	Code	Meaning
X	1969	N	1981	1	January
A	1970	P	1982	2	February
B	1971	R	1983	3	March
C	1972	S	1984	4	April
D	1973	T	1985	5	May
E	1974	U	1986	6	June
F	1975	V	1987	7	July
H	1976	W	1988	8	August
J	1977	X	1989	9	September
K	1978			O	October
L	1979			N	November
M	1980			D	December

For example, R8 means manufactured in August 1983.

Four-character system

The first two characters are the year of manufacture and the second two characters are the week of manufacture. For example, 8507 means week 7 of 1985.

A2.2.2 **Capacitors**

The value is defined by a sequence of numbers, as for resistors, with a multiplier code letter representing the decimal point.

Code	Multiplier
p	10^{-12}
n	10^{-9}
μ(u)	10^{-6}
m	10^{-3}
F	1

For example:

p10	0.1 pF
333p	333 pF
15n	15 nF

The tolerance is indicated by the same one-letter code as for resistors. For example, 333 pK means a 333 pF capacitor with $\mp$10% tolerance.

Appendix 3
Thévenin's theorem

Thévenin's theorem is extremely useful in the analysis of complex electrical circuits. It states that any network which has two accessible terminals A and B can be replaced, as far as its external behaviour is concerned, by a single e.m.f. acting in series with a single resistance between A and B. The single equivalent e.m.f. is that e.m.f. which is measured across A and B when the circuit external to the network is disconnected. The single equivalent resistance is the resistance of the network when all current and voltage sources within it are reduced to zero. To calculate this internal resistance of the network, all current sources within it are treated as open circuits and all voltage sources as short circuits. The proof of Thévenin's theorem can be found in Skilling (1967).

Figure A3.1 shows part of a network consisting of a voltage source and four resistances. As far as its behaviour external to the terminals A and B is concerned, this can be regarded as a single voltage source V_t and a single resistance R_t. Applying Thévenin's theorem, R_t is found first of all by treating V_1 as a short circuit, as shown in Figure A3.2. This is simply two resistances, R_1 and $(R_2 + R_4 + R_5)$ in parallel. The equivalent resistance R_t is thus given by:

$$R_t = \frac{R_1(R_2 + R_4 + R_5)}{R_1 + R_2 + R_4 + R_5}$$

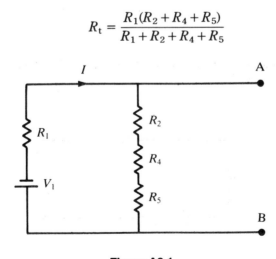

Figure A3.1

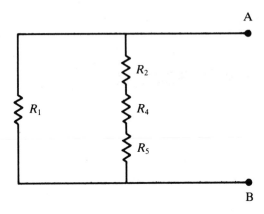

Figure A3.2

V_t is the voltage drop across AB. To calculate this, it is necessary to carry out an intermediate step of working out the current flowing, I. Referring to Figure A3.1, this is given by:

$$I = \frac{V_1}{R_1 + R_2 + R_4 + R_5}$$

Now, V_t can be calculated from:

$$V_t = I(R_2 + R_4 + R_5)$$
$$= \frac{V_1(R_2 + R_4 + R_5)}{R_1 + R_2 + R_4 + R_5}$$

The network of Figure A3.1 has thus been reduced to the simpler equivalent network shown in Figure A3.3.

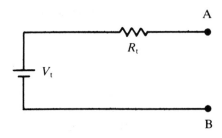

Figure A3.3

Let us now proceed to the typical network problem of calculating the current flowing in the resistor R_3 of Figure A3.4. R_3 can be regarded as an external circuit or load on the rest of the network consisting of V_1, R_1, R_2, R_4 and R_5, as shown in Figure A3.5. This network of V_1, R_1, R_2, R_4 and R_5 is that shown in Figure A3.6. This can be rearranged to the network shown in Figure A3.1, which is equivalent to the single voltage source and resistance, V_t and R_t, calculated above. The whole

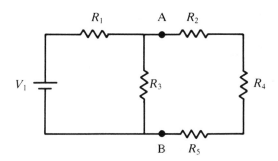

Figure A3.4

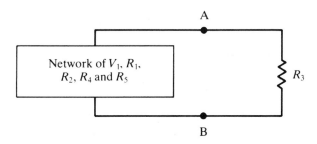

Figure A3.5

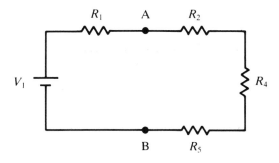

Figure A3.6

circuit is then equivalent to that shown in Figure A3.7, and the current flowing through R_3 can be written as:

$$I_{AB} = \frac{V_t}{R_t + R_3}$$

Thévenin's theorem can be applied successively to solve ladder networks of the form shown in Figure A3.8. Suppose in this network that it is required to calculate the current flowing in branch XY.

The first step is to imagine two terminals in the circuit A and B and regard the network to the right of AB as a load on the circuit to the left of AB. The circuit to the left of AB can be reduced to a single equivalent voltage source, E_{AB}, and resistance, R_{AB}, by Thévenin's theorem. If the 50 V source is replaced by its zero internal resistance (i.e. by a short circuit), then R_{AB} is given by:

$$\frac{1}{R_{AB}} = \frac{1}{100} + \frac{1}{2000} = \frac{2000 + 100}{200\,000}$$

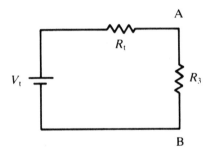

Figure A3.7

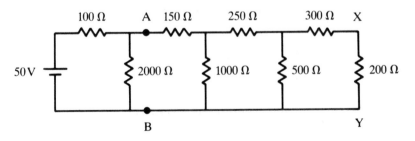

Figure A3.8

Hence:

$$R_{AB} = 95.24 \, \Omega$$

When AB is open circuit, the current flowing round the loop to the left of AB is given by:

$$I = \frac{50}{100 + 2000}$$

Hence, E_{AB}, the open-circuit voltage across AB, is given by:

$$E_{AB} = I \times 2000 = 47.62 \, \text{V}$$

We can now replace the circuit shown in Figure A3.8 by the simpler equivalent circuit shown in Figure A3.9.

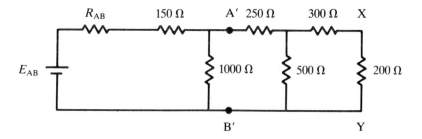

Figure A3.9

The next stage is to apply an identical procedure to find an equivalent circuit consisting of voltage source $E_{A'B'}$ and resistance $R_{A'B'}$ for the network to the left of points A' and B' in Figure A3.9:

$$\frac{1}{R_{A'B'}} = \frac{1}{R_{AB} + 150} + \frac{1}{1000} = \frac{1}{245.24} + \frac{1}{1000} = \frac{1245.24}{245\,240}$$

Hence:

$$R_{A'B'} = 196.94 \, \Omega$$

$$E_{A'B'} = \frac{1000}{R_{AB} + 150 + 1000} E_{AB} = 38.24 \, \text{V}$$

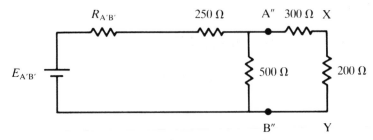

Figure A3.10

The circuit can now be represented in the yet simpler form shown in Figure A3.10. Proceeding as before to find an equivalent voltage source and resistance, $E_{A''B''}$ and $R_{A''B''}$, for the circuit to the left of A″B″ in Figure A3.10:

$$\frac{1}{R_{A''B''}} = \frac{1}{R_{A'B'} + 250} + \frac{1}{500} = \frac{500 + 446.94}{223\,470}$$

Hence:

$$R_{A''B''} = 235.99\ \Omega$$

$$E_{A''B''} = \frac{500}{R_{A'B'} + 250 + 500} E_{A'B'} = 20.19\ \text{V}$$

The circuit has now been reduced to the form shown in Figure A3.11, where the current through branch XY can be calculated simply as:

$$I_{XY} = \frac{E_{A''B''}}{R_{A''B''} + 300 + 200} = \frac{20.19}{735.99} = 27.43\ \text{mA}$$

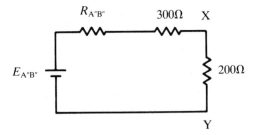

Figure A3.11

References and further reading

Skilling, H. H. (1967) *Electrical Engineering Circuits*, Wiley: New York.

Appendix 4
Thermocouple tables

Type E: chromel–constantan
Type J: iron–constantan
Type K: chromel–alumel
Type N: nicrosil–nisil
Type S: platinum/10% rhodium–platinum
Type T: copper–constantan

Temp. (°C)	Type E	Type J	Type K	Type N	Type S	Type T
−270	−9.834		−6.458	−4.345		
−260	−9.795		−6.441	−4.336		
−250	−9.719		−6.404	−4.313		
−240	−9.604		−6.344	−4.277		−6.105
−230	−9.456		−6.262	−4.227		−6.003
−220	−9.274		−6.158	−4.162		−5.891
−210	−9.063	−8.096	−6.035	−4.083		−5.753
−200	−8.824	−7.890	−5.891	−3.990		−5.603
−190	−8.561	−7.659	−5.730	−3.884		−5.438
−180	−8.273	−7.402	−5.550	−3.766		−5.261
−170	−7.963	−7.122	−5.354	−3.634		−5.070
−160	−7.631	−6.821	−5.141	−3.491		−4.865
−150	−7.279	−6.499	−4.912	−3.336		−4.648
−140	−6.907	−6.159	−4.669	−3.170		−4.419
−130	−6.516	−5.801	−4.410	−2.994		−4.177
−120	−6.107	−5.426	−4.138	−2.807		−3.923
−110	−5.680	−5.036	−3.852	−2.612		−3.656
−100	−5.237	−4.632	−3.553	−2.407		−3.378
−90	−4.777	−4.215	−3.242	−2.193		−3.089
−80	−4.301	−3.785	−2.920	−1.972		−2.788
−70	−3.811	−3.344	−2.586	−1.744		−2.475
−60	−3.306	−2.892	−2.243	−1.509		−2.152
−50	−2.787	−2.431	−1.889	−1.268	−0.236	−1.819
−40	−2.254	−1.960	−1.527	−1.023	−0.194	−1.475
−30	−1.709	−1.481	−1.156	−0.772	−0.150	−1.121
−20	−1.151	−0.995	−0.777	−0.518	−0.103	−0.757

Temp. (°C)	Type E	Type J	Type K	Type N	Type S	Type T
−10	−0.581	−0.501	−0.392	−0.260	−0.053	−0.383
0	0.000	0.000	0.000	0.000	0.000	0.000
10	0.591	0.507	0.397	0.261	0.055	0.391
20	1.192	1.019	0.798	0.525	0.113	0.789
30	1.801	1.536	1.203	0.793	0.173	1.196
40	2.419	2.058	1.611	1.064	0.235	1.611
50	3.047	2.585	2.022	1.339	0.299	2.035
60	3.683	3.115	2.436	1.619	0.365	2.467
70	4.329	3.649	2.850	1.902	0.432	2.908
80	4.983	4.186	3.266	2.188	0.502	3.357
90	5.646	4.725	3.681	2.479	0.573	3.813
100	6.317	5.268	4.095	2.774	0.645	4.277
110	6.996	5.812	4.508	3.072	0.719	4.749
120	7.683	6.359	4.919	3.374	0.795	5.227
130	8.377	6.907	5.327	3.679	0.872	5.712
140	9.078	7.457	5.733	3.988	0.950	6.204
150	9.787	8.008	6.137	4.301	1.029	6.702
160	10.501	8.560	6.539	4.617	1.109	7.207
170	11.222	9.113	6.939	4.936	1.190	7.718
180	11.949	9.667	7.338	5.258	1.273	8.235
190	12.681	10.222	7.737	5.584	1.356	8.757
200	13.419	10.777	8.137	5.912	1.440	9.286
210	14.161	11.332	8.537	6.243	1.525	9.820
220	14.909	11.887	8.938	6.577	1.611	10.360
230	15.661	12.442	9.341	6.914	1.698	10.905
240	16.417	12.998	9.745	7.254	1.785	11.456
250	17.178	13.553	10.151	7.596	1.873	12.011
260	17.942	14.108	10.560	7.940	1.962	12.572
270	18.710	14.663	10.969	8.287	2.051	13.137
280	19.481	15.217	11.381	8.636	2.141	13.707
290	20.256	15.771	11.793	8.987	2.232	14.281
300	21.033	16.325	12.207	9.340	2.323	14.860
310	21.814	16.879	12.623	9.695	2.414	15.443
320	22.597	17.432	13.039	10.053	2.506	16.030
330	23.383	17.984	13.456	10.412	2.599	16.621
340	24.171	18.537	13.874	10.772	2.692	17.217
350	24.961	19.089	14.292	11.135	2.786	17.816
360	25.754	19.640	14.712	11.499	2.880	18.420
370	26.549	20.192	15.132	11.865	2.974	19.027
380	27.345	20.743	15.552	12.233	3.069	19.638
390	28.143	21.295	15.974	12.602	3.164	20.252
400	28.943	21.846	16.395	12.972	3.260	20.869
410	29.744	22.397	16.818	13.344	3.356	
420	30.546	22.949	17.241	13.717	3.452	

Temp. (°C)	Type E	Type J	Type K	Type N	Type S	Type T
430	31.350	23.501	17.664	14.091	3.549	
440	32.155	24.054	18.088	14.467	3.645	
450	32.960	24.607	18.513	14.844	3.743	
460	33.767	25.161	18.938	15.222	3.840	
470	34.574	25.716	19.363	15.601	3.938	
480	35.382	26.272	19.788	15.981	4.036	
490	36.190	26.829	20.214	16.362	4.135	
500	36.999	27.388	20.640	16.744	4.234	
510	37.808	27.949	21.066	17.127	4.333	
520	38.617	28.511	21.493	17.511	4.432	
530	39.426	29.075	21.919	17.896	4.532	
540	40.236	29.642	22.346	18.282	4.632	
550	41.045	30.210	22.772	18.668	4.732	
560	41.853	30.782	23.198	19.055	4.832	
570	42.662	31.356	23.624	19.443	4.933	
580	43.470	31.933	24.050	19.831	5.034	
590	44.278	32.513	24.476	20.220	5.136	
600	45.085	33.096	24.902	20.609	5.237	
610	45.891	33.683	25.327	20.999	5.339	
620	46.697	34.273	25.751	21.390	5.442	
630	47.502	34.867	26.176	21.781	5.544	
640	48.306	35.464	26.599	22.172	5.648	
650	49.109	36.066	27.022	22.564	5.751	
660	49.911	36.671	27.445	22.956	5.855	
670	50.713	37.280	27.867	23.348	5.960	
680	51.513	37.893	28.288	23.740	6.064	
690	52.312	38.510	28.709	24.133	6.169	
700	53.110	39.130	29.128	24.526	6.274	
710	53.907	39.754	29.547	24.919	6.380	
720	54.703	40.382	29.965	25.312	6.486	
730	55.498	41.013	30.383	25.705	6.592	
740	56.291	41.647	30.799	26.098	6.699	
750	57.083	42.283	31.214	26.491	6.805	
760	57.873	42.922	31.629	26.885	6.913	
770	58.663	43.563	32.042	27.278	7.020	
780	59.451	44.207	32.455	27.671	7.128	
790	60.237	44.852	32.866	28.063	7.236	
800	61.022	45.498	33.277	28.456	7.345	
810	61.806	46.144	33.686	28.849	7.454	
820	62.588	46.790	34.095	29.241	7.563	
830	63.368	47.434	34.502	29.633	7.672	
840	64.147	48.076	34.908	30.025	7.782	
850	64.924	48.717	35.314	30.417	7.892	
860	65.700	49.354	35.718	30.808	8.003	

Temp. (°C)	Type E	Type J	Type K	Type N	Type S	Type T
870	66.473	49.989	36.121	31.199	8.114	
880	67.245	50.621	36.524	31.590	8.225	
890	68.015	51.249	36.925	31.980	8.336	
900	68.783	51.875	37.325	32.370	8.448	
910	69.549	52.496	37.724	32.760	8.560	
920	70.313	53.115	38.122	33.149	8.673	
930	71.075	53.729	38.519	33.538	8.786	
940	71.835	54.341	38.915	33.926	8.899	
950	72.593	54.949	39.310	34.315	9.012	
960	73.350	55.553	39.703	34.702	9.126	
970	74.104	56.154	40.096	35.089	9.240	
980	74.857	56.753	40.488	35.476	9.355	
990	75.608	57.349	40.879	35.862	9.470	
1000	76.357	57.942	41.269	36.248	9.585	
1010		58.533	41.657	36.633	9.700	
1020		59.121	42.045	37.018	9.816	
1030		59.708	42.432	37.402	9.932	
1040		60.293	42.817	37.786	10.048	
1050		60.877	43.202	38.169	10.165	
1060		61.458	43.585	38.552	10.282	
1070		62.040	43.968	38.934	10.400	
1080		62.619	44.349	39.315	10.517	
1090		63.199	44.729	39.696	10.635	
1100		63.777	45.108	40.076	10.754	
1110		64.355	45.486	40.456	10.872	
1120		64.933	45.863	40.835	10.991	
1130		65.510	46.238	41.213	11.110	
1140		66.087	46.612	41.590	11.229	
1150		66.664	46.985	41.966	11.348	
1160		67.240	47.356	42.342	11.467	
1170		67.815	47.726	42.717	11.587	
1180		68.389	48.095	43.091	11.707	
1190		68.963	48.462	43.464	11.827	
1200		69.536	48.828	43.836	11.947	
1210			49.192	44.207	12.067	
1220			49.555	44.577	12.188	
1230			49.916	44.947	12.308	
1240			50.276	45.315	12.429	
1250			50.633	45.682	12.550	
1260			50.990	46.048	12.671	
1270			51.344	46.413	12.792	
1280			51.697	46.777	12.913	
1290			52.049	47.140	13.034	
1300			52.398	47.502	13.155	
1310			52.747		13.276	

Temp. (°C)	Type E	Type J	Type K	Type N	Type S	Type T
1320			53.093		13.397	
1330			53.438		13.519	
1340			53.782		13.640	
1350			54.125		13.761	
1360			54.467		13.883	
1370			54.807		14.004	
1380					14.125	
1390					14.247	
1400					14.368	
1410					14.489	
1420					14.610	
1430					14.731	
1440					14.852	
1450					14.973	
1460					15.094	
1470					15.215	
1480					15.336	
1490					15.456	
1500					15.576	
1510					15.697	
1520					15.817	
1530					15.937	
1540					16.057	
1550					16.176	
1560					16.296	
1570					16.415	
1580					16.534	
1590					16.653	
1600					16.771	
1610					16.890	
1620					17.008	
1630					17.125	
1640					17.243	
1650					17.360	
1660					17.477	
1670					17.594	
1680					17.711	
1690					17.826	
1700					17.942	
1710					18.056	
1720					18.170	
1730					18.282	
1740					18.394	
1750					18.504	
1760					18.612	

Appendix 5
Error function tables ($F(z)$, where $z = (x - \mu)/\sigma$) (area under a normalized Gaussian curve)

	0.00	0.01	0.02	0.03	0.04	0.05	0.06	0.07	0.08	0.09
0.0	0.5000	0.5040	0.5080	0.5120	0.5160	0.5199	0.5239	0.5279	0.5319	0.5359
0.1	0.5398	0.5438	0.5478	0.5517	0.5557	0.5596	0.5636	0.5675	0.5714	0.5753
0.2	0.5793	0.5832	0.5871	0.5910	0.5948	0.5987	0.6026	0.6064	0.6103	0.6141
0.3	0.6179	0.6217	0.6255	0.6293	0.6331	0.6368	0.6406	0.6443	0.6480	0.6517
0.4	0.6554	0.6591	0.6628	0.6664	0.6700	0.6736	0.6772	0.6808	0.6844	0.6879
0.5	0.6915	0.6950	0.6985	0.7019	0.7054	0.7088	0.7123	0.7157	0.7190	0.7224
0.6	0.7257	0.7291	0.7324	0.7357	0.7389	0.7422	0.7454	0.7486	0.7517	0.7549
0.7	0.7580	0.7611	0.7642	0.7673	0.7703	0.7734	0.7764	0.7793	0.7823	0.7852
0.8	0.7881	0.7910	0.7939	0.7967	0.7995	0.8023	0.8051	0.8078	0.8106	0.8133
0.9	0.8159	0.8186	0.8212	0.8238	0.8264	0.8289	0.8315	0.8340	0.8365	0.8389
1.0	0.8413	0.8438	0.8461	0.8485	0.8508	0.8531	0.8554	0.8577	0.8599	0.8621
1.1	0.8643	0.8665	0.8686	0.8708	0.8729	0.8749	0.8770	0.8790	0.8810	0.8830
1.2	0.8849	0.8869	0.8888	0.8906	0.8925	0.8943	0.8962	0.8980	0.8997	0.9015
1.3	0.9032	0.9049	0.9066	0.9082	0.9099	0.9115	0.9131	0.9147	0.9162	0.9177
1.4	0.9192	0.9207	0.9222	0.9236	0.9251	0.9265	0.9279	0.9292	0.9306	0.9319
1.5	0.9332	0.9345	0.9357	0.9370	0.9382	0.9394	0.9406	0.9418	0.9429	0.9441
1.6	0.9452	0.9463	0.9474	0.9484	0.9495	0.9505	0.9515	0.9525	0.9535	0.9545
1.7	0.9554	0.9564	0.9573	0.9582	0.9591	0.9599	0.9608	0.9616	0.9625	0.9633
1.8	0.9641	0.9648	0.9656	0.9664	0.9671	0.9678	0.9686	0.9693	0.9699	0.9706
1.9	0.9713	0.9719	0.9726	0.9732	0.9738	0.9744	0.9750	0.9756	0.9761	0.9767
2.0	0.9772	0.9778	0.9783	0.9788	0.9793	0.9798	0.9803	0.9808	0.9812	0.9817
2.1	0.9821	0.9826	0.9830	0.9834	0.9838	0.9842	0.9846	0.9850	0.9854	0.9857
2.2	0.9861	0.9864	0.9868	0.9871	0.9875	0.9878	0.9881	0.9884	0.9887	0.9890
2.3	0.9893	0.9896	0.9898	0.9901	0.9904	0.9906	0.9909	0.9911	0.9913	0.9916
2.4	0.9918	0.9920	0.9922	0.9924	0.9927	0.9929	0.9930	0.9932	0.9934	0.9936
2.5	0.9938	0.9940	0.9941	0.9943	0.9945	0.9946	0.9948	0.9949	0.9951	0.9952
2.6	0.9953	0.9955	0.9956	0.9957	0.9959	0.9960	0.9961	0.9962	0.9963	0.9964
2.7	0.9965	0.9966	0.9967	0.9968	0.9969	0.9970	0.9971	0.9972	0.9973	0.9974
2.8	0.9974	0.9975	0.9976	0.9977	0.9977	0.9978	0.9979	0.9979	0.9980	0.9981
2.9	0.9981	0.9982	0.9982	0.9983	0.9984	0.9984	0.9985	0.9985	0.9986	0.9986
3.0	0.9986	0.9987	0.9987	0.9988	0.9988	0.9989	0.9989	0.9989	0.9990	0.9990

	0.00	0.01	0.02	0.03	0.04	0.05	0.06	0.07	0.08	0.09
3.1	0.9990	0.9991	0.9991	0.9991	0.9992	0.9992	0.9992	0.9992	0.9993	0.9993
3.2	0.9993	0.9993	0.9994	0.9994	0.9994	0.9994	0.9994	0.9995	0.9995	0.9995
3.3	0.9995	0.9995	0.9995	0.9996	0.9996	0.9996	0.9996	0.9996	0.9996	0.9996
3.4	0.9997	0.9997	0.9997	0.9997	0.9997	0.9997	0.9997	0.9997	0.9997	0.9998
3.5	0.9998	0.9998	0.9998	0.9998	0.9998	0.9998	0.9998	0.9998	0.9998	0.9998
3.6	0.9998	0.9998	0.9998	0.9999	0.9999	0.9999	0.9999	0.9999	0.9999	0.9999

Appendix 6
Solutions to exercises

Chapter 2

2.5   0.0175 mV/°C

2.7   (a) 2.62;   (b) 2.94, 0.32

2.8   (a) 20 μm/kg, 22 μm/kg;   (b) 200 μm, 2 μm/kg;   (c) 14.3 μm/°C, 0.143 (μm per kg)/°C

2.9   (a)

Time	Depth	Temp. reading	Temp. error
0	0	20.0	0.0
100	50	19.716	0.216
200	100	19.245	0.245
300	150	18.749	0.249
400	200	18.250	0.250
500	250	17.750	0.250

(b) 10.25 °C

Chapter 3

3.3   3.9%

3.5   5.0%, 24 750 Ω

3.6   10.0%

3.9   Mean 31.1, median 30.5, standard deviation 3.0

3.10 Mean 1.537, standard deviation 0.021, accuracy of mean value = $\pm$0.007, i.e. mean value = 1.537 $\pm$ 0.007; accuracy improved by a factor of ten

3.11 86.6%

3.12 97.7%

3.13 $\pm$0.7%

3.14 $\pm$4.7%

3.15 $\pm$3%

3.16 46.7 Ω $\pm$1.3%

3.17 4.5%

3.18 (a) 0.31 m^3/min; (b) $\pm$5.6%

Chapter 6

6.1 (a) $V_0 = V_s \left(\dfrac{R_1}{R_1 + R_4} - \dfrac{R_2}{R_2 + R_3} \right)$

(b) 81.9 mV

6.2 378 mV

6.3 (a) $V_0 = V_i \left(\dfrac{R_u}{R_u + R_3} - \dfrac{R_1}{R_1 + R_2} \right)$

(b) 0.82 mV/°C
(c) indicated temperature 101.9 °C, error 1.9 °C

6.4 24 V, 1.2 W

6.6 85.9 mV

6.7 (a) 69.6 Ω, 930.4 Ω; (b) 110.3 Ω

6.8 (a) $R_u = R_2 R_3 / R_1$, $L_u = R_2 R_3 C$
(b) 1.57 Ω, 100 mH
(c) 20

6.9 2.538 V_{rms}

6.10 50 μF

6.11 (a) At balance:

$$\frac{R_1 + j\omega L}{R_3} = \frac{R_2}{R_4 - j/\omega C}$$

By taking real and imaginary parts and manipulating:

$$L = R_1 / \omega^2 R_4 C \tag{A}$$

$$R_1 = \frac{R_2 R_3}{R_4(1 + 1/\omega^2 R_4^2 C^2)}$$

Hence, at balance:

$$L = \frac{R_2 R_3 C}{1 + \omega^2 R_4^2 C^2} \tag{B}$$

(b) $Q = \omega L / R_1 = 1/\omega R_4 C$ using Equation (A) above. For large Q, $\omega^2 R_4^2 C^2 \ll 1$, and Equation (B) above becomes:

$$L = R_2 R_3 C$$

This is independent of frequency because there is no ω-term in the expression.

(c) 20 mH

Chapter 7

7.2 (a) Moving-coil (mean) 0.685 A
(b) Moving-iron (r.m.s.) 6.455 A
7.3 (a) 3.18 A; (b) 7.07 A, 6.37 A
7.4 (a) (i) Mean 6.27 V, (ii) r.m.s. 6.46 V
7.5 (a) 1.73 V; (b) 1.67 V; (c) 2.12 V
7.6 (i) Dynamometer (r.m.s.) 0.583 A, moving-coil (mean) 0.267 A
(ii) 68 W

Chapter 8

8.1 (b) $\omega_n = (K_S/J)^{1/2}$, $\beta = K_I/[2R(JK_S)]^{1/2}$, sensitivity $= K_I/(K_S R)$
(d) 0.7
(e) Typical bandwidth 100 Hz, maximum frequency 30 Hz
8.3 $a = 12.410$, $b = 40.438$
8.4 9.8 Ω
8.5 $a = 1.12$, $b = 2.00$
8.6 (a) $C = 5.77 \times 10^{-7}$, $T_0 = 11\,027$
(b) $T = 428$ K

Chapter 10

10.2

	One's compl.	Two's compl.
(a)	01010000	01010001
(b)	10001000	10001001
(c)	10011010	10011011
(d)	00101001	00101010
(e)	00010011	00010100

10.3 (a) 111001, 71, 39
(b) 1100101, 145, 65
(c) 10101111, 257, AF
(d) 100000011, 403, 103
(e) 1111100111, 1747, 3E7
(f) 10011010010, 2322, 4D2

10.4 (a) 7515; (b) F4D
10.5 (a) 130645; (b) B1A5
10.6 (a) 1214; (b) 28C; (c) 3352

Chapter 12

12.1 300 °C
12.2 147.1 °C
12.3 700 °C
12.4 610 °C, 678.4 °C
12.5 15.55 mV, 228.5 °C

Index

Coventry University

β345 sc

Principles of Measurement and Instrumentation

SECOND EDITION

ALAN S. MORRIS

Following the success of the first edition, this new edition has been thoroughly revised and expanded with new material and five new chapters covering, calibration principles and practice, fiber optic sensors and transmission systems, instrumentation networks, dimension measurement, and measurement of other miscellaneous quantities, to reflect the growing importance of this expanding field.

Presenting the subject of instrumentation, and its use within measurement systems as an integrated and coherent subject, the text is divided into two parts. Part one covers the principles of measurement, and part two presents the range of instruments available for measuring various physical quantities.

THE BOOK

Presents integrated treatment of systematic and random errors, statistical data analysis and calibration procedures

Includes important recent developments such as the use of fiber optics and instrumentation networks

Contains encyclopaedic overview of measuring instruments and transducers

Provides numerous worked examples and student exercises

This readable text will provide the student with the theory and knowledge of different types of instruments necessary for a professional career in this field. It will also be a valuable source of reference to all practising engineers and instrumentation technologists.

DR ALAN MORRIS is a senior lecturer at Sheffield University, and has 20 years experience in measurement and control research.

ISBN 0-13-489709-9

PRENTICE HALL

9 780134 897097